Finite Element Methods for Computational Fluid Dynamics

A Practical Guide

Computational Science & Engineering

The SIAM series on Computational Science and Engineering publishes research monographs, advanced undergraduate- or graduate-level textbooks, and other volumes of interest to an interdisciplinary CS&E community of computational mathematicians, computer scientists, scientists, and engineers. The series includes both introductory volumes aimed at a broad audience of mathematically motivated readers interested in understanding methods and applications within computational science and engineering and monographs reporting on the most recent developments in the field. The series also includes volumes addressed to specific groups of professionals whose work relies extensively on computational science and engineering.

SIAM created the CS&E series to support access to the rapid and far-ranging advances in computer modeling and simulation of complex problems in science and engineering, to promote the interdisciplinary culture required to meet these large-scale challenges, and to provide the means to the next generation of computational scientists and engineers.

Series Volumes

Kuzmin, Dmitri and Hämäläinen, Jari, *Finite Element Methods for Computational Fluid Dynamics: A Practical Guide*

Rostamian, Rouben, *Programming Projects in C for Students of Engineering, Science, and Mathematics*

Smith, Ralph C., *Uncertainty Quantification: Theory, Implementation, and Applications*

Dankowicz, Harry and Schilder, Frank, *Recipes for Continuation*

Mueller, Jennifer L. and Siltanen, Samuli, *Linear and Nonlinear Inverse Problems with Practical Applications*

Shapira, Yair, *Solving PDEs in C++: Numerical Methods in a Unified Object-Oriented Approach, Second Edition*

Borzì, Alfio and Schulz, Volker, *Computational Optimization of Systems Governed by Partial Differential Equations*

Ascher, Uri M. and Greif, Chen, *A First Course in Numerical Methods*

Layton, William, *Introduction to the Numerical Analysis of Incompressible Viscous Flows*

Ascher, Uri M., *Numerical Methods for Evolutionary Differential Equations*

Zohdi, T. I., *An Introduction to Modeling and Simulation of Particulate Flows*

Biegler, Lorenz T., Ghattas, Omar, Heinkenschloss, Matthias, Keyes, David, and van Bloemen Waanders, Bart, Editors, *Real-Time PDE-Constrained Optimization*

Chen, Zhangxin, Huan, Guanren, and Ma, Yuanle, *Computational Methods for Multiphase Flows in Porous Media*

Shapira, Yair, *Solving PDEs in C++: Numerical Methods in a Unified Object-Oriented Approach*

DMITRI KUZMIN
TU Dortmund University
Dortmund, Germany

JARI HÄMÄLÄINEN
Lappeenranta University of Technology
Lappeenranta, Finland

Finite Element Methods for Computational Fluid Dynamics

A Practical Guide

siam.
Society for Industrial and Applied Mathematics
Philadelphia

This SIAM edition is partly based on the Finnish-language book *Elementtimenetelmä virtauslaskennassa* (in English: The Finite Element Method in Computational Fluid Dynamics) by J. Hämäläinen and J. Järvinen, CSC Press, 1994.

10 9 8 7 6 5 4 3 2 1

Library of Congress Cataloging-in-Publication Data

Kuzmin, Dmitri, author.
Finite element methods for computational fluid dynamics : a practical guide / Dmitri Kuzmin, Dortmund University of Technology, Dortmund, Germany, Jari Hämäläinen, Lappeenranta University of Technology, Lappeenranta, Finland.
pages cm. – (Computational science and engineering series)
"This SIAM edition is partly based on the Finnish-language book, Elementtimenetelmä virtauslaskennassa (in English: The finite element method in computational fluid dynamics) by J. Hämäläinen and J. Järvinen, CSC Press, 1994."
Includes bibliographical references and index.
ISBN 978-1-611973-60-0
1. Fluid dynamics–Mathematics. 2. Finite element method–Data processing. I. Hämäläinen, Jari, 1965- author. II. Title.
TA357.5.D37.K89 2014
620.1'0640151825–dc23

2014029504

Contents

Preface

Computational fluid dynamics (CFD) is the art of solving partial differential equations that model the motion of fluids, as well as mass and heat transfer phenomena. This book describes one the most powerful numerical techniques for CFD, the finite element method. The finite element method is readily applicable to domains of complex geometrical shape and provides a great freedom in the choice of numerical approximations. Regrettably, only finite difference and/or finite volume methods are usually discussed in introductory courses, and most CFD codes are based on these methods. One of the reasons for this state of affairs is the lack of self-contained elementary textbooks on finite elements for fluids.

In 1994, the second author and his colleague Jari Järvinen wrote a Finnish-language guide to finite elements for CFD (*Elementtimenetelmä virtauslaskennassa* [95]) which became a popular textbook at Finnish universities. The revised and extended second edition appeared in 2006. The present text brings [95] up to date and adds a lot of new material.

The mission of our guidebook is to provide an informal introduction to finite element methods for CFD applications without burdening the reader with the underlying mathematical theory. Instead of tedious proofs of convergence for the Poisson equation in a unit square, the reader will find a clear and detailed presentation of state-of-the-art numerical algorithms for convection-dominated transport problems and the incompressible Navier–Stokes equations. The methods to be presented have been chosen among many others for their accuracy, robustness, and simplicity. Moreover, the book will equip the reader with a collection of fail-safe tools for enforcing physical constraints such as nonnegativity.

We anticipate that the readers of this book will have very diverse backgrounds. The first chapters are written for beginners and present elementary concepts assuming no prior knowledge of CFD or finite elements. The material covered in these chapters may be used in undergraduate-level courses for students majoring in mathematics or computational engineering. In the chapters that follow, basic methods are extended to increasingly complex flow problems. Following a practice-oriented approach, numerical algorithms for representative models are discussed in great detail. The best way to gain the full understanding of these methods is to implement them. Simple problems can be solved using just a few lines of MATLAB code. As a software development kit for more advanced applications, we recommend the open-source finite element library *Elmer* developed at Finnish center for scientific computing CSC - IT Center for Science (www.csc.fi/elmer). The sections describing *Elmer* in Chapters 2 and 5 were contributed by its chief developers Peter Råback, Mika Malinen, and Juha Ruokolainen (CSC). The authors gratefully acknowledge this contribution.

The book will guide the reader through all the steps that are involved in solving the equations of fluid mechanics using continuous finite elements. The text is organized in nine chapters. Chapter 1 introduces the basic notation and mathematical foundations. Chapter 2 describes the general CFD philosophy and acquaints the reader with numerical methods for partial differential equations. In Chapter 3, some basic models of fluid flow and heat transfer are derived from physical conservation principles. An introduction to the Galerkin finite element method is given in Chapter 4 using the one-dimensional heat equation as a model problem. Chapter 5 extends the presented concepts to the two-dimensional heat equation and to the steady incompressible Navier–Stokes equations. In Chapters 6 and 7, we present a self-contained review of stabilization techniques and high-resolution schemes for convection-dominated transport problems. Chapter 8 addresses the numerical solution of the time-dependent incompressible Navier–Stokes equations and the implementation of the $k-\varepsilon$ turbulence model. In addition to numerous citations throughout the book, the reader will find a list of books for further reading in Chapter 9. To some extent, the eclectic choice of topics reflects the authors' personal research interests and teaching philosophy. Nevertheless, we hope that this excursion into the realm of finite element methods for CFD will be a useful experience for the reader and the first step toward solving real-life problems.

Our own interest in finite element methods for CFD goes back to our graduate studies at the University of Jyväskylä (Finland) in the 1990s. This book is an outcome of teaching and research activities which have been greatly influenced by the feedback of our students and colleagues. The methods and software that we develop are based on the work of many other people. We express our sincere gratitude to all individuals with whom we had the privilege to work during the last two decades. Special thanks go to Suncica Canic, Roland Glowinski, Heikki Haario, Jari Järvinen, Raino Mäkinen, Pekka Neittaanmäki, John Shadid, Friedhelm Schieweck, Mikhail Shashkov, Timo Tiihonen, Jari Toivanen, Stefan Turek, and to our late advisor Valery Rivkind. We also wish to thank Markus Hämäläinen, the son of the second author, for bringing the figures from the 20-year-old book [95] up to present-day standards. Last but not least, we thank our wives, Antonia and Taija, for patience and understanding.

Dmitri Kuzmin
Dortmund, Germany

Jari Hämäläinen
Lappeenranta, Finland

Chapter 1

Notation and Preliminaries

The mathematical models to be considered in this text are based on partial differential equations (PDEs) in $d \in \{1,2,3\}$ space dimensions. In this chapter, we introduce the basic mathematical notation that will be employed throughout the book. We also review some useful theorems of vector calculus and the basic definitions of functional analysis.

Blackboard bold capitals are commonly used to denote the sets of natural numbers $\mathbb{N}$, integers $\mathbb{Z}$, rational numbers $\mathbb{Q}$, real numbers $\mathbb{R}$, and complex numbers $\mathbb{C}$.

Lowercase roman, italic, or Greek letters refer to scalar quantities. For example, we denote the time variable by t and use the standard notation x_1, x_2, x_3 or x, y, z for the Cartesian coordinates of a point in the three-dimensional Euclidean space $\mathbb{R}^3$. Further real-valued variables or constants may be named, e.g., a, b, c or α, β, γ. The letters i, j, k, l, m, n are commonly reserved for integers (subscripts, superscripts, dimensions, etc.).

Lowercase boldface symbols denote vectors, i.e., elements of $\mathbb{R}^d$. The individual components of a given vector field are denoted by lowercase letters with subscripts. For example, the boldface symbol $\mathbf{x}$ is just shorthand notation for

$$\mathbf{x} = (x_1, \ldots, x_d)^T.$$

The subscript notation may also be used to pick out the nodal values of an approximate solution defined on a computational mesh. Superscripts usually refer to time levels or iteration steps. For example, $f_i^n \approx f(\mathbf{x}_i, t^n)$ may stand for the numerical value of a scalar function f at some point $\mathbf{x}_i \in \mathbb{R}^d$ and time instant $t^n \in \mathbb{R}$. If the computation of f_i^n requires several iterations, the provisional value calculated at the mth step is denoted by $f_i^{(m)}$.

Uppercase letters are reserved for matrices. The components of the square matrix

$$A = \begin{pmatrix} a_{11} & \cdots & a_{1d} \\ \vdots & \ddots & \vdots \\ a_{d1} & \cdots & a_{dd} \end{pmatrix}$$

are denoted by a_{ij}, where i is the row number and j is the column number.

The above subscript notation is also adopted for second-rank tensors denoted by boldface symbols with double overbars. The jth column of the tensor

$$\bar{\bar{\boldsymbol{\sigma}}} = \begin{pmatrix} \sigma_{11} & \cdots & \sigma_{1d} \\ \vdots & \ddots & \vdots \\ \sigma_{d1} & \cdots & \sigma_{dd} \end{pmatrix}$$

is given by the vector $\boldsymbol{\sigma}_j = (\sigma_{1j},\ldots,\sigma_{dj})^T$. Using this notation, we can write

$$\bar{\bar{\boldsymbol{\sigma}}} = (\boldsymbol{\sigma}_1,\ldots,\boldsymbol{\sigma}_d).$$

The *dot product* and *dyadic product* of two vectors $\mathbf{u},\mathbf{v} \in \mathbb{R}^d$ are defined by

$$\mathbf{u}\cdot\mathbf{v} = u_1 v_1 + \cdots + u_d v_d = \sum_{i=1}^{d} u_i v_i$$

and

$$\mathbf{u}\otimes\mathbf{v} = \begin{pmatrix} u_1 v_1 & \ldots & u_1 v_d \\ \vdots & \ddots & \vdots \\ u_d v_1 & \ldots & u_d v_d \end{pmatrix},$$

respectively. The dot product of a tensor $\bar{\bar{\boldsymbol{\sigma}}} \in \mathbb{R}^{d\times d}$ and a vector $\mathbf{n} \in \mathbb{R}^d$ is a vector

$$\bar{\bar{\boldsymbol{\sigma}}}\cdot\mathbf{n} = (\boldsymbol{\sigma}_1\cdot\mathbf{n},\ldots,\boldsymbol{\sigma}_d\cdot\mathbf{n})^T,$$

while the *double dot product* of two tensors $\bar{\bar{\boldsymbol{\sigma}}},\bar{\bar{\boldsymbol{\varepsilon}}} \in \mathbb{R}^{d\times d}$ is a scalar quantity

$$\bar{\bar{\boldsymbol{\sigma}}} : \bar{\bar{\boldsymbol{\varepsilon}}} = \sum_{i=1}^{d}\sum_{j=1}^{d} \sigma_{ij}\varepsilon_{ji}.$$

The first and second derivative of a function $f : \mathbb{R} \to \mathbb{R}$ are denoted by $f'(x)$ and $f''(x)$, respectively. The first partial derivative of $f : \mathbb{R}^d \to \mathbb{R}$ with respect to (w.r.t.) the independent variable x_i is denoted by $\frac{\partial f}{\partial x_i}$. The vector of first partial derivatives is denoted by

$$\nabla = \left(\frac{\partial}{\partial x_1},\ldots,\frac{\partial}{\partial x_d}\right)^T.$$

The *gradient* of a scalar-valued function $f : \mathbb{R}^d \to \mathbb{R}$ is defined as

$$\nabla f = \left(\frac{\partial f}{\partial x_1},\ldots,\frac{\partial f}{\partial x_d}\right)^T.$$

The *Laplacian* of f represents the sum of second partial derivatives

$$\Delta f = \sum_{j=1}^{d} \frac{\partial^2 f}{\partial x_j^2}.$$

The *divergence* of a vector field $\mathbf{u} : \mathbb{R}^d \to \mathbb{R}^d$ is defined as the dot product

$$\nabla\cdot\mathbf{u} = \frac{\partial u_1}{\partial x_1} + \cdots + \frac{\partial u_d}{\partial x_d} = \sum_{i=1}^{d} \frac{\partial u_i}{\partial x_i}.$$

Note that the gradient of a scalar is a vector, while the divergence of a vector is a scalar. In particular, it is easy to verify that $\nabla\cdot\nabla f = \Delta f$, i.e., $\Delta = \nabla^2$.

The gradient of a vector field $\mathbf{u}$ is a tensor containing all partial derivatives:

$$\nabla\mathbf{u} = \begin{pmatrix} \frac{\partial u_1}{\partial x_1} & \cdots & \frac{\partial u_1}{\partial x_d} \\ \vdots & \ddots & \vdots \\ \frac{\partial u_d}{\partial x_1} & \cdots & \frac{\partial u_d}{\partial x_d} \end{pmatrix} = (\nabla u_1, \ldots, \nabla u_d)^T.$$

The divergence of a second-rank tensor $\bar{\bar{\boldsymbol{\sigma}}}$ is a vector field defined by the formula

$$\nabla \cdot \bar{\bar{\boldsymbol{\sigma}}} = (\nabla \cdot \boldsymbol{\sigma}_1, \ldots, \nabla \cdot \boldsymbol{\sigma}_d)^T.$$

A composition of two functions can be differentiated by the *chain rule* of calculus,

$$(f \circ g)' = (f' \circ g) \cdot g',$$

where the prime mark is used to denote the total derivatives, i.e., the Jacobian matrices.

For example, let $f = f(\mathbf{x}, t)$ be a scalar-valued function of the space coordinates $\mathbf{x} \in \mathbb{R}^3$ and time $t \in \mathbb{R}$. The time rate of change along the path $(\mathbf{x}(t), t)$ is given by

$$\frac{\mathrm{d}f}{\mathrm{d}t} = \nabla f \cdot \frac{\mathrm{d}\mathbf{x}}{\mathrm{d}t} + \frac{\partial f}{\partial t}.$$

In fluid dynamics, the trajectory $\mathbf{x}(t)$ of a fluid particle moving with a prescribed velocity $\mathbf{u}$ is determined by integrating the ordinary differential equation (ODE)

$$\frac{\mathrm{d}\mathbf{x}}{\mathrm{d}t} = \mathbf{u}.$$

In this case, the so-called *substantial derivative* (alias *material derivative*)

$$\frac{Df}{Dt} = \frac{\partial f}{\partial t} + \mathbf{u} \cdot \nabla f \tag{1.1}$$

coincides with the total time derivative of $f(\mathbf{x}, t)$ along the path defined by $\mathbf{x}(t)$.

The integral of $f(\mathbf{x}, t)$ over a time-dependent bounded domain $\Omega(t)$ with a moving boundary $\partial\Omega(t)$ evolves in accordance with the *Reynolds transport theorem*

$$\frac{\mathrm{d}}{\mathrm{d}t}\int_{\Omega(t)} f \,\mathrm{d}\mathbf{x} = \int_{\Omega(t)} \frac{\partial f}{\partial t}\,\mathrm{d}\mathbf{x} + \int_{\partial\Omega(t)} f\mathbf{u} \cdot \mathbf{n}\,\mathrm{d}s, \tag{1.2}$$

where $\mathbf{n}$ denotes the outward-pointing unit normal vector. The surface integral can be transformed into a volume integral using *Green's formula*:

$$\int_{\partial\Omega} f\mathbf{u} \cdot \mathbf{n}\,\mathrm{d}s = \int_{\Omega} \mathbf{u} \cdot \nabla f \,\mathrm{d}\mathbf{x} + \int_{\Omega} f \nabla \cdot \mathbf{u}\,\mathrm{d}\mathbf{x}. \tag{1.3}$$

This useful identity represents a generalization of the integration-by-parts formula for functions of a single real variable. Assuming that the flux $\mathbf{g} = f\mathbf{u}$ is differentiable, Green's formula can be easily inferred from the *divergence theorem*

$$\int_{\Omega} \nabla \cdot \mathbf{g}\,\mathrm{d}\mathbf{x} = \int_{\partial\Omega} \mathbf{g} \cdot \mathbf{n}\,\mathrm{d}s \tag{1.4}$$

and the generalized *product rule*

$$\nabla \cdot (f\mathbf{u}) = \mathbf{u} \cdot \nabla f + f \nabla \cdot \mathbf{u}. \tag{1.5}$$

Knowledge of the above theorems and transformation rules is essential for the derivation of mathematical models and numerical algorithms to be presented in this guidebook.

In our presentation, we tacitly assumed that all integrals and partial derivatives exist, at least in a certain generalized sense. This assumption relies on the right choice of functional spaces, so we close this chapter with some definitions from functional analysis.

Let $\Omega \subset \mathbb{R}^d$ be an open set. The space of functions that are square integrable in Ω is denoted by $L^2(\Omega)$. The formal mathematical definition of this space is as follows:

$$L^2(\Omega) = \left\{ v : \Omega \to \mathbb{R} \,\middle|\, \int_\Omega |v(\mathbf{x})|^2 \, d\mathbf{x} < \infty \right\}. \tag{1.6}$$

The space $L^2(\Omega)$ is a Hilbert space. The L^2 scalar product and norm are defined by

$$(w, v) = \int_\Omega w(\mathbf{x}) v(\mathbf{x}) \, d\mathbf{x}, \qquad w, v \in L^2(\Omega), \tag{1.7}$$

$$||v|| = \sqrt{\int_\Omega |v(\mathbf{x})|^2 \, d\mathbf{x}}, \qquad v \in L^2(\Omega). \tag{1.8}$$

Note that the value of $||v||$ is finite by definition of the functional space $L^2(\Omega)$, while the value of (w, v) is finite by the *Cauchy-Schwarz inequality*

$$|(w, v)| \leq ||w|| \, ||v||, \qquad w, v \in L^2(\Omega). \tag{1.9}$$

The space of functions that are continuous on Ω is denoted by $C(\Omega)$ or $C^0(\Omega)$. If all partial derivatives of order $k \in \mathbb{N}$ exist and are continuous, the function belongs to the space $C^k(\Omega)$. A function v is said to have *compact support* if $v \equiv 0$ outside a bounded subset $\Omega' \subset \Omega$ which has a positive distance from the boundary $\partial\Omega$. The space of infinitely differentiable functions that have compact supports is denoted by $C_0^\infty(\Omega)$.

Generalized partial derivatives of a function $v \in L^2(\Omega)$ can be defined using multiplication by test functions $w \in C_0^\infty(\Omega)$ and integration by parts. Let $\mathbf{e}_i \in \mathbb{R}^d$, $i \in 1, \ldots, d$, denote the ith unit vector in $\mathbb{R}^d$. If v is differentiable in the classical sense, the application of Green's formula (1.3) to the vector field $\mathbf{u} = v\mathbf{e}_i$ and scalar field $f = w$ yields

$$\int_\Omega w \frac{\partial v}{\partial x_i} \, d\mathbf{x} = -\int_\Omega \frac{\partial w}{\partial x_i} v \, d\mathbf{x}. \tag{1.10}$$

The surface integral vanishes due to the assumption that $w \in C_0^\infty(\Omega)$ has a compact support.

In light of the above, it is natural to define a generalized (weak) partial derivative of $v \in L^2(\Omega)$ w.r.t. x_i as a function $\partial_i v \in L^2(\Omega)$ such that

$$\int_\Omega w \partial_i v \, d\mathbf{x} = -\int_\Omega v \frac{\partial w}{\partial x_i} \, d\mathbf{x} \qquad \forall w \in C_0^\infty(\Omega). \tag{1.11}$$

Higher-order generalized partial derivatives are defined in a similar way. Let

$$\alpha = (\alpha_1, \ldots, \alpha_d)$$

be a vector of nonnegative integers. The sum of its components is denoted by

$$|\alpha| = \alpha_1 + \cdots + \alpha_d.$$

In analogy to definition (1.11), a square integrable function

$$D^\alpha v = \frac{\partial^{|\alpha|} v}{\partial x_1^{\alpha_1} \dots \partial x_d^{\alpha_d}} \in L^2(\Omega)$$

represents a generalized (weak) partial derivative of order $|\alpha|$ if

$$\int_\Omega w D^\alpha v \, d\mathbf{x} = (-1)^{|\alpha|} \int_\Omega v D^\alpha w \, d\mathbf{x} \qquad \forall w \in C_0^\infty(\Omega). \tag{1.12}$$

If the partial derivative $D^\alpha v$ exists in the usual sense of differential calculus, this formula can be derived using repeated integration by parts. Thus, a classical partial derivative (if it exists) is always equal to its generalized counterpart defined in the sense of distributions.

Since we are interested in solving PDEs, we will need to assume the existence of (at least) first generalized partials. The *Sobolev space*

$$H^1(\Omega) = \{v \in L^2(\Omega) \mid D^\alpha v \in L^2(\Omega), \ |\alpha| = 1\}$$

contains all square integrable functions with square integrable first generalized partials.

If the function v is continuous on $\bar{\Omega}$, its boundary values are defined by the restriction $v|_{\partial\Omega}$. However, two functions $v, w \in L^2(\Omega)$ are equivalent if they differ on a set of measure zero. By the *trace theorem* of functional analysis, there exists a bounded linear operator

$$\gamma : H^1(\Omega) \to L^2(\partial\Omega) : \quad \gamma v = v|_{\partial\Omega} \quad \forall v \in H^1(\Omega) \cap C(\bar{\Omega}).$$

Hence, the trace γv represents a well-defined generalization of $v|_{\partial\Omega}$ to $v \in H^1(\Omega)$. The space of H^1 functions with vanishing traces on $\partial\Omega$ is denoted by

$$H_0^1(\Omega) = \{v \in H^1(\Omega) \mid \gamma v = 0\}.$$

Generalized derivatives and Sobolev spaces are widely used in texts on functional analysis and finite element methods for PDEs. We will also refer to them occasionally in this guidebook and introduce additional definitions in the course of presentation.

List of Variables

α	multi-index
α_{ij}	correction factor
β	thermal expansion coefficient
Γ	boundary of a domain
δ_{ij}	Kronecker's delta
ε	emissivity, diffusion coefficient, turbulent dissipation rate
$\bar{\bar{\varepsilon}}$	strain rate tensor,
$\varkappa$	von Karman constant
μ	dynamic viscosity
μ_T	eddy viscosity
ν	kinematic viscosity
Φ	limiter function
ρ	density

List of Variables, continued

σ	Stefan–Boltzmann constant
$\sigma_k, \sigma_\varepsilon$	constants in the k–ε turbulence model
$\bar{\bar{\boldsymbol{\sigma}}}$	Cauchy stress tensor
τ	stabilization parameter
$\bar{\bar{\tau}}$	Reynolds stress tensor
τ_w	wall shear stress
Ω	bounded domain
c_p	specific heat capacity at constant pressure
C_μ, C_1, C_2	constants in the k–ε turbulence model
e	internal energy
E	roughness coefficient
f_{ij}	numerical flux
$\mathbf{f}$	flux, body force
g	gravitational constant
$\mathbf{g}$	gravitational force
h	heat source, mesh size
h_c	heat transfer coefficient
h_r	effective heat transfer coefficient
$\mathbf{I}$	identity tensor
k	heat conductivity, turbulent kinetic energy
$\mathbf{n}$	unit outward normal
p	static pressure
Pe	Péclet number
$\mathbf{q}$	heat flux
Re	Reynolds number
s	source term
S	control surface
T	temperature
$\mathbf{u}$	velocity
u_τ	friction (shear) velocity
$\mathbf{v}$	velocity
V	control volume
y	dimensional wall distance
y^+	dimensionless wall distance

List of Abbreviations

1D	one-dimensional
2D	two-dimensional
3D	three-dimensional
ADI	alternating direction implicit
ALE	arbitrary Lagrangian Eulerian
AMG	algebraic multigrid
BDF	backward difference formula
BiCGStab	biconjugate gradient stabilized
BW	Beam–Warming
CCS	compressed column storage
CDR	convection-diffusion-reaction

List of Abbreviations, continued

CES	compressed edge storage
CFD	computational fluid dynamics
CFL	Courant–Friedrichs–Lewy
CG	conjugate gradient
CGS	conjugate gradient squared
CPU	central processing unit
CRS	compressed row storage
CV	control volume
DAE	differential-algebraic equation
DC	discontinuity capturing
DG	discontinuous Galerkin
DMP	discrete maximum principle
DNS	direct numerical simulation
DOF	degree of freedom
FCT	flux-corrected transport
FDM	finite difference method
FEM	finite element method
FSI	fluid-structure interaction
FVM	finite volume method
GLS	Galerkin least squares
GMG	geometric multigrid
GMRES	generalized minimal residual
GPU	graphics processing unit
GUI	graphical user interface
ILU	incomplete lower-upper
LBB	Ladyzhenskaya–Babuška–Brezzi
LED	local extremum diminishing
LES	large eddy simulation
LP	linearity preserving
LW	Lax–Wendroff
MC	monotonized centered
MPI	message passing interface
MOL	method of lines
ODE	ordinary differential equation
PDE	partial differential equation
PGP	pressure gradient projection
PPE	pressure Poisson equation
PSC	pressure Schur complement
RANS	Reynolds-averaged Navier–Stokes
SD	streamline diffusion
SGS	subgrid scale
SI	system of units
SOR	successive overrelaxation
SU	streamline upwind
SUPG	streamline upwind Petrov–Galerkin
TDMA	tridiagonal matrix algorithm
TFQMR	transpose-free quasi-minimal residual
TG	Taylor–Galerkin
TVD	total variation diminishing

Chapter 2

Introduction to Numerical Methods for PDEs

Fluid dynamics is a science that studies transport processes in gases and liquids. Unlike solid materials, fluids offer little resistance to shear forces and are characterized by their ability to flow. Examples of fluid flows include winds, rains, floods, hurricanes, and many other natural phenomena. The associated heat and mass transfer processes are responsible for temperature fluctuations and for the transport of pollutants in air, water, or soil. The hydrodynamic behavior of fluids also plays an important role in many industrial applications such as aerodynamic shape design, oil recovery from an underground reservoir, and chemical reactor engineering. Hence, the ability to understand and predict the physics of fluid flows offers substantial economic benefits and contributes to human well-being.

The experimental approach to investigation of fluid flows is based on observations and measurements for an existing physical system or a laboratory-scale model. The amount of information that can be obtained in this way is limited, and its quality depends on the precision of the measuring instruments. Experimental studies are often time-consuming, costly, dangerous, or simply impossible. In many cases, the use of mathematical models and numerical simulation techniques is an attractive alternative to experimentation.

A mathematical model consists of differential and/or algebraic equations which describe how the quantities of interest evolve and interact with one another. The equations of fluid dynamics are derived from physical conservation laws for mass, momentum, and energy. The Navier–Stokes equations which embody these fundamental principles have been known for almost two centuries, but the complexity and nonlinearity of this model makes it impossible to obtain closed-form solutions to most problems of practical interest. For this reason, the mathematical theory of fluid flows relies on drastic simplifications, and there is little hope that we will witness a major breakthrough in the foreseeable future.

The invention of digital computers and the rapidly increasing computing power have stimulated a growing interest in numerical methods for PDEs. By the end of the 20th century, *computational fluid dynamics* (CFD) has become one of the most important research fields in scientific computing. Although the results of a computer simulation contain numerical errors, they may provide the desired information at a fraction of the cost that a full-scale experimental study would require. Another significant advantage of the computational approach is its universality. An amazing variety of fluid flows can be studied using CFD simulations in actual-size domains under realistic operating conditions. High pressures, toxic chemicals or hot temperatures pose no hazard to a CFD practitioner. The instantaneous values of all quantities of interest can be calculated at any point without disturbing the flow field. However, experimental data are still required to determine the parameters of the mathematical model and validate the numerical results.

In this introductory chapter, we will explain the CFD design philosophy using a scalar transport equation as a toy problem. In the chapters that follow, we will introduce more sophisticated flow models and numerical algorithms based on the finite element method (FEM). Throughout the book, we will focus on computational aspects rather than mathematical theory. We encourage the reader to perform numerical experiments using his/her own implementation of the methods described in the book or the finite element library *Elmer* to be presented at the end of this chapter. The "learning by doing" approach is the best way to understand what hides behind the colorful pictures and what may cause the code to crash or produce nonphysical solutions, such as negative densities or concentrations.

2.1 ▪ Convection and Diffusion

The equations of fluid mechanics are based on conservation principles. Let $u(\mathbf{x},t)$ denote the concentration of a scalar conserved quantity u at point $\mathbf{x}$ and time $t \geq 0$. The evolution of u in a bounded domain Ω is governed by a PDE of the form

$$\frac{\partial u}{\partial t} + \nabla \cdot \mathbf{f}(u) = s \qquad \text{in } \Omega, \tag{2.1}$$

where $\mathbf{f}(u)$ is a vector-valued *flux function* and s is the source term density. This model is derived from an integral conservation law for a small *control volume*. We will use this technique to derive the Navier–Stokes equations of fluid dynamics in the next chapter.

The physical significance of the terms that appear in (2.1) is as follows. The local time derivative $\frac{\partial u}{\partial t}$ describes the temporal variations of u at a fixed point $\mathbf{x} \in \Omega$. The flux $\mathbf{f}(u)$ is the instantaneous rate of transport (per unit area and time) at this point, and s is the rate at which u is produced or destroyed by processes like chemical reactions.

In a typical CFD model, the flux function $\mathbf{f}(u)$ describes the transport of u by *convection* (also called *advection*) and *diffusion*. Since these transport phenomena play an important role in fluid dynamics, let us discuss the underlying physics in some detail.

Convective effects arise when fluids flow and transport the quantities of interest downstream. For example, consider a horizontal pipe filled with water which flows from left to right at constant speed. In experimental studies, the flow pattern is commonly visualized by tracking a set of small tracer particles which are convected with the flow as time goes. Suppose that some particles are white, while others are black. The distribution of particles at time $t = 0$ is displayed in Fig. 2.1(a). Since the water is in motion, it carries

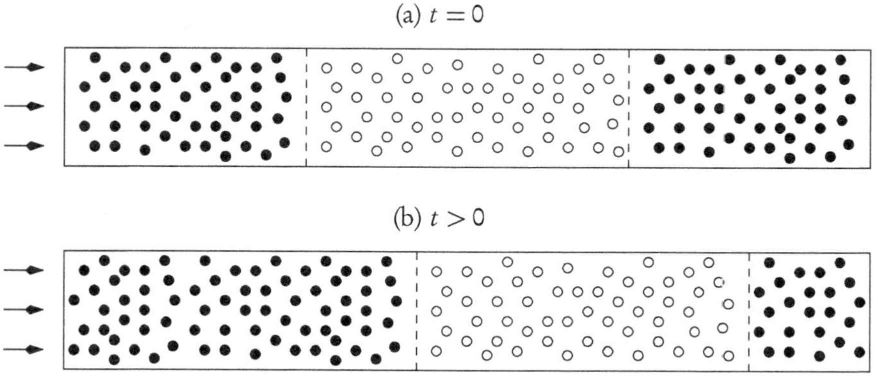

Figure 2.1. *Transport of tracer particles in a pipe filled with moving water.*

the suspension of tracer particles towards the outlet. If we keep injecting black particles at the left end of the pipe, a snapshot of the particle distribution at a later time ($t > 0$) might look like the image depicted in Fig. 2.1(b). On the other hand, if we change the temperature of the water at the inlet, this will affect the temperature distribution inside the pipe and, eventually, at the outlet. This effect is utilized in many heating and cooling devices that we use in everyday life.

Mass and heat may also be transported from one place to another by diffusion or heat conduction, respectively. Random molecular motion gives rise to diffusive fluxes even if the fluid is at rest. To illustrate this process, consider a tank filled with liquid in which two distinct chemical species are dissolved. In Fig. 2.2, the black and white circles represent the molecules of species A and B, respectively. Initially, these species are separated by a diaphragm that divides the tank into two parts (Fig. 2.2(a)). When the diaphragm is removed, some white molecules cross the interface and end up in the left half of the tank. At the same time, some black molecules travel in the opposite direction and end up in the right half (Fig. 2.2(b)). In the long run, the mixture will become homogeneous, and each half of the tank will contain the same number of black and white molecules (Fig. 2.2(c)). In the context of central heating, convection transports hot water into the radiator, but heat transfer inside the room is a diffusive phenomenon since it is driven by the temperature gradients.

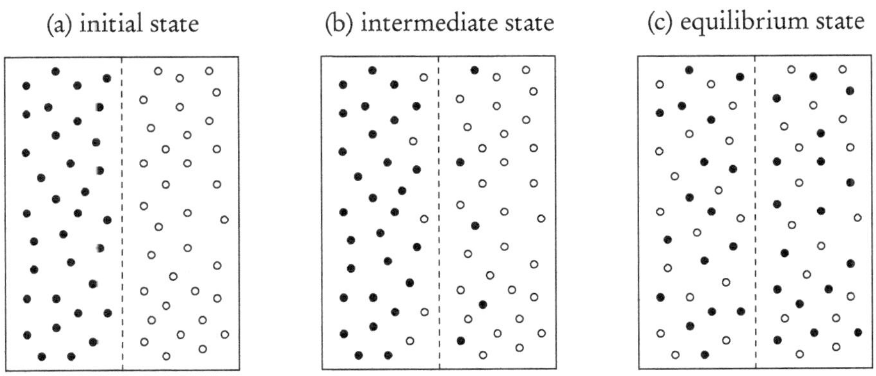

Figure 2.2. *Random motion of molecules across an interface in a stationary liquid.*

In general, the transport of conserved quantities from regions of high concentration into regions of low concentration may be caused by random molecular motion or turbulence. Molecular diffusion represents the natural tendency of a physical system toward an equilibrium, whereas turbulent dispersion is used to model convective transport by unresolved eddies that enhance the macroscopic mixing rate. The mathematical models of diffusion and dispersion look alike but the coefficients differ by orders of magnitude.

A flux function that incorporates the convective and diffusive effects is given by

$$\mathbf{f}(u) = \mathbf{v}u - d\nabla u, \tag{2.2}$$

where $\mathbf{v}$ is a given velocity field and d is a nonnegative diffusion coefficient. The relative strength of convection and diffusion can be expressed in terms of the *Péclet number*,

$$\mathrm{Pe} = \frac{v_0 l_0}{d_0}, \tag{2.3}$$

where v_0 is a reference velocity, l_0 is a geometric length scale (e.g., the width of the domain), and d_0 is a diffusion coefficient. The dimensionless Péclet number is infinite in the limit of pure convection ($d_0 = 0$) and zero in the limit of pure diffusion ($v_0 = 0$).

2.2 ▪ Generic Transport Equation

The common structure of mathematical models based on conservation laws of the form (2.1) suggests a unified approach to their analysis and numerical treatment. Substituting the flux function (2.2) into the conservation law (2.1), one obtains the *generic transport equation*

$$\frac{\partial u}{\partial t} + \nabla \cdot (\mathbf{v}u) - \nabla \cdot (d \nabla u) = s \qquad \text{in } \Omega, \tag{2.4}$$

also known as the *convection-diffusion-reaction (CDR) equation*. This model problem is a good starting point for learning the basics of CFD and testing numerical methods.

By virtue of the vector identity $\nabla \cdot (\mathbf{v}u) = \mathbf{v} \cdot \nabla u + (\nabla \cdot \mathbf{v})u$, the convective term reduces to $\mathbf{v} \cdot \nabla u$ in the case $\nabla \cdot \mathbf{v} = 0$. The diffusive term reduces to $d \Delta u$ if d is constant. That is why the transport equation is sometimes written in the *nonconservative form*

$$\frac{\partial u}{\partial t} + \mathbf{v} \cdot \nabla u - d \Delta u = s \qquad \text{in } \Omega. \tag{2.5}$$

The two formulations are equivalent for $\nabla \cdot \mathbf{v} = 0$ and $d = const$, but only (2.4) is conservative for arbitrary velocity fields and diffusion coefficients.

2.2.1 ▪ Initial and Boundary Conditions

To obtain a well-posed problem, an equation like (2.4) or (2.5) must be supplemented by suitable initial and boundary conditions. The initial values of $u(\mathbf{x}, t)$ are defined by

$$u(\mathbf{x}, 0) = u_0(\mathbf{x}) \qquad \forall \mathbf{x} \in \Omega, \tag{2.6}$$

where $u_0 : \Omega \to \mathbb{R}$ is a given function. Since Ω is a bounded domain, convection and diffusion may transport u across the boundary $\Gamma = \partial\Omega$. If the values of u are known on the boundary part $\Gamma_D \subseteq \Gamma$, they may be imposed using the *Dirichlet boundary condition*

$$u(\mathbf{x}, t) = g(\mathbf{x}, t) \qquad \forall \mathbf{x} \in \Gamma_D. \tag{2.7}$$

The rate of transport across $\Gamma_N \subseteq \Gamma$ may be prescribed using the boundary condition

$$(\alpha \mathbf{v} u - d \nabla u) \cdot \mathbf{n} = f_n(\mathbf{x}, t) \qquad \forall \mathbf{x} \in \Gamma_N, \tag{2.8}$$

where $\mathbf{n}$ is the unit outward normal vector and α is a parameter that determines the type of the boundary condition (Neumann for $\alpha = 0$, Newton for $\alpha = 1$). The Dirichlet boundary condition (2.7) may be imposed weakly using (2.8) with $\alpha = 1$ and $f_n = g\mathbf{v} \cdot \mathbf{n}$.

In the presence of convective transport, the boundary Γ may be decomposed into an inlet $\Gamma_- = \{\mathbf{x} \in \Gamma \,|\, \mathbf{v} \cdot \mathbf{n} < 0\}$, an outlet $\Gamma_+ = \{\mathbf{x} \in \Gamma \,|\, \mathbf{v} \cdot \mathbf{n} > 0\}$, and a solid wall $\Gamma_0 = \{\mathbf{x} \in \Gamma \,|\, \mathbf{v} \cdot \mathbf{n} = 0\}$. A possible choice of boundary conditions is as follows:

$$u = g \quad \text{or} \quad (\mathbf{v}u - d\nabla u) \cdot \mathbf{n} = g\mathbf{v} \cdot \mathbf{n} \quad \text{on } \Gamma_-, \tag{2.9}$$

$$(-d \nabla u) \cdot \mathbf{n} = 0 \quad \text{on } \Gamma_+ \cup \Gamma_0. \tag{2.10}$$

The so-defined boundary value problem remains well-posed even in the limit of vanishing diffusion ($d \to 0$) in which no boundary condition is required on $\Gamma_+ \cup \Gamma_0$.

The behavior of analytical and numerical solutions to (2.4) and (2.5) depends on the interplay of the four terms that appear in these equations. The time derivative may be large as long as the other terms are out of balance but vanish in the steady-state limit. The interplay of the convective and diffusive effects depends on the Péclet number (2.3). The source term s may be equal to zero in some problems but play a dominant role in other applications. It is instructive to analyze the models that arise when certain terms in the generic transport equation are neglected. The omission of these terms may affect the type of the PDE, the choice of initial and boundary conditions, qualitative properties of the exact solutions, and the performance of numerical schemes. Below we classify the existing reduced models and discuss the ramifications of each simplification.

2.2.2 ▪ Elliptic Transport Problems

If the convective and diffusive fluxes are in equilibrium with the source term, then the time derivative of the transported quantity vanishes, and equation (2.4) reduces to

$$\nabla \cdot (\mathbf{v}u - d\nabla u) = s \qquad \text{in } \Omega. \tag{2.11}$$

This stationary CDR equation is of *elliptic* type provided the diffusion coefficient d is strictly positive. In elliptic problems, information propagates in all directions at infinite speeds. The variation of u at some point may immediately influence the solution value at any other point. Boundary conditions must be prescribed on $\Gamma = \Gamma_D \cup \Gamma_N$. Of course, no initial conditions need to be imposed for stationary problems.

Elliptic transport models describe an equilibrium state of the system that may represent the steady-state limit of an evolutionary process. If the velocity field, the diffusion coefficient, and the boundary conditions do not change, the solution to the time-dependent transport equation (2.4) will eventually become stationary and satisfy (2.11).

If the fluid is at rest, equation (2.11) simplifies to the generalized *Poisson equation*

$$-\nabla \cdot (d\nabla u) = s \qquad \text{in } \Omega. \tag{2.12}$$

In the case $d = const$ and $s = 0$, this PDE becomes the classical *Laplace equation*

$$\Delta u = 0 \qquad \text{in } \Omega. \tag{2.13}$$

For example, the solution to this equation can be used to calculate the velocity field $\mathbf{v} = -\nabla u$ for an incompressible irrotational flow model ($\nabla \cdot \mathbf{v} = 0,\ \nabla \times \mathbf{v} = 0$).

2.2.3 ▪ Hyperbolic Transport Problems

In the case of vanishing diffusion, equation (2.4) reduces to the hyperbolic conservation law

$$\frac{\partial u}{\partial t} + \nabla \cdot (\mathbf{v}u) = s \qquad \text{in } \Omega. \tag{2.14}$$

The homogeneous version of this equation ($s = 0$) describes the evolution of fluid density in compressible flow models. This well-known *continuity equation* will be derived in Chapter 3 using the principle of mass conservation and the control volume method.

The steady-state counterpart of the unsteady transport equation (2.14) is given by

$$\nabla \cdot (\mathbf{v}u) = s \qquad \text{in } \Omega. \tag{2.15}$$

The nature of hyperbolic problems requires that boundary conditions be specified at the inlet Γ_- only. It would be incorrect to prescribe any boundary conditions elsewhere.

In hyperbolic problems, information travels at finite speeds along the characteristic curves that may be interpreted as trajectories of tracer particles injected into the flow field. In the stationary case, the characteristics coincide with the streamlines. Due to the lack of diffusive effects, hyperbolic conservation laws may admit discontinuous weak solutions. The qualitative behavior of these solutions is well understood, but numerical methods may fail to reproduce this behavior in the presence of discontinuities or steep fronts.

2.2.4 ▪ Parabolic Transport Problems

Unsteady PDEs of second order are of *parabolic* type if their stationary counterparts are elliptic. For example, the full CDR equation (2.4) is parabolic. In the absence of convection, the evolution of u is governed by the simplified parabolic equation

$$\frac{\partial u}{\partial t}-\nabla\cdot(d\nabla u)=s \qquad \text{in } \Omega. \tag{2.16}$$

This reduced model describes unsteady transport processes such as mass diffusion or heat conduction. The redistribution of u continues until the time derivative vanishes and the solution to the elliptic Poisson equation (2.12) is obtained in the steady-state limit.

In unsteady parabolic models, information propagates forward in time. The distribution of u at any time instant $\bar{t}$ depends on the previous evolution history, but only the solution at a later time may be influenced by what happens at $t=\bar{t}$. Steady CDR equations can also be parabolic if there is a predominant flow direction, and the diffusive transport in this direction is neglected. For example, the steady three-dimensional (3D) parabolic PDE

$$\frac{\partial u}{\partial x_1}-d\left(\frac{\partial^2 u}{\partial x_2^2}+\frac{\partial^2 u}{\partial x_3^2}\right)=s \qquad \text{in } \Omega \tag{2.17}$$

exhibits the same structure as the unsteady heat conduction equation (2.16) in two dimensions. The parabolic nature of (2.17) is due to the existence of a *time-like coordinate* x_1 such that information is convected downstream and no recirculation may occur.

2.2.5 ▪ Summary of Reduced Models

As shown above, the generic transport equation (2.4) represents the whole range of PDE types. If the time derivative, convection, or diffusion is neglected, the resulting equation looks simpler. Ironically, it may be more difficult to solve numerically. It is also important to understand what happens if some coefficients are small compared to others. A general-purpose numerical scheme must be able to handle all possible parameter constellations. Therefore, reduced models are a useful tool for the analysis and design of new methods.

PDEs are easier to analyze and solve in one space dimension. In the one-dimensional (1D) case, the gradient and divergence operators reduce to the partial derivative with respect to x, and the unsteady CDR equation (2.4) becomes

$$\frac{\partial u}{\partial t}+\frac{\partial}{\partial x}\left(vu-d\frac{\partial u}{\partial x}\right)=s. \tag{2.18}$$

If the velocity v and diffusion coefficient d are constant, this equation can be written as

$$\frac{\partial u}{\partial t}+v\frac{\partial u}{\partial x}-d\frac{\partial^2 u}{\partial x^2}=s. \tag{2.19}$$

Again, the generic transport equation can be used to generate 1D model problems for all important limiting cases. The taxonomy of reduced transport models is summarized in Table 2.1. The PDE type listed in the first column remains unchanged in the case $s = 0$.

Table 2.1. *Summary and classification of reduced transport models.*

PDE	multidimensional	one-dimensional
elliptic	$\nabla\cdot(\mathbf{v}u - d\nabla u) = s$ $-\nabla\cdot(d\nabla u) = s$	$\frac{\partial}{\partial x}\left(vu - d\frac{\partial u}{\partial x}\right) = s$ $-\frac{\partial}{\partial x}\left(d\frac{\partial u}{\partial x}\right) = s$
hyperbolic	$\frac{\partial u}{\partial t} + \nabla\cdot(\mathbf{v}u) = s$ $\nabla\cdot(\mathbf{v}u) = s$	$\frac{\partial u}{\partial t} + \frac{\partial}{\partial x}(vu) = s$ $\frac{\partial}{\partial x}(vu) = s$
parabolic	$\frac{\partial u}{\partial t} + \nabla\cdot(\mathbf{v}u - d\nabla u) = s$ $\frac{\partial u}{\partial t} - \nabla\cdot(d\nabla u) = s$	$\frac{\partial u}{\partial t} + \frac{\partial}{\partial x}\left(vu - d\frac{\partial u}{\partial x}\right) = s$ $\frac{\partial u}{\partial t} - \frac{\partial}{\partial x}\left(d\frac{\partial u}{\partial x}\right) = s$

2.2.6 ▪ Maximum Principles

The theory of PDEs makes it possible to prove existence and uniqueness of solutions to (initial-)boundary value problems for all models listed in Table 2.1. Moreover, it is sometimes possible to prove that the exact solution is bounded by the initial and/or boundary conditions. Such *maximum principles* are important for several reasons. On the one hand, they usually represent certain physical constraints such as nonnegativity of densities and concentrations. On the other hand, maximum principles are used to obtain a priori estimates, uniqueness proofs, and comparison principles. Last but not least, discrete maximum principles play an important role in the development of numerical methods.

All transport equations that we considered so far can be written in the generic form

$$\sigma\frac{\partial u}{\partial t} + \mathscr{L}u = s \quad \text{in } \Omega, \tag{2.20}$$

where $\sigma = 1$ for unsteady transport models and $\sigma = 0$ for stationary PDEs. The spatial differential operator $\mathscr{L}$ is defined by $\mathscr{L}u = \nabla\cdot\mathbf{f}(u)$, where $\mathbf{f}$ is the net flux due to convection and/or diffusion. Let Σ denote the set of points and time instants at which the initial or boundary conditions are prescribed. Using elementary calculus, one can prove that

$$s \geq 0, \quad u|_\Sigma \geq 0 \quad \Rightarrow \quad u \geq 0 \quad \text{in } D, \tag{2.21}$$

where D is the entire domain on which the function u is defined, i.e., $D = \bar{\Omega}$ in the steady case and $D = \bar{\Omega}\times[0,\tau]$ for an unsteady PDE to be solved up to the final time $\tau > 0$.

By virtue of (2.21), the solution cannot become negative if the data (the source term and initial/boundary values) are nonnegative. This important result holds for all transport equations in Table 2.1 and is known as the principle of *positivity preservation*. To prove sign preservation for nonpositive data, one should multiply (2.20) by -1. This gives

$$s \leq 0, \quad u|_\Sigma \leq 0 \quad \Rightarrow \quad u \leq 0 \quad \text{in } D. \tag{2.22}$$

If the fluid is at rest or the flow is incompressible ($\nabla \cdot \mathbf{v} = 0$), the differential operator $\mathscr{L}$ has the property that $\mathscr{L}(u+c) = \mathscr{L}u$ for any constant $c \in \mathbb{R}$. This property makes it possible to prove the following maximum/minimum principles for transport equations:

$$s \leq 0, \quad \nabla \cdot \mathbf{v} = 0 \quad \Rightarrow \quad \max_D u = \max_\Sigma u, \tag{2.23}$$

$$s \geq 0, \quad \nabla \cdot \mathbf{v} = 0 \quad \Rightarrow \quad \min_D u = \min_\Sigma u. \tag{2.24}$$

In the case $s = 0$, the solution u is bounded by its initial and/or boundary values:

$$\min_\Sigma u \leq u \leq \max_\Sigma u \quad \text{in } D. \tag{2.25}$$

This two-sided estimate holds for all transport models with $\nabla \cdot \mathbf{v} = 0$ and $s = 0$ but the definition of Σ and D is problem-dependent; see Table 2.2. Note that in the hyperbolic case boundary conditions are prescribed at the inlet $\Gamma_- = \{\mathbf{x} \in \Gamma \,|\, \mathbf{v} \cdot \mathbf{n} < 0\}$ only.

Table 2.2. *Domain and initial/boundary conditions for reduced transport models.*

	PDE type	D	Σ
unsteady	parabolic hyperbolic	$\bar{\Omega} \times [0,T]$ $\bar{\Omega} \times [0,T]$	$\Gamma \times [0,T]$ $\Gamma_- \times [0,T]$
steady	elliptic hyperbolic	$\bar{\Omega}$ $\bar{\Omega}$	Γ Γ_-

2.3 ▪ Numerical Approximations

Computers are of little help in obtaining closed-form solutions to a PDE model. However, they can be programmed to solve large systems of algebraic equations. Replacing calculus with algebra, we will construct numerical approximations to the generic transport equation and more advanced mathematical models to be presented in the next chapter.

The design of numerical methods for PDEs begins with a subdivision of the domain Ω into small subdomains (cells, elements, control volumes) which constitute a *computational mesh* (also called *grid*). The governing equation or an equivalent integral formulation also needs to be discretized. On each subdomain, a given function can be approximated by a polynomial of certain degree. Such polynomial approximations are used to replace the derivatives or integrals by linear combinations of the function values at certain points. The result is a system of algebraic equations for a finite number of *degrees of freedom*. For example, the unknowns of the resultant *discrete problem* may represent the values of the numerical solution at the vertices of the mesh or cell/face averages.

2.3.1 ▪ Computational Meshes

The complexity of mesh generation depends on the shape of the domain. While it is easy to cut a rectangle or a box into a regular array of smaller rectangles/boxes, the presence of curvilinear boundaries, internal obstacles, and small-scale features may require the use of more sophisticated meshes and data structures. The process of automatic mesh generation and adaptation is described in many excellent texts [29, 76, 78, 153, 225]. A typical computational mesh consists of simplices (intervals in one dimension, triangles in two dimensions, tetrahedra in three dimensions), quadrilaterals, or hexahedra. Depending on

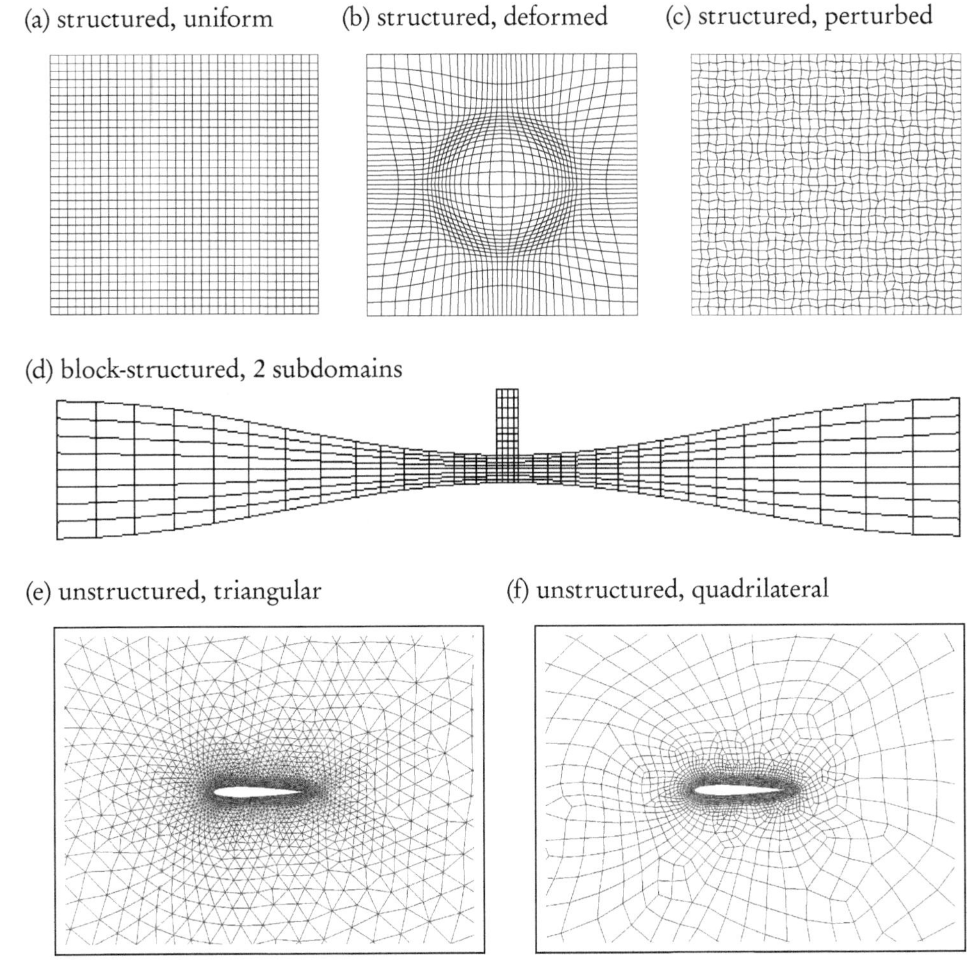

Figure 2.3. *Examples of computational meshes for 2D domains.*

the geometric complexity of the domain, a structured, block-structured, or unstructured mesh may be employed (Fig. 2.3).

In the 1D case, the computational domain is an interval $\Omega = (a, b)$. Let us subdivide this interval into N_{el} subintervals $\Omega_k = (x_{k-1}, x_k)$ of equal length:

$$h = \frac{b-a}{N_{\text{el}}}.$$

The result is the simplest representative of structured meshes. The $N_{\text{el}} + 1$ grid points

$$x_i = ih, \qquad i = 0, 1, \dots, N_{\text{el}}, \tag{2.26}$$

are numbered from left to right. Each interior grid point x_i has two nearest neighbors whose indices $i \pm 1$ and coordinates $x_{i\pm 1}$ are known. In general, nonuniform spacing may be used to achieve higher mesh resolution in regions of particular interest.

A multidimensional structured mesh is a net of grid lines (Fig. 2.3(a)–(c)) which can be numbered consecutively. In the simplest case, each coordinate axis is discretized using a uniform 1D mesh, as defined by (2.26). Again, the search for nearest neighbors is easy, and their number is the same for each interior point. The data structures and numerical algorithms are almost as simple as in one dimension. A nonuniform body-fitted mesh can be generated using a mapping onto a uniform Cartesian grid [3, 225]. However, all grid lines must begin and end on the boundary. Thus, increasing the mesh resolution in some subdomain may give rise to unintended mesh refinements in other parts of the domain [69].

Block-structured mesh generation involves a subdivision of the domain into overlapping or nonoverlapping subdomains (blocks) to be discretized using structured meshes [29, 69]. Fig. 2.3(d) shows a mesh which consists of two nonoverlapping components. The use of multiple blocks makes it easier to deal with nonrectangular geometries and moving objects. Domain decomposition methods [29, 196] may be employed to distribute the work between multiple processors in a parallel computing environment. Local mesh refinement and solution algorithms tailored to structured meshes can be applied blockwise, but special care is required to transfer information between the blocks in a conservative manner. It is also possible to use structured grids in some subdomains (e.g., to generate a *Cartesian core* [165] or resolve a boundary layer) and unstructured meshes elsewhere.

Domains of particularly complex geometric shapes call for the use of a fully unstructured mesh. Recent advances in computational geometry make it possible to generate such meshes automatically for 2D and 3D problems as complex as flow around a car, a submarine, or a space shuttle [163]. Unstructured mesh methods are very flexible and ideally suited for mesh adaptation. An arbitrary number of elements are allowed to meet at a single vertex (Fig. 2.3(e)–(f)), so it is easy to insert extra grid points in regions of insufficient mesh resolution. It is also easy to remove points in regions where a coarser mesh would suffice. However, sophisticated data structures are required to handle the irregular connectivity pattern, and data access is rather slow. Moreover, efficient solution methods are more difficult to develop and program than in the case of (block-)structured grids. Nevertheless, many successful general-purpose CFD codes support the use of unstructured meshes, since the high complexity and overhead costs are often compensated for by a gain of accuracy.

In numerical methods for time-dependent problems, the nodes of the mesh may be fixed or move in a prescribed fashion. The available algorithms can be classified into Eulerian, Lagrangian, and arbitrary Lagrangian Eulerian (ALE). An Eulerian method employs a fixed mesh, i.e., the positions of mesh points do not change as time evolves. The nodes of a Lagrangian mesh move with the flow velocity, so that the convective transport is built into the mesh motion and only diffusive fluxes need to be discretized. In a pure Lagrangian algorithm, the shape and size of mesh cells may change in an uncontrollable manner. In particular, mesh tangling may occur, unless global or local remeshing is performed. Within the ALE framework, some nodes may remain fixed, while others may move with prescribed velocities. This is the most general formulation which lends itself to simulating flows with free interfaces and solving fluid-structure interaction (FSI) problems [50, 163].

2.3.2 ▪ Space Discretization Techniques

In the process of *space discretization*, the exact solution u is approximated by a finite number of nodal values $\{u_j\}$ which may be associated with vertices, edges, faces, cells, or control volumes. Depending on the chosen approximation method, these degrees of freedom may represent pointwise function values, cell averages, or coefficients of piecewise-polynomial basis functions. If the governing equation models an evolutionary process, then the nodal values are generally time-dependent and must be updated step by step.

Each space discretization technique produces a system of algebraic or differential-algebraic equations for the nodal values of the approximate solution. For example, the structure of the semidiscrete problem associated with equation (2.4) is as follows:

$$\sum_{j=1}^{N}\left(m_{ij}\frac{\mathrm{d}u_j}{\mathrm{d}t}\right)+\sum_{j=1}^{N}k_{ij}u_j=r_i, \qquad i=1,\ldots,N, \tag{2.27}$$

where N is the number of degrees of freedom. The only derivative that still appears in each semidiscrete equation is the local time derivative (hence the use of the total derivative notation). The size of the system and the values of the coefficients m_{ij}, k_{ij}, and r_i depend on the computational mesh and on the type of approximation (see below).

Invoking the initial condition (2.6), one obtains an initial-value problem of the form

$$M\frac{\mathrm{d}u}{\mathrm{d}t}+Ku=r, \qquad u(0)=u_0, \tag{2.28}$$

where u is the vector of nodal values, u_0 is the initial data, $M=\{m_{ij}\}$ is the *mass matrix*, $K=\{k_{ij}\}$ is the discrete transport operator, and $r=\{r_i\}$ is the vector of source terms.

The matrices M and K are usually sparse. That is, most of their entries are equal to zero and do not need to be stored. The sparsity pattern of the discrete operators depends on the type of the underlying mesh (structured or unstructured) and on the numbering of nodes. The mass matrix M is diagonal or symmetric positive definite. The discrete diffusion operator K is generally nonsymmetric due to the presence of convective terms.

The steady-state counterpart of the differential-algebraic system (2.28) is given by

$$Ku=r. \tag{2.29}$$

In particular, all stationary transport problems lead to sparse linear systems of this form.

Polynomial Approximations

Polynomials play an important role in the discretization process since they are easy to differentiate and integrate. The most popular discretization techniques based on polynomial approximations are the finite difference, finite volume, and finite element methods. In this chapter, we briefly introduce these methods in the context of a scalar transport equation. An in-depth presentation of the finite element method will follow in subsequent chapters.

Finite Difference Method

The finite difference method (FDM) is the oldest discretization technique for PDEs which is still widely used due to its simplicity and efficiency. Finite difference approximations are particularly easy to construct on structured meshes which are topologically equivalent to a uniform Cartesian grid. Each degree of freedom

$$u_i(t)\approx u(\mathbf{x}_i,t) \tag{2.30}$$

represents a pointwise approximation to the exact solution of the PDE at a grid point $\mathbf{x}_i$.

Taylor series expansions or polynomial fitting techniques are used to approximate the space derivatives in terms of u_i and degrees of freedom associated with a number of neighboring nodes. On a uniform 1D mesh, the second-order accurate *central difference* approximations to the first and second derivatives at the grid point $x_i = ih$ are given by

$$\left(\frac{\partial u}{\partial x}\right)_i \approx \frac{u_{i+1} - u_{i-1}}{2h}, \tag{2.31}$$

$$\left(\frac{\partial^2 u}{\partial x^2}\right)_i \approx \frac{u_{i+1} - 2u_i + u_{i-1}}{h^2}. \tag{2.32}$$

On a nonuniform mesh, the coefficients are different and must be derived individually for each grid point or obtained using a mapping onto a regular Cartesian grid [3].

In finite difference methods for 1D conservation laws, the derivative of the flux function $f(u)$ is usually discretized using approximations of the form [156]

$$\left(\frac{\partial f}{\partial x}\right)_i \approx \frac{f_{i+1/2} - f_{i-1/2}}{h}, \tag{2.33}$$

where $f_{i+1/2}$ and $f_{i-1/2}$ are the *numerical fluxes* defined on the boundaries of the control volume $[x_{i-1/2}, x_{i+1/2}]$ around node i. The convective and diffusive components of $f_{i\pm1/2}$ are expressed in terms of the primary unknowns using a flux function F such that

$$f_{i+1/2} = F(u_{i-k}, \dots, u_{i+m}), \tag{2.34}$$

where k and m are nonnegative integers. For example, the central difference approximation of the flux $f(u) = vu - d\frac{\partial u}{\partial x}$ at the point $x_{i+1/2}$ is defined by the formula

$$f_{i+1/2} = v\frac{u_{i+1} + u_i}{2} - d\frac{u_{i+1} - u_i}{h}. \tag{2.35}$$

The resulting scheme is second-order accurate but tends to produce nonphysical oscillations around steep fronts. This problem may be cured using a first-order *upwind* approximation of the convective flux. The corresponding flux function is given by

$$f_{i+1/2} = v\frac{u_{i+1} + u_i}{2} - \left(d + \frac{|v|h}{2}\right)\frac{u_{i+1} - u_i}{h}. \tag{2.36}$$

Note that the convective part of the so-defined flux $f_{i+1/2}$ reduces to vu_i for $v > 0$ and to vu_{i+1} for $v < 0$. This upwind-biased approximation is guaranteed to be nonoscillatory but has the side effect of artificially increasing the diffusion coefficient, which leads to strong smearing of the solution profiles on coarse meshes. The amount of numerical diffusion may be reduced using a weighted average of (2.35) and (2.36). The design of nonlinear high-resolution schemes based on this idea will be discussed in Chapter 6.

Any finite difference discretization of the unsteady CDR equation (2.4) can be written in the generic form (2.27) which determines the relationship between u_i and the approximate solution values at neighboring nodes. The set of all grid points that make a nonzero contribution to the ith equation is called the *stencil* of the finite difference scheme. The matrix form of the semidiscrete problem is given by (2.28), where $M = I$ is the identity matrix. The coefficients of the discrete transport operator K depend on the parameters of the model, on the choice of finite difference approximations, and on the mesh size h.

Global mesh refinement results in higher accuracy but increases the number of unknowns and, consequently, the computational cost. On structured meshes, the stencil of each interior point has the same size, and nearest neighbors are easy to identify if the grid lines are numbered consecutively. In the case of an unstructured mesh, the number of neighbors may vary, and the distribution of grid points is irregular. Polynomial fitting for scattered data is possible but computationally intensive and difficult to implement. For this reason, unstructured meshes are rarely used in the context of finite difference methods.

Finite Volume Method

The finite volume method (FVM) yields a finite difference–like approximation on a uniform Cartesian grid but is readily applicable to unstructured meshes. The majority of general-purpose CFD codes are based on this class of discretization techniques.

FVMs for the generic transport equation are based on the integral form of (2.4) which can be recovered by integrating this conservation law over a control volume V with a control surface S. By the divergence theorem, this gives

$$\frac{\mathrm{d}}{\mathrm{d}t}\int_V u\,\mathrm{d}\mathbf{x} + \int_S (\mathbf{v}u - d\nabla u)\cdot\mathbf{n}\,\mathrm{d}s = \int_V s\,\mathrm{d}\mathbf{x}, \tag{2.37}$$

where $\mathbf{n}$ is the unit outward normal. This integral conservation law implies that the amount of u contained in V may only change due to the net flux across S and internal sources.

When it comes to the design of numerical approximations, a local conservation law of the form (2.37) is formulated for each element V_i of the computational mesh (Fig. 2.4) or of the associated *dual tessellation*. In the 2D case, the dual cell around the vertex $\mathbf{x}_i$ can be constructed by joining the midpoints of mesh edges and the centroids of the neighboring cells, as shown in Fig. 2.5. Depending on the definition of V_i, finite volume schemes are classified into *cell-centered* and *vertex-centered* approximations.

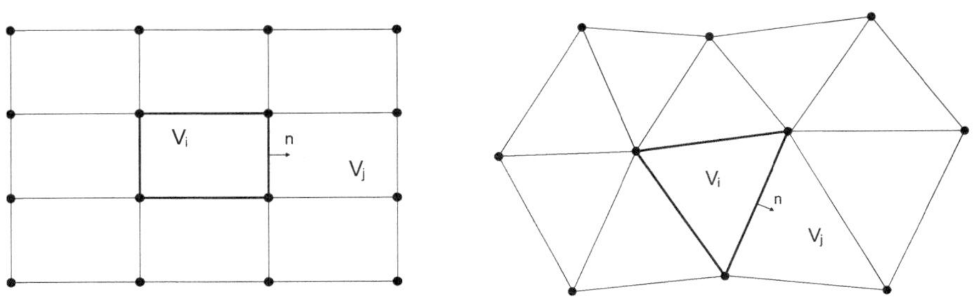

Figure 2.4. *Control volumes for a cell-centered FVM in two dimensions.*

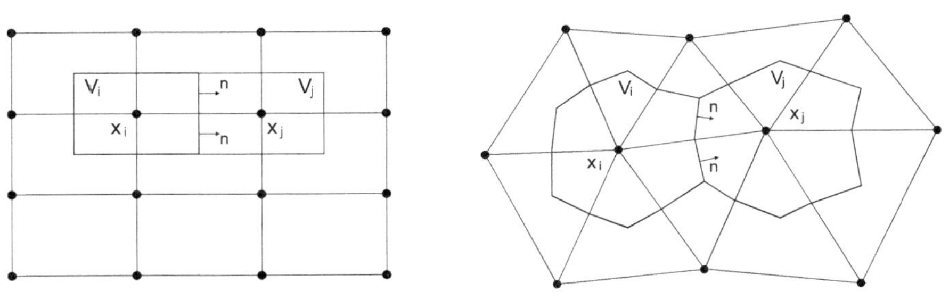

Figure 2.5. *Control volumes for a vertex-centered FVM in two dimensions.*

Averaging the solution and the source term over V_i, we define the mean values

$$u_i(t) = \frac{1}{|V_i|}\int_{V_i} u(\mathbf{x},t)\,\mathrm{d}\mathbf{x} \tag{2.38}$$

and

$$s_i(t) = \frac{1}{|V_i|}\int_{V_i} s(\mathbf{x},t)\,\mathrm{d}\mathbf{x}. \tag{2.39}$$

Using this notation, the integro-differential equation (2.37) can be written as

$$|V_i|\frac{\mathrm{d}u_i}{\mathrm{d}t} + \int_{S_i} (\mathbf{v}u - d\nabla u)\cdot\mathbf{n}\,\mathrm{d}s = |V_i|s_i. \tag{2.40}$$

The boundary S_i of a polygonal or polyhedral control volume V_i consists of several patches,

$$S_i = \bigcup_j S_{ij}. \tag{2.41}$$

For any index $j \neq i$, the patch $S_{ij} = S_i \cap S_j$ is the interface between V_i and one of its neighbors V_j. The index $j = i$ is reserved for boundary patches $S_{ii} = S_i \cap \Gamma$.

Splitting the surface integral into a sum over patches, one can write (2.40) as

$$|V_i|\frac{\mathrm{d}u_i}{\mathrm{d}t} + \sum_j \int_{S_{ij}} (\mathbf{v}u - d\nabla u)\cdot\mathbf{n}\,\mathrm{d}s = |V_i|s_i. \tag{2.42}$$

If the control volumes for a cell-centered or vertex-centered FVM are defined as shown in Figs. 2.4–2.5, then the normal to S_{ij} is constant or piecewise constant.

The total flux across a single patch S_{ij} of the control surface S_i is given by

$$f_{ij} = \int_{S_{ij}} (\mathbf{v}u - d\nabla u)\cdot\mathbf{n}\,\mathrm{d}s. \tag{2.43}$$

Since the internal and external normals to the patch S_{ij} have opposite signs, we have

$$f_{ji} = -f_{ij} \quad \forall j \neq i,$$

i.e., the mass leaving V_i across S_{ij} is equal to the mass received by V_j. The integral over a boundary patch S_{ii} is evaluated using flux boundary conditions of the form (2.8).

To express the semidiscrete problem (2.42) in the form (2.27), the numerical fluxes f_{ij} must be defined using a combination of the mean values u_i in surrounding cells. In FVMs, this process involves two levels of approximation [69]:

- Surface and volume integrals are approximated using numerical quadrature rules.
- Interpolation techniques are used to get the function values at quadrature points.

By the midpoint rule, the control volume averages $u_i(t)$ and $s_i(t)$ represent second-order accurate approximations to the values of u and s at the center of mass defined by

$$\bar{\mathbf{x}}_i = \frac{1}{|V_i|}\int_{V_i} \mathbf{x}\,\mathrm{d}\mathbf{x}.$$

The surface integral in the right-hand side of (2.43) can also be evaluated by the midpoint rule. If the normal to S_{ij} is not constant, numerical integration is performed using a mapping to a reference patch. Interpolation is required to approximate the flux function at the quadrature points since only $u_i(t) \approx u(\bar{\mathbf{x}}_i, t)$ are available. If the same interpolation formula is used on both sides of the interface, then the discretization is conservative, both locally and globally. Indeed, summation over i yields the integral conservation law for $\bar{\Omega} = \bigcup_i V_i$.

Most introductory courses and textbooks on numerical methods for CFD focus on the finite volume approach. The underlying concepts are easy to explain and implement, especially in the context of first- and second-order approximations on structured meshes. Higher-order schemes are rather difficult to derive within the framework of the above two-level approximation strategy, which relies on a well-balanced choice of numerical algorithms for integration and interpolation. Moreover, the discretization of diffusive fluxes involves numerical differentiation, which makes it difficult to achieve optimal accuracy and preserve the symmetry of the diffusion operator on general meshes. Nevertheless, the market for commercial CFD software is dominated by unstructured mesh finite volume codes which produce reasonably accurate results in many applications of practical importance.

Finite Element Method

The finite element method (FEM) is a relative newcomer to the field of CFD. Unlike finite differences and finite volumes, it is backed by rigorous mathematical theory and provides the *best approximation property* when applied to elliptic and parabolic problems at relatively low Péclet numbers. The development of high-resolution finite element schemes for hyperbolic and convection-dominated transport equations is a topic of active research. A review of existing algorithms is included in this guidebook.

Finite element approximations are based on a variational form of the PDE which is derived using the *method of weighted residuals*. The residual of (2.4) is given by

$$\mathcal{R}(u) = \frac{\partial u}{\partial t} + \nabla \cdot (\mathbf{v}u - d\nabla u) - s \tag{2.44}$$

and equals zero if u is the exact solution of equation (2.4). Thus, the magnitude of the residual $\mathcal{R}(u_h)$ measures the accuracy of an approximate solution $u_h \approx u$.

To convert the PDE into a general integral form, the residual is multiplied by a suitable *weighting function* (a.k.a. *test function*) and integrated over Ω. This gives

$$\int_\Omega w\mathcal{R}(u)\,\mathrm{d}\mathbf{x} = 0 \qquad \forall w \in \mathcal{W}, \tag{2.45}$$

where $\mathcal{W}$ denotes the space of admissible test functions (e.g., the Sobolev space $H_0^1(\Omega)$).

Substituting (2.44) into (2.45), one obtains the weighted residual formulation

$$\int_\Omega w\left[\frac{\partial u}{\partial t} + \nabla \cdot (\mathbf{v}u - d\nabla u) - s\right] \mathrm{d}\mathbf{x} = 0 \qquad \forall w \in \mathcal{W}. \tag{2.46}$$

To get rid of the second derivatives, integration by parts is performed using Green's formula

$$\int_\Omega w\nabla \cdot (\mathbf{v}u - d\nabla u)\,\mathrm{d}\mathbf{x} = -\int_\Omega \nabla w \cdot (\mathbf{v}u - d\nabla u)\,\mathrm{d}\mathbf{x} + \int_\Gamma w(\mathbf{v}u - d\nabla u)\cdot \mathbf{n}\,\mathrm{d}s. \tag{2.47}$$

Note that the volume integral on the right-hand side vanishes for $w \equiv 1$. This proves that convective and diffusive terms cannot produce or destroy u inside the domain.

If the Dirichlet boundary condition (2.7) is imposed on a boundary part $\Gamma_D \subseteq \Gamma$, then the trace of w is required to vanish on Γ_D. Boundary conditions of the form (2.8) are imposed weakly by substituting the given normal flux f_n into the surface integral

$$\int_{\Gamma_N} w(\mathbf{v}u - d\nabla u)\cdot \mathbf{n}\,ds = \int_{\Gamma_N} w[(1-\alpha)\mathbf{v}u + f_n]\cdot \mathbf{n}\,ds, \tag{2.48}$$

where α is the boundary condition selector: $\alpha = 1$ if f_n is the total flux and $\alpha = 0$ if the Neumann boundary condition $f_n = -(d\nabla u)\cdot \mathbf{n}$ is prescribed on Γ_N.

The *weak solution* u to the CDR equation (2.4) belongs to the space $\mathscr{V}$ of sufficiently smooth functions satisfying the Dirichlet boundary conditions (if any) and the variational formulation which can be written using the convenient shorthand notation

$$\left(w, \frac{\partial u}{\partial t}\right) + a(w,u) = b(w) \qquad \forall w \in \mathscr{W}, \tag{2.49}$$

where

$$\left(w, \frac{\partial u}{\partial t}\right) = \int_\Omega w \frac{\partial u}{\partial t}\,d\mathbf{x},$$

$$a(w,u) = (1-\alpha)\int_{\Gamma_N} w(\mathbf{v}u)\cdot \mathbf{n}\,ds - \int_\Omega \nabla w \cdot (\mathbf{v}u - d\nabla u)\,d\mathbf{x},$$

$$b(w) = \int_\Omega ws\,d\mathbf{x} - \int_{\Gamma_N} wf_n\,ds.$$

In finite element methods, an approximate solution u_h to (2.49) is defined as a linear combination of basis functions $\{\varphi_i\}$ spanning a finite-dimensional space $\mathscr{V}_h$,

$$u_h(\mathbf{x},t) = \sum_j u_j(t)\varphi_j(\mathbf{x}). \tag{2.50}$$

The standard basis functions are required to possess the following properties [44]:

- there exists a set of nodes $\mathbf{x}_i \in \Omega$ such that $\varphi_i(\mathbf{x}_i) = 1 \wedge \varphi_i(\mathbf{x}_j) = 0\ \forall j \neq i$;
- the restriction of φ_i to each cell is a polynomial function of local coordinates.

By the first property and (2.50), the nodal values of the approximate solution are given by $u_h(\mathbf{x}_i,t) = u_i(t)$. In the simplest case, the nodal points $\mathbf{x}_i$ are located at the vertices of the mesh. Other possible locations are the midpoints/barycenters of edges, faces, and cells [44]. A typical piecewise-linear basis function φ_i is depicted in Fig. 2.6.

Let $\mathscr{W}_h$ be the space of test functions spanned by $\{\psi_i\}$. Using $u = u_h$ and $w = \psi_i$ in the integral form (2.49), one obtains a semidiscrete equation of the form (2.27), where

$$m_{ij} = (\psi_i, \varphi_j), \qquad k_{ij} = a(\psi_i, \varphi_j), \qquad r_i = b(\psi_i).$$

For the discrete problem to be solvable, the number of equations must be equal to the number of unknowns, which is clearly the case if we take $\psi_i = \varphi_i$. This setting is used in the standard *Galerkin method*, a finite element scheme in which the numerical solution and the test functions live in finite-dimensional spaces spanned by the same basis

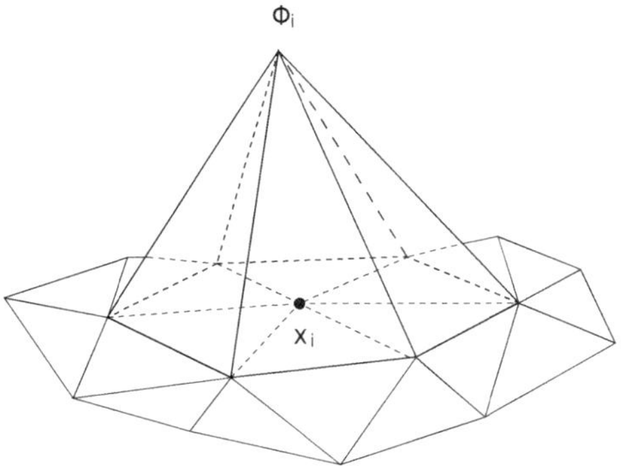

Figure 2.6. *A piecewise-linear finite element basis function on a triangular mesh.*

functions. Finite element approximations with $\psi_i \neq \varphi_i$ are called *Petrov–Galerkin methods* and offer certain advantages, e.g., in the context of convection-dominated transport problems.

Finite element and finite volume discretizations of convection-diffusion equations have a lot in common and are largely equivalent in the case of low-order polynomials [163, 213, 214]. The traditional strengths of the FVM and FEM are complementary. Convective terms call for the use of an upwind-biased discretization which is easier to construct within the finite volume framework. On the other hand, the FEM is better suited for the discretization of diffusive terms. The *discontinuous Galerkin* FEM is a generalization of FVM which offers many attractive features, including local conservation. However, the treatment of diffusive fluxes is not straightforward and requires special care, as in the case of FVM. In this guidebook, we will focus on continuous finite element approximations.

2.3.3 ▪ Time Discretization Techniques

Discretization of a stationary PDE leads to an algebraic system of the form (2.29) which needs to be solved just once since the degrees of freedom are independent of time. If the time derivative cannot be neglected, the discretization of space derivatives using finite differences, finite volumes, or finite elements gives rise to an initial-value problem of the form (2.27) which must be integrated in time using a suitable numerical method.

In nonstationary flow models, the time represents an additional independent variable which is usually discretized using a 1D finite difference mesh to define the time levels

$$0 = t^0 < t^1 < \cdots < t^{N_T} = T.$$

If the time step Δt is constant, the nth time level is given by the simple formula

$$t^n = n\Delta t, \qquad \Delta t = \frac{T}{N_T}.$$

The value of the approximate solution at node i and time level t^n is denoted by

$$u_i^n \approx u_i(t^n).$$

The value of $u_i^0 = u_i(0)$ is determined using the initial condition (2.6), which may also be interpreted as an inflow boundary condition for the space-time domain $D = \Omega \times (0, T)$.

In principle, finite difference, finite volume, and finite element methods are readily applicable to functions defined in D. However, simultaneous computation of u_i^n for all nodes and time levels is usually impractical. Since information propagates forward in time, it is makes more sense to advance the numerical solution in time step by step.

Time-stepping (or *marching*) methods initialize the vector of discrete nodal values by $u^0 = \{u_i^0\}$ and use $u^n = \{u_i^n\}$ as initial data for the computation of $u^{n+1} = \{u_i^{n+1}\}$. This solution strategy is faster and requires less memory than a coupled space-time discretization. The size of the algebraic systems to be solved at each time step depends only on the number of spatial degrees of freedom and not on the number of time levels. In essence, a very large system is replaced with a sequence of smaller ones, which results in considerable savings of computer time. Intermediate solutions are usually overwritten to save computer memory. Last but not least, a nonphysical dependence of u^n on u^{n+1} is ruled out by construction.

A time-stepping technique for a semidiscrete system of the form (2.28) can be designed using any numerical method for solving initial-value problems of the form

$$\frac{du}{dt} = R(u,t), \qquad t \in (t^n, t^{n+1}), \qquad u(t^n) = u^n. \tag{2.51}$$

Our semidiscrete system (2.28) can be decomposed into subproblems of this form with

$$R(u,t) = M^{-1}[r(t) - Ku(t)]. \tag{2.52}$$

The use of numerical solution techniques for ordinary differential equations (ODEs) as time integrators for PDEs is called the *method of lines* (MOL) since the nodal value $u_i(t)$ is advanced in time along the line $x = x_i$.

The simplest representative of MOL time discretizations is the two-level θ-scheme

$$\frac{u^{n+1} - u^n}{\Delta t} = \theta R(u^{n+1}) + (1-\theta)R(u^n), \qquad 0 \leq \theta \leq 1. \tag{2.53}$$

In general, the right-hand side is a weighted average of $R(u^{n+1}, t^{n+1})$ and $R(u^n, t^n)$. The resulting time discretization is *explicit* for $\theta = 0$ and *implicit* for $\theta \in (0,1]$.

The trivial choice $\theta = 0$ corresponds to the first-order accurate forward Euler method

$$u^{n+1} = u^n + \Delta t R(u^n), \tag{2.54}$$

whereas $\theta = \frac{1}{2}$ corresponds to the second-order accurate Crank–Nicolson scheme

$$u^{n+1} = u^n + \frac{\Delta t}{2}[R(u^{n+1}) + R(u^n)], \tag{2.55}$$

and $\theta = 1$ corresponds to the first-order accurate backward Euler method

$$u^{n+1} = u^n + \Delta t R(u^{n+1}). \tag{2.56}$$

In the explicit Euler method (2.54), each nodal value u_i^{n+1} can be calculated independently using the data from the previous time level. In the implicit methods (2.55) and (2.56), the value of u_i^{n+1} depends on the new data at other points, so the equations of the linear system must be solved simultaneously. Explicit methods are easier to implement, but their implicit counterparts are more stable and may be operated safely with larger time steps.

The application of the θ-scheme to (2.51)–(2.52) yields a linear system of the form

$$[M+\theta\Delta tK]u=[M-(1-\theta)\Delta tK]u^n+\Delta t[\theta r^{n+1}+(1-\theta)r^n]. \qquad (2.57)$$

If M is not a diagonal matrix, then the equations of this linear system are coupled even for case $\theta=0$. Finite difference and finite volume methods produce diagonal mass matrices. In finite element methods, the sparsity pattern of M depends on the choice of the basis functions. A diagonal mass matrix is obtained only in the case of an orthogonal basis or if the entries of M are calculated using an inexact numerical quadrature rule.

The generic θ-scheme (2.53) is a *two-level* method since only u^n is involved in the computation of u^{n+1}. Such time-stepping methods can be at most second-order accurate. Higher-order time discretizations must use information from additional time levels. In Runge–Kutta methods, all time levels $t^{n+\alpha}$ with $0\le\alpha\le1$ belong to the time interval $[t^n,t^{n+1}]$, and a predictor-corrector strategy is used to compute u^{n+1}. Another possibility is to integrate a polynomial fitted to the values of $R(u,t)$ at $t^{n+\alpha},\dots,t^{n-\beta}$, where α and β are nonnegative integers. Multipoint methods of Adams–Bashforth (explicit) and Adams–Moulton (implicit) type correspond to $\alpha=0$ and $\alpha=1$, respectively. For a concise introduction to Adams and Runge–Kutta schemes, we refer to Ferziger and Perić [69].

2.3.4 ▪ Solvers for Algebraic Systems

The fully discrete problem is an algebraic system which can be written in the generic form

$$Au=b, \qquad (2.58)$$

where $A=\{a_{ij}\}$ is a sparse matrix, u is the vector of unknowns, and b is a vector that depends on the initial/boundary conditions and on the previously computed data.

If A is a diagonal matrix, each degree of freedom is given by $u_i=b_i/a_{ii}$. In the case of a general sparse matrix, an efficient numerical algorithm must be used to solve system (2.58). The choice of the solution strategy is largely independent of the underlying approximations, but the size and sparsity pattern of A should be taken into account.

Direct Methods

Direct methods for solving linear systems of the form (2.58) accomplish this task in one step. The input parameters are the matrix A and the right-hand side b. In the absence of roundoff errors, the result is the exact solution u of the discrete problem. Direct solvers perform well for linear systems of moderate size, but the memory requirements and the CPU time increase rapidly with the number of unknowns. If the sparsity pattern of the matrix A exhibits some regular structure, then this structure can sometimes be exploited to design a fast direct solver, such as the Thomas (TDMA) algorithm for tridiagonal matrices [69]. This algorithm can also be embedded in alternating-direction-implicit (ADI) solvers for banded matrices that result from 2D and 3D discretizations on structured meshes. Efficient multifrontal direct solvers for nonsymmetric sparse linear systems are implemented in the software library UMFPACK [235], which is available under a GNU LGPL license and provides interfaces to Fortran, C, and MATLAB. The rapid increase in computer memory and advances in parallel supercomputing have revived the interest in direct solvers for large-scale applications.

Iterative Methods

Iterative methods solve linear systems using a sequence of explicit updates starting with an initial guess which must be supplied as an additional input parameter. An iterative algorithm is said to be *convergent* if the approximate solution to the linear system becomes exact after sufficiently many iterations. In practice, it is neither necessary nor advisable to iterate until full convergence. The iterative process is terminated as soon as certain stopping criteria are satisfied. As a rule, these criteria boil down to monitoring the differences between two successive iterates and/or the residuals measured in a certain norm. Stopping too early gives rise to large iteration errors; stopping too late results in a waste of CPU time.

Iterative solvers require a relatively small amount of computer memory. Many of them do not even require that the matrix A be assembled and stored. A subroutine that evaluates matrix-vector products of the form Au is usually provided instead of A. The most efficient iterative algorithms are based on *multigrid methods* [93, 230, 242]. If properly configured, they can solve a system of N equations using as few as $\mathcal{O}(N)$ arithmetic operations, as compared to $\mathcal{O}(N^3)$ for Gaussian elimination and $\mathcal{O}(N^2)$ for forward/backward substitution with a precomputed LU factorization. Thus, the iterative approach pays off when it comes solving sparse linear systems for a large number ($N > 10^5$) of degrees of freedom.

If the algebraic system corresponds to the discretization of a stationary problem, a good initial guess is rarely available. The default is zero, and many iterations are typically required to achieve the prescribed tolerance. The computational effort associated with advancing the solution of an unsteady problem from one time level to the next is much smaller. Since u^n provides a relatively good guess for u^{n+1}, just a few iterations per time step are normally required to obtain an accurate approximation. Thus, each update is typically much cheaper than that performed with a direct solver. If the time step is very small, a single iteration may suffice, so the cost of solving the linear system is similar to that of a fully explicit update. As the time step increases, so does the number of iterations for the linear solver.

Due to the above, stationary PDEs are often solved using time-stepping methods for the associated unsteady problems. Given an arbitrary initial condition, a long time simulation will produce the desired steady-state solution. This approach is called *pseudo time-stepping*. In essence, it represents an iterative solver in which the time step serves as a relaxation parameter. In this context, the accuracy of the time discretization is not important, and the artificial time step should be chosen so as to reach the steady state as fast as possible.

Nonlinear Solvers

If the coefficients of the numerical scheme depend on the unknown solution, then the discrete problem is nonlinear and an iterative solution strategy must be employed. Starting with a suitable initial guess, the available solution values are used to update the matrix entries and obtain the next guess by solving a linearized algebraic system. This process is repeated until the changes and/or the residual of the nonlinear system become small enough. In principle, a direct or iterative method may be used to solve the involved linear systems. However, the use of direct methods is usually impractical since they spend a large amount of CPU time on solving linear systems with coefficients that may change at the next iteration.

Within a fully iterative approach, the coefficients are calculated using the solution values from the previous *outer iteration*, and a few *inner iterations* are performed to obtain an improved solution. Since the coefficients are tentative, the amount of work spent on the solution of each linear system should not be inordinately large. As soon as the residuals have decreased by a factor of 10 or so, one may stop the inner iteration process and proceed to the calculation of the coefficients for the next outer iteration. Well-balanced stopping criteria make it possible to minimize the total number of inner and outer iterations.

If the nonlinearity is too strong, the residuals may exhibit oscillatory behavior or even increase, causing the iterative process to diverge. A possible remedy to this problem is the use of underrelaxation and convergence acceleration techniques. The basic idea is to construct a weighted average of two or more iterates, so as to limit the changes of the solution [69, 185] or speed up convergence by choosing optimal weights [239].

2.4 ▪ Properties of Numerical Methods

The quality of numerical solutions to transport equations depends on the underlying mesh, on the properties of the discretization techniques, and on the Péclet number. The mesh size and time step should also be chosen carefully, especially in the case of explicit schemes. The choice of numerical algorithms determines the coefficients of the algebraic system (2.58) for the nodal values of the approximate solution. The properties of the matrix A and of the discrete operators built into the right-hand side b may make or break the entire simulation run. The type of PDE and the smoothness of the solution may also play an important role. Some algorithms are tailored for equations of a certain type or cannot handle nonsmooth solutions properly. Other methods do not work at all or produce reasonably looking results which have little in common with the exact solution of the governing equation.

The difference between the exact solution u of the mathematical model and an approximate solution u_h produced by a bug-free computer code is caused by numerical errors which can be classified into discretization errors, roundoff errors, and iteration errors. The discretization error $\epsilon_h = u - u_h$ depends on the mesh size h and, possibly, on the time step Δt. The magnitude of ϵ_h can be estimated using Taylor series expansions or the tools of functional analysis. Roundoff errors are due to the finite precision of computer arithmetics. They are usually much smaller than ϵ_h and iteration errors which depend on the prescribed tolerances and stopping criteria for linear solvers. It is essential to ensure that u_h converges to the exact solution of the differential equation as h and Δt go to zero. To control the magnitude of the total error, the design and evaluation of new discretization techniques must always be accompanied by analysis of *consistency*, *stability*, and *convergence*.

A good numerical method must also provide a realistic description of the physical system and preserve important qualitative properties of the exact solution of the mathematical model, such as conservation and boundedness. The above criteria guarantee a certain level of accuracy and robustness for a sufficiently broad range of applications. In this section, we summarize the most important design principles and explain their implications.

2.4.1 ▪ Consistency

A numerical method is said to be consistent if the discretization error ϵ_h vanishes in the limit $h \to 0$ and $\Delta t \to 0$. Consistency is a relationship between the exact solutions to the continuous and discrete problems. A consistent discretization is "asymptotically correct." In practice, finite values of h and Δt are used to keep simulations affordable. Since the computational cost increases rapidly with the number of unknowns, it is natural to require that the discretization errors become smaller as we refine the mesh and/or the time step. Moreover, we wish to know how fast the numerical error decreases when we do so.

Let u_{ex} and u_{num} denote the vectors of nodal values for the exact solutions to the continuous and discrete problems, respectively. By definition, u_{num} satisfies an algebraic system of the form $Au_{\text{num}} = b$, while the residual $b - Au_{\text{ex}}$ is generally nonzero. The magnitude of the consistency error can be measured using a suitable vector norm $||\cdot||$, e.g., the Euclidean norm or the maximum norm. The discretization method is consistent if

$$||b - Au_{\text{ex}}|| \leq C(h, \Delta t), \tag{2.59}$$

where $C(h, \Delta t)$ is nonnegative and goes to zero as $h \to 0$ and $\Delta t \to 0$. The underlying discretization is said to be of order p in space and of order q in time if

$$C(h, \Delta t) = \mathcal{O}(h^p, \Delta t^q). \tag{2.60}$$

We remark that numerical analysis of finite element approximations is usually performed using functions $u_{\text{num}}, u_{\text{ex}} : \bar{\Omega} \to \mathbb{R}$ defined in terms of their discrete nodal values.

The formal order of accuracy is not always a good indicator of quality. The errors produced by two schemes of the same order may differ significantly, and a low-order scheme might perform better than a high-order one on a coarse mesh. Strictly speaking, the Taylor series analysis is not applicable to discontinuous solutions, and the contribution of higher-order terms may become nonnegligible if the corresponding derivatives are large. As a consequence, numerical errors may decrease much slower than expected. Moreover, even a consistent scheme may fail to converge if it is not immune to numerical instabilities.

2.4.2 ▪ Stability

Stability is a relationship between the exact solution of the discrete problem and the actual solution which is influenced by roundoff and iteration errors. Some numerical algorithms enjoy unconditional stability; others are stable under certain restrictions on the choice of input parameters. Unstable schemes may be fixed by adding extra terms, provided the discretization remains consistent. This idea leads to a rich variety of stabilized schemes.

It is instructive to investigate what happens if we solve the system $A\tilde{u}_{\text{num}} = \tilde{b}$ instead of $Au_{\text{num}} = b$. Given a vector norm $||\cdot||$ and a compatible matrix norm, the error

$$u_{\text{num}} - \tilde{u}_{\text{num}} = A^{-1}(b - \tilde{b})$$

induced by the perturbation of the right-hand side b can be estimated as follows:

$$||u_{\text{num}} - \tilde{u}_{\text{num}}|| \leq ||A^{-1}|| \, ||b - \tilde{b}||. \tag{2.61}$$

This estimate implies that the discretization is stable if $||A^{-1}|| \leq C_A$, where $C_A > 0$ is independent of h. The proof of this property is based on problem-dependent arguments.

Detailed stability analysis for numerical approximations to unsteady transport problems is usually based on the *von Neumann method* or on the method of *modified equations* [3, 50, 104] as applied to linear model problems with constant coefficients. The theory of *discrete maximum principles* makes it possible to obtain maximum-norm stability estimates under restrictive assumptions on the coefficients of the numerical scheme (see below).

2.4.3 ▪ Convergence

A numerical method is said to be convergent if the numerical solution of the discrete problem approaches the exact solution of the differential equation as the mesh size and time step go to zero. High-order methods converge to smooth solutions faster than low-order approximations. For linear problems, consistency and stability are the necessary and sufficient conditions of convergence. Furthermore, the boundedness of $||A^{-1}||$ implies that

$$\begin{aligned} ||u_{\text{num}} - u_{\text{ex}}|| &= ||A^{-1}A(u_{\text{num}} - u_{\text{ex}})|| = ||A^{-1}(b - Au_{\text{ex}})|| \\ &\le ||A^{-1}||C(h^p, \Delta t^q) \le C_A C(h^p, \Delta t^q). \end{aligned} \tag{2.62}$$

Hence, the approximation error goes to zero as fast as the consistency error $||b - Au_{\text{ex}}||$.

In the context of finite difference approximations to well-posed linear initial-value problems, the relationship between consistency and convergence is known as the *Lax equivalence theorem*. In the nonlinear case, a proof of convergence can be obtained under the assumption that the *total variation* of the numerical solution is uniformly bounded [156].

It is a good practice to verify convergence numerically by running the same simulation on at least three successively refined meshes and varying the time step size. The results of this grid convergence study may be used to estimate the order of accuracy [69, 158]:

$$p = \log_2 \frac{||u_{2h} - u||}{||u_h - u||} \approx \log_2 \frac{||u_{2h} - u_{4h}||}{||u_h - u_{2h}||}. \tag{2.63}$$

If the discrepancy between the solutions obtained with several different meshes and time steps is insignificant, this indicates that the numerical errors are small, and the results are close enough to the exact solution of the PDE. Otherwise, the refinement process should be continued to make sure that the method converges to a grid-independent solution.

2.4.4 ▪ Conservation

If the mathematical model is derived from conservation principles, the same principles should apply to the approximate solution. If the quantity of interest is conserved, then numerical errors can only distribute it improperly. Conservation property forces the numerical algorithm to respect an important qualitative property of the physical system. The *Lax–Wendroff theorem* states that if a consistent and conservative scheme converges, then it converges to a weak solution of the conservation law. Convergence to nonphysical weak solutions, such as shocks moving at wrong speeds, is ruled out by this theorem.

A consistent and stable nonconservative scheme will produce correct results on sufficiently fine meshes. Unfortunately, it is usually difficult to tell a priori how fine the mesh and time step must be chosen to keep conservation errors small enough. Thus, even solutions to rather simple problems may behave in an unpredictable manner.

A finite difference scheme proves conservative if it can be written in terms of numerical fluxes. Finite volume and discontinuous Galerkin methods are conservative by construction, both globally and locally. The continuous Galerkin FEM is globally conservative, while its local conservation property is a matter of definition [108, 113]. Conservation may be lost if inexact quadrature rules or nonstandard approximations are employed. Any modification of a conservative scheme is potentially harmful and must be examined critically.

2.4.5 ▪ Boundedness

In many applications, the transported quantity is known to lie within certain bounds. For example, volume fractions and concentrations are always bounded by 0 and 1. The exact solution to the mathematical model satisfies such physically motivated constraints by the maximum principle and by the principle of positivity preservation (see Section 2.2.6). However, a numerical approximation may produce undershoots and overshoots, which cannot be tolerated in many cases. It is a dangerous practice to eliminate nonphysical solution values using pointwise corrections since this may result in the loss of conservation property. Instead, the numerical scheme should be designed to produce bounded solutions.

The discretization of a steady or unsteady transport equation to be solved subject to Dirichlet boundary conditions (2.7) prescribed on Γ_D can be written in the form [137]

$$Au = Bg + f, \tag{2.64}$$

where u is the vector of nodal values, f is the source term, and g is the vector of previously calculated solution values and prescribed boundary values. The number of equations in the extended system equals the total number of nodes (including those on the Dirichlet boundary). A boundary condition of the form $u_i = g_i$ is enforced as a constraint by setting $f_i = 0$ and overwriting the ith rows of A and B by the ith row of the identity matrix.

The above discretization satisfies the *discrete maximum principle* if

$$f \le 0 \quad \Rightarrow \quad u \le \max g, \qquad f \ge 0 \quad \Rightarrow \quad u \ge \min g \tag{2.65}$$

and the principle of *discrete positivity preservation* if

$$g \ge 0 \quad \Rightarrow \quad u \ge 0, \qquad g \le 0 \quad \Rightarrow \quad u \le 0.$$

To prove the latter property, it is sufficient to show that $A^{-1} \ge 0$ and $B \ge 0$ componentwise. The discrete maximum principle holds if, additionally, the row sums of A and B are equal [135]. The only condition which cannot be verified directly is the *monotonicity constraint* $A^{-1} \ge 0$. A typical proof of monotonicity is based on the following sufficient conditions [238]:

- A is strictly or irreducibly diagonally dominant;
- all diagonal entries of A are strictly positive;
- all off-diagonal entries of A are nonnegative.

A matrix with these properties is called an *M-matrix*. Unfortunately, the above conditions are very restrictive, especially in the case of high-order approximations and general meshes. Some low-order approximations are guaranteed to produce matrices with the desired properties, but their accuracy is very poor on coarse meshes. High-order schemes perform well for smooth data but tend to produce undershoots/overshoots around steep fronts.

A conservative, bounded, and convergent high-order scheme may be difficult to design and expensive to run but the results are usually rewarding. Indeed, if the total amount of the conserved quantity is correct, if its distribution is accurate, and if all relevant constraints are satisfied, then the numerical solution must be a good approximation to the exact one.

2.4.6 ▪ Causality

The nature of the physical and mathematical problem may impose further constraints. In some applications, it is important to ensure that information travels in the right direction (downstream and forward in time) and at the right speed. This design criterion is known as *causality* [182]. Some algorithms transmit information too far or fast; others produce unrealistic patterns of wave propagation. To prevent such side effects, the difference between the analytical and numerical domains of dependence should be as small as possible.

2.4.7 ▪ Efficiency

The overall efficiency of the simulation software depends on the properties of the algebraic system, on the way in which this system is solved, and on the practical implementation of numerical algorithms. The best choice of the numerical method is always problem-dependent. For example, an evolutionary process can be simulated efficiently using an explicit scheme, but the use of large pseudo-time steps in steady-state computations calls for the use of an unconditionally stable implicit algorithm. The cost-accuracy ratio also depends on the geometric properties of the mesh, on the local mesh size, and on the time step. In addition to robust and accurate discretization techniques, one needs a feedback mechanism for a posteriori error estimation. In adaptive methods, error estimators are used to adjust the mesh, the time step, or the order of numerical approximations automatically in the process of simulation. Adaptive mesh refinement is a particularly useful tool for improving the accuracy of numerical solutions and reducing the computational cost.

The optimization of iterative solvers for linear and nonlinear systems involves the design of Newton-like methods, convergence acceleration techniques, problem-specific preconditioners, monolithic multigrid solvers, and domain decomposition methods.

In addition to the above mathematical challenges, the growing demands of the CFD industry require a further investment in the development of hardware-oriented implementation techniques for modern computer architectures. The main bottleneck to achieving high performance is not the actual data processing but slow memory access. For this reason, the actual MFLOP/s rates are typically very low compared to the theoretical peak performance. A major gain of efficiency can be achieved, for example, by using cache-based implementation techniques and exploiting the tensor product structure of stencils for block-structured grids. Such a hardware-oriented approach may yield an overall speedup factor of up to 1000 even on a single processor [231, 232]. On top of that, the use of optimal parallelization strategies [163] may boost the performance of the code by further orders of magnitude.

In light of the above, the key to achieving optimal performance lies in shifting the distribution of CPU times from costly memory access tasks (assembly of matrices / right-hand sides / residuals, adaptive mesh refinement / coarsening) toward more arithmetic-intensive work (solution of sparse linear systems). High-performance computing techniques based on this philosophy are already available and prove remarkably efficient [232]. In recent years, graphics processing units (GPUs) have become a popular tool for scientific computing. In fact, the practical implementation of a numerical algorithm may be as important as the choice of its mathematical components. Hence, the methods of scientific computing will continue to evolve following the technological trends in computer architecture.

2.5 ▪ Software Libraries

Numerical methods for scalar transport equations can be implemented from scratch using MATLAB or a high-level programming language like Fortran, C++, or Java. The implementation of more sophisticated CFD models to be presented in the next chapter is a time-consuming task which requires excellent programming skills. The process of software development can be accelerated by using existing commercial or research codes. Some of the most popular software libraries for general PDE-based models and various CFD applications are listed in Table 2.3. The companion software to this guidebook is the open source library *Elmer* developed at the Finnish CSC-IT Center for Science. *Elmer*

Table 2.3. *Commercial (C) and open-source (O) software libraries based on finite volume (FV) and finite element (FE) discretizations (in alphabetical order).*

	Type	URL
ALBERTA	O, FE	http://www.alberta-fem.de
ANSYS CFX / Fluent	C, FV	http://www.ansys.com
ANSYS Polyflow	C, FE	http://www.ansys.com
Clawpack	O, FV	http://www.clawpack.org
COMSOL Multiphysics	C, FE	http://www.comsol.com
deal.II	O, FE	http://www.dealii.org
Diffpack	C, FE	http://www.diffpack.com
DUNE	O, FE	http://www.dune-project.org
Hermes	O, FE	http://hpfem.org/hermes
Elmer	O, FE	http://www.csc.fi/elmer
FeatFlow	O, FE	http://www.featflow.de
FEniCS	O, FE	http://fenicsproject.org
FreeFem++	O, FE	http://www.freefem.org/ff++
OpenFOAM	O, FV	http://www.openfoam.com
STAR-CD	C, FV	http://www.cd-adapco.com

provides a general-purpose finite element toolbox for solving PDEs as well as state-of-the-art numerical algorithms for advanced applications in the fields of fluid dynamics, structural mechanics, electromagnetics, heat transfer, and acoustics. The next section presents the functionality of *Elmer* and describes selected features in some detail.

2.6 ▪ Introduction to *Elmer*

The process of computer simulation by using finite element software is traditionally based on the use of three software components. A preprocessor component is mainly utilized to create a mesh that enables the construction of spatial discretization over the continuum. Based on additional PDE model data to be specified by the analyst in conjunction with preprocessing, an actual computational PDE model is then created by using a software component which is referred to as the (finite element) solver. This program also carries out the actual computational work, i.e., updates the numerical solution via the time-stepping algorithm, nonlinear iterations, and linear solves needed in these actions. Finally a third software component, called postprocessor, is used for analyzing the solver output and displaying the computational solution in graphical form (visualization).

2.6.1 ▪ Software Components and Workflow

The design of the *Elmer* software package is based on the conventional three-component division of the finite element modeling process. The key part of the *Elmer* package is its solver program, which can also be used independently. The workflow relating to the usage of *Elmer* can thus be adapted such that other software may also be used for preprocessing and postprocessing tasks. The greatest versatility of modeling options is in fact available when this mode of usage is adopted. However, a beginner may well start learning the usage by employing the native software components provided by the *Elmer* package. Then the graphical user interface program *ElmerGUI* contained in the *Elmer* package offers modern ways to unite the entire workflow within a single program.

In connection with *Elmer*, the desired output of the preprocessing phase is principally a collection of data files that define a mesh in a specific manner. The *Elmer* package contains an independent program, having the name *ElmerGrid*, which can be used to convert other mesh files into the *Elmer* format. In addition, the *ElmerGrid* program can create *Elmer* mesh files directly in the case of simple geometries which can be discretized using structured meshes. This option offers a quick way to start learning *Elmer* usage without a need to import meshes generated with other preprocessing programs.

When suitable mesh files are available, the execution of the solver program of *Elmer* can be controlled simply by providing a special text file which contains sets of commands. If the graphical user interface program of the *Elmer* package is utilized, this data file, which is referred to as the solver input file, may be generated automatically based on a collection of model specifications given via the *ElmerGUI* program. Learning the syntax of the solver input file is not difficult, so a model specification can easily be altered afterwards by using some standard text editor. Alternatively, a solver input file can also be created directly by using just a text editor. *Elmer* is delivered with a large selection of built-in example cases which may then give a useful starting point for creating input data for the solver. Some tutorials are also included in Chapter 5 of this textbook.

The usual output of the *Elmer* solver is a result file that can be read by a postprocessor program. The postprocessor of the *Elmer* package is called *ElmerPost*. It is expected that using *ElmerPost* is enough for beginner's needs. The *Elmer* solver can also create output in some other file formats. For example, visualizing the results with *ParaView* is possible.

The *Elmer* software is being actively updated with the current efforts devoted mostly to the development of the *Elmer* solver. Hence the solver program is by no doubt the most advanced part of the package. Its relative merits are, for example, related to supporting computations in a parallel computer, handling interactions of several PDE models (occurring in multiphysics applications), and adaptability via modifying the source code. While the *Elmer* package contains all necessary components for performing the process of finite element simulation, advanced users are likely to adapt their workflows eventually for other software to be employed in combination with the solver of *Elmer*, especially for preprocessing purposes. Furthermore, understanding the solver input file syntax is typically required so that advanced features of the solver, which may not be accessible simply by using the current version of the graphical user interface program, can be utilized.

The principal design paradigm of the *Elmer* solver is modularity. Consequently, the current version of *Elmer* offers a large selection of physical models which can be combined flexibly to produce advanced computational models describing interactions of multiple physical phenomena in order to meet the modern needs of multiphysics simulations. The implemented basic models include equations which are related, for example, to fluid mechanics, structural mechanics, acoustics, electromagnetism, heat transfer, and species transport. In addition to utilizing these built-in models, the user may also employ a wide collection of numerical tools and algorithms contained in *Elmer* to create completely new models which can then be interfaced easily with the existing models.

2.6.2 ▪ Obtaining *Elmer* Software and Documentation

Elmer is being actively developed, mainly by CSC-IT Center for Science in Finland, and the newest version of the software source code, which is maintained under Subversion version control system, may be obtained via the project repository site [63], where the executable files for Windows operating systems are also provided. To obtain the latest revision, the executable files must be compiled from the source codes. To this end, the user is supplied with accessory configuration management aids and automated tests which help in compiling the software from the source codes and testing the executables.

The available documentation of the software and compiling instructions are best accessed via the software's main web site [62]. Additional references such as links to the discussion forum may also be found there. The following list describes the most important parts of the software documentation that are available at the software's main web site.

"Overview of Elmer" This document gives an overview of the *Elmer* software package presenting general information on the capabilities of the software, its usage, and the way in which the available features are organized.

"ElmerGUI Manual" This manual gives a detailed description of the graphical user interface of the *Elmer* package. Walk-through tutorials and some technical details are also included.

"Elmer GUI Tutorials" This document describes example problems which are handled by using the graphical user interface. It contains walk-through cases that teach the user to choose the right program menus and to write the right values.

"ElmerGrid Manual" This document is the manual of the *ElmerGrid* program. The examples related to the *ElmerGrid* documentation are provided in an additional file which can be found on the same site where the manuals are located.

"ElmerSolver Manual" This manual gives an overview of the general capabilities of the solver, focusing on the utilities that are of common use to several physical models. It follows from this organization of material that information specific to a certain physical PDE model is not included in this manual.

"Elmer Models Manual" The *Elmer* Models Manual describes the variety of physical models which the solver of *Elmer* can handle. The specific options for controlling the corresponding solution procedures are also documented. In addition, this manual describes certain utilities for other tasks, such as the computation of derived quantities from the primary solution.

Additional documents and related material are provided on the *Elmer* software web site. They may be particularly useful for advanced users.

2.6.3 ▪ Solution Procedures: Basics

Computational techniques have traditionally been designed for individual models which had originally emerged within a single branch of science or engineering. Therefore, different codes have been developed independently for different fields of study, such as fluid mechanics and structural mechanics. However, modern applications of computational simulation are increasingly multidisciplinary and have led to the concepts of multiphysics and coupled problems. A defining factor behind these concepts is that the principal intention is to model the interaction of several (physical) phenomena that occur simultaneously.

In order to fully understand the way in which the solver of *Elmer* operates and is controlled, it must be said that *Elmer* has been particularly developed with multiphysics simulations in mind. That is, its applicability is by no means limited to the conventional type of CFD models. In the following, we shall describe some basic concepts relating to simulations with *Elmer*. The standard solution procedure which the solver of *Elmer* uses to handle the discrete version of a coupled problem is also introduced. However, it is expected that a beginner is likely to start with applications of *Elmer* to single-physics models, and the full utilization of multimodel concepts may be unnecessary at this stage.

The models handled by *Elmer* may generally be stationary or evolutionary, with nonlinearities possible in both cases. Starting from a weak formulation of the governing equations, a computational version of the model is typically constructed using a finite element approximation and an implicit time-stepping method (for evolutionary problems). In the simplest case of single-physics models, we end up solving systems of the form

$$F(u) = 0, \tag{2.66}$$

where u represents either the vector of coefficients in the finite element expansion of the stationary solution or the coefficient vector containing the nodal values of the time-dependent finite element solution at a given time $t = t^n$. Thus, in the evolutionary case, the solution is advanced in time by solving problems of type (2.66) repeatedly.

For linear models, problem (2.66) reduces to solving a linear system with residual

$$F(u) = b - Au,$$

where the coefficient matrix A is often referred to as the stiffness matrix, and b corresponds to the right-hand-side vector in the linear system. Otherwise F is a nonlinear mapping, and an iterative method is needed to obtain the solution of problem (2.66). The nonlinear iteration methods available in *Elmer* generally depend on the model, as the definition of the linearization strategy is a part of the computational description of each physical model.

It is worth mentioning that many single-physics models offer the possibility of using the Newton method. At each iteration step, an approximation $u^{(m)}$ to the solution u of the nonlinear system is given, and the next iterate is calculated as follows:

$$\begin{aligned} DF(u^{(m)})[\delta u^{(m)}] &= -F(u^{(m)}), \\ u^{(m+1)} &= u^{(m)} + \delta u^{(m)}, \end{aligned} \tag{2.67}$$

where $DF(u^{(m)})$ is the derivative (Jacobian matrix) of F at $u^{(m)}$. We note that the preferred way to obtain the derivative is to appeal directly to the definition. By the Taylor theorem, the derivative matrix defines a linear transformation given by the formula

$$F(u+v) = F(u) + DF(u)[v] + o(v), \tag{2.68}$$

with $o(v)$ indicating terms that approach zero faster than v. It is noted that the nonlinear solution update involves the assembly of $DF(u^{(m)})$ and solution of a linear system at each iteration step. As an alternative to the Newton method, linearization strategies based on lagged-value approximations are also often available. In addition, relaxation is conventionally offered as a way to enable convergence in situations when the basic nonlinear iteration fails to produce convergence. Given the current nonlinear iterate $u^{(m)}$ and a computed correction $\delta u^{(m)}$ to the approximation, the new nonlinear iterate is then defined by

$$u^{(m+1)} = u^{(m)} + \lambda^{(m)} \delta u^{(m)},$$

where $\lambda^{(m)}$ is an adjustable parameter referred to as the relaxation parameter.

Having introduced the basic concepts in the context of single-physics models, we now proceed to describe how the modularity employed in the design of *Elmer* allows us to create models which represent interactions of multiple (physical) phenomena. To this end, we assume that the complete model describes an interaction of N constituent models, the computational versions of which lead to algebraic systems for the coefficient vectors u_i, $i = 1, 2, \ldots, N$. As before, the components of u_i are usually associated with the finite element expansion of a stationary solution to the corresponding subproblem or the nodal values of an evolutionary solution at a time level $t = t^n$.

The fully discrete version of the coupled model leads to a problem of the form

$$\begin{aligned} F_1(u_1, u_2, \ldots, u_N) &= 0, \\ F_2(u_1, u_2, \ldots, u_N) &= 0, \\ &\cdots \\ F_N(u_1, u_2, \ldots, u_N) &= 0. \end{aligned} \tag{2.69}$$

If all constituent models are linear, problem (2.69) corresponds to a linear system where the coefficient matrix is an $N \times N$ block matrix. Otherwise (2.69) describes a nonlinear problem. Although (2.69) could in principle be solved in the same way as explained in the context of single-physics models, i.e., by performing either a coupled linear solve or a Newton iteration, coupled problems of this form are usually handled differently in order to enable the reuse of solvers developed for single-physics models and facilitate extensions of the code to new applications. To this end, the nonlinear Gauss–Seidel iteration is usually employed. Instead of solving the coupled subproblems simultaneously, the components of the iterate $u^{(j)} = (u_1^{(j)}, u_2^{(j)}, \ldots, u_N^{(j)})$ are updated in the following segregated manner:

$$\begin{aligned} F_1(u_1^{(j)}, u_2^{(j-1)}, u_3^{(j-1)}, \ldots, u_N^{(j-1)}) &= 0, \\ F_2(u_1^{(j)}, u_2^{(j)}, u_3^{(j-1)}, \ldots, u_N^{(j-1)}) &= 0, \\ &\cdots \\ F_N(u_1^{(j)}, u_2^{(j)}, \ldots, u_N^{(j)}) &= 0. \end{aligned} \tag{2.70}$$

It is noted that the kth discrete model description in (2.70) depends implicitly only on its primary variable u_k, while the dependencies on the other constituent model variables are treated explicitly. This brings us to solving a nonlinear single-field problem,

$$F(u_k^{(j)}) = F_k(v_1, \ldots, v_{k-1}, u_k^{(j)}, v_{k+1}, \ldots, v_N) = 0, \text{ with all } v_l \text{ given}, \tag{2.71}$$

which is handled using the methods already described in the context of single-physics models. We also note that (2.70) reduces to the block Gauss–Seidel iteration for linear systems if all the constituent models are linear. Relaxation may again be applied as an attempt to improve the convergence behavior of the basic iteration (2.70).

The main advantage of using the nonlinear Gauss–Seidel scheme is its support for the modular software design. Also, it brings us to handling coupled problems via solving linear systems which are smaller than those which would result from treating all constraints in (2.69) simultaneously. Despite these merits, the suitability of the loosely coupled iteration (2.70) is generally case-dependent as convergence problems may occur in cases where a strong coupling is involved. Such problems are often best handled by methods which treat all the constituent models in (2.69) simultaneously. Certain physical models available in *Elmer* indeed employ this alternative strongly coupled solution strategy. However, these models have initially been developed independently, as common high-abstraction *Elmer* utilities for creating strongly coupled iteration methods in a general manner are less developed.

The following pseudocode presentation describes the basic loosely coupled iteration scheme employed by the solver of *Elmer*. This rough description illustrates what needs to be done to create a working computational solution procedure for a coupled problem.

! *The time integration loop*
for $k = 1 : M$
  Generate an initial guess $u^{(0)} = (u_1^{(0)}, u_2^{(0)}, \dots, u_N^{(0)})$ at $t = t^n$
  ! *The nonlinear Gauss–Seidel iteration*
  for $j = 1, 2, \dots$
    ! *Generate the next coupled system iterate* $u^{(j)}$ *by performing single-field updates*
    for $i = 1 : N$
      Set $v_l = u_l^{(j)}$ for $l = 1, 2, \dots, i-1$
      Set $v_l = u_l^{(j-1)}$ for $l = i+1, \dots, N$
      Perform the nonlinear solve of $F_i(v_1, \dots, v_{i-1}, u_i^{(j)}, v_{i+1}, \dots, v_N) = 0$
      Apply a relaxation to set $u_i^{(j)} := u_i^{(j-1)} + \alpha_i (u_i^{(j)} - u_i^{(j-1)})$
    end
  end
end

Here the descriptions of the stopping criteria for the iterative solver have been omitted for brevity. It is also noted that, obviously, the time integration loop is not needed in the case of a stationary problem. On the other hand, the use of a marching method based on implicit pseudo-time integration may be advantageous when it comes to efficient computation of multiple stationary solutions for a range of model parameter values.

2.6.4 • The Key Capabilities of the Solver of *Elmer*

In the following, we shall focus on describing the key capabilities of the solver of *Elmer* which enable the user to create new models and configure the solver for different purposes.

Extendability by Modular Design

A module of the *Elmer* software which makes it possible to create a discrete model description of type (2.71) and update the corresponding solution component keeping the other variables fixed is generally called a solver (this notion of solver does not thus refer to an independent program which we have already called the finite element solver). The solvers of *Elmer* are truly modular in this manner and have a standard interface. Thus, each solver usually contains an implementation of the nonlinear iteration, instructions to assemble the corresponding linear systems from elementwise contributions, and standard subroutine calls to set up constraints and to actually solve the linear systems assembled. It follows that enabling an interaction with another field, designated by v_l in (2.71), is simply a matter of solver-level implementation. Therefore, interactions which have not been implemented yet can be enabled by making modifications which are localized to the solvers. Moreover, a completely new physical model may be added by introducing a new solver which represents a separate software module and can be developed independently from the main program. As a result, applying the loosely coupled solution procedure to a coupled problem based on the new physical model again requires making only solver-level modifications.

Approximation by Various Finite Element Formulations

Elmer has traditionally employed the Galerkin finite element approximation of weak formulations. A standard abstraction of linearized subproblems can usually be given. A typical problem statement is as follows: Find a finite element solution $u_h \in \mathcal{V}_h$ such that

$$a(u_h, v_h) = b(v_h)$$

for any test function $v_h \in \mathcal{W}_h$. In the most typical case, the bilinear form $a : \mathcal{V} \times \mathcal{W} \to \mathbb{R}$ and the linear functional $b : \mathcal{W} \to \mathbb{R}$ are well-defined when

$$\mathcal{V} = \mathcal{W} = H^1(\Omega),$$

where Ω denotes the domain where the equation is posed and $H^1(\Omega)$ is the space of square integrable functions whose first derivatives are also all square integrable in Ω. Traditionally, the Lagrange interpolation basis functions defined for various element shapes have been used to construct the finite-dimensional spaces $\mathcal{V}_h$ and $\mathcal{W}_h$. In this connection, the piecewise polynomial approximation of degree $1 \leq p \leq 3$ is possible for 2D domains, while 3D domains may be discretized by using elements of degree $1 \leq p \leq 2$. Curved element shapes can also be generated with these traditional elements.

Discrete models based on more recent versions of the Galerkin finite element approximation are also possible. As an alternative to the standard Lagrange interpolation basis functions, the Galerkin approximation based on hierarchic high-degree piecewise polynomials can be employed. In this case, the degree of polynomial approximation can also be defined elementwise, so that, in effect, the use of the hp-version of the FEM is enabled. We note that *Elmer* provides a built-in mechanism to guarantee the continuity of such solutions. The H^1-regularity of discrete solutions is thus ensured. However, generic ways to describe curved surfaces accurately in connection with high-degree finite elements have not been implemented yet, which may limit the current utility of these elements. Anyhow, discretizations that capture localized solution details in the interior of the domain can generally be created without addressing the question of the geometry representation.

The way to define a high-degree approximation is based on the idea that a background mesh for representing the standard lowest-degree continuous finite element expansion is first provided, so that a specific element-type definition in relation to elements present in the background mesh may then be given in order to enrich the approximation. The same idea has been adopted to create other alternative finite element formulations. For example, finite element formulations which enrich the approximation defined on the background mesh by a subscale approximation spanned by elementwise bubble basis functions can be obtained in this way. We note that this strategy is widely used in *Elmer* to stabilize otherwise unstable formulations. Another example of the use of a user-supplied element definition relates to creating approximations based on the discontinuous Galerkin method.

As a final example, we mention that it is also possible to enrich the approximation on the background mesh by introducing additional degrees of freedom associated with either faces or edges of elements and then omit the original nodal degrees of freedom. This approach leads to a suitable set of unknowns for creating discretizations based on the face or edge element interpolation. If $\Omega \subset \mathbb{R}^d$ and $L_2(\Omega)$ is the set of square integrable scalar functions, the space of square integrable vector fields is denoted by $L_2(\Omega)^d$. Instead of interpolating each component of a vector using scalar basis functions, one may design bases for approximating vector fields in finite-dimensional versions of the spaces

$$X = H(\mathrm{div}, \Omega) = \{\mathbf{v} \in L_2(\Omega)^d \mid \nabla \cdot \mathbf{v} \in L_2(\Omega)\}$$

or

$$X = H(\mathrm{curl}, \Omega) = \{\mathbf{v} \in L_2(\Omega)^d \mid \nabla \times \mathbf{v} \in L_2(\Omega)^d\},$$

where $\nabla \times \mathbf{v}$ is the curl of $\mathbf{v}$. A physical motivation for using these finite element spaces is that vector fields with only either normal or tangential continuity on interelement boundaries can then be approximated in a natural and realistic manner.

Model-Specific Finite Element Approximation

In the most basic setting, all constituent model variables u_i, $i = 1, \dots, N$, of a coupled problem are approximated by using the same piecewise polynomial basis functions defined on a single mesh. In addition, the solver of *Elmer* offers a built-in functionality to perform a coupled problem simulation using solver-specific finite element meshes. The solver description is then augmented by the specification of the independent mesh to be used by the solver. To make this functional in connection with the solution of coupled problems, *Elmer* has the capability of performing data transfer which is needed to exchange information between the solvers in the loosely coupled solution procedure, even when the meshes are nonmatching. It must be understood, however, that the loss of fine-scale details is unavoidable when a high-resolution approximation is represented using a coarser finite element mesh.

Monolithic Discretizations

In the case of a strongly coupled problem, the standard segregated solution procedure of *Elmer* may become ineffective due to the need for using small values of relaxation parameters to avoid the divergence of the iteration. Using monolithic discretizations, i.e., handling all constituent components of the solution simultaneously, may then have a relative merit, at least in terms of obtaining robustness. Monolithic discretizations may also be created using *Elmer* as, in principle, it does not pose any restriction on how much physics may be included into a solver module definition. Some basic physical models, such as the system of compressible Navier–Stokes equations of fluid mechanics, may indeed be thought of as intrinsically multiphysical, and handling them in this way may actually appear as the most natural way. It should be noted, however, that high-level abstractions relating to implementing this alternative solution strategy do not necessarily exist, and the design of an efficient solver for the resulting linear systems is likely to become an additional burden. We shall return to this issue when discussing the linear algebra capabilities below.

Discretization in Time

For handling evolutionary problems, the solver of *Elmer* offers implementations of many standard time-stepping algorithms for first-order and second-order problems. The usage of these time integration routines is done in a standardized manner so that just a few modifications of an existing stationary solver suffice to enable transient simulations.

Linear Algebra Tools

The ability to solve large linear systems efficiently is a central aspect of the simulation process with *Elmer*. As already explained, in the basic setting a linear solve is needed to perform the solution update at each step of the nonlinear iteration. In practice, linear solves are usually done iteratively, which adds an extra iteration level to the pseudocode presentation at the end of Section 2.6.3.

The solver of *Elmer* offers a large selection of strategies for configuration of linear solvers. The majority of them are directly implemented into *Elmer* software, but interfaces to external linear algebra libraries are also available. When it comes to the use of linear solvers, the most critical decision typically relates to identifying an effective preconditioning strategy for the linear system at hand. Traditionally, *Elmer* has employed generic preconditioning strategies based on the fully algebraic approach. Highly efficient alternatives to these standard preconditioners may also be obtained by using a two-level iteration procedure where the preconditioner is derived by applying multigrid methods.

If the linear system at hand stems from a monolithic discretization of a coupled problem, the use of basic options in the linear solver may become a hindrance since standard preconditioners may not be effective. A coupled linear system of this kind usually has a typical block structure, whence standard segregation strategies may be employed as natural preconditioners for the iterative solution procedure. The utilities provided by *Elmer* have been used to generate sophisticated block-preconditioned linear solvers for specific models [172, 173], but an encapsulation of common abstraction in certain versions of these solvers has also been pursued recently. We conclude that although the monolithic discretization may ultimately be the best way of handling some strongly coupled problems, the numerical treatment of the resulting linear systems often needs to be addressed at the early stages of program development when it comes to efficient implementation of the model.

Parallel Computations

One of the major strengths of the solver of *Elmer* is its parallel computing capability. The employed message-passing library approach is based on the Message Passing Interface (MPI) standard. The parallelization of the code significantly widens the range of problem sizes which can be considered.

After a parallel version of the *Elmer* solver has been made available in the given parallel computing environment, the main challenge is the efficient distribution of work between the central processing units (they may be cores in the case of modern multicore computer architectures) to be used in the simulation run. This task is accomplished using domain decomposition methods to generate subproblems to be solved on each processor/core. In practice, this decomposition is created prior to running the *Elmer* solver by partitioning the mesh files accordingly, so that each computing unit may primarily work on its own data as long as the associated piece of computation does not necessitate communication between the units. The *Elmer* package offers preprocessing tools for performing the mesh partitioning, including the possibility of utilizing the external METIS library. It also provides postprocessing tools for merging parallel simulation results stored in separate files.

Chapter 3

Equations of Fluid Dynamics and Heat Transfer

3.1 ▪ Conservation Laws

The governing equations of heat transfer and fluid dynamics are derived from *conservation laws* and provide a mathematical description of these laws. In physics, a certain quantity is said to be conserved if the total amount of this quantity does not change in the absence of external supply. Examples of conserved quantities include mass, momentum, and energy. The volume of an incompressible fluid is also conserved. In order to derive the mathematical equations that describe the evolution of a conserved fluid property, the conservation principle must be applied to the fluid confined within a closed *control volume*. Each control volume represents a small region of space, and a local balance equation of the form

$$\textit{time rate of change} = (\textit{inflow} - \textit{outflow}) + (\textit{production} - \textit{consumption})$$

can be formulated for each quantity of interest. The use of small but finite control volumes leads to equations containing surface and volume integrals. The use of infinitesimally small control volumes leads to PDEs which embody the same conservation principles but are only valid under additional regularity assumptions. In particular, integral conservation laws may have discontinuous solutions, whereas solutions to PDE models must be continuous for the partial derivatives to exist in the classical sense.

In addition to the choice between finite and infinitesimal control volumes, one has a choice between fixed and moving ones. The fixed control volume approach leads to the *Eulerian form* of the governing equations. This flow model corresponds to the viewpoint of a standing observer watching the changes of fluid properties in a fixed region of space. The moving control volume approach leads to the equivalent *Lagrangian form* which corresponds to the viewpoint of an observer moving with the velocity of the fluid. While the ambiguity in the definition of control volumes may seem confusing, it is an easy exercise to show that the resulting models are actually the same. In this chapter, we will derive the Navier–Stokes equations of fluid dynamics using the moving finite and fixed infinitesimal control volume models. For a detailed presentation of other models and a proof of equivalence, we refer the reader to Anderson [3].

3.1.1 ▪ Conservation of Mass

The mass of the fluid contained in a control volume $\Omega \subset \mathbb{R}^3$ is given by the volume integral

$$\int_\Omega \rho \, d\mathbf{x},$$

where $\rho = \rho(\mathbf{x}, t)$ is the density of the fluid at the point $\mathbf{x} = (x_1, x_2, x_3)^T$ and time $t \geq 0$.

Adopting the Lagrangian model of fluid motion, we consider a finite volume of fluid $\Omega(t)$ transported by a given velocity field $\mathbf{u} = (u_1, u_2, u_3)^T$. The shape of $\Omega(t)$ may change as time evolves, but the total mass remains unchanged because the moving control volume is always composed of the same molecules. By the law of mass conservation, we have

$$\frac{d}{dt} \int_{\Omega(t)} \rho \, d\mathbf{x} = 0.$$

By the Reynolds transport theorem (1.2) and Green's formula (1.3), this gives

$$\begin{aligned} 0 &= \int_{\Omega(t)} \frac{\partial \rho}{\partial t} \, d\mathbf{x} + \int_{\partial\Omega(t)} \rho \mathbf{u} \cdot \mathbf{n} \, ds \\ &= \int_{\Omega(t)} \frac{\partial \rho}{\partial t} \, d\mathbf{x} + \int_{\Omega(t)} (\mathbf{u} \cdot \nabla \rho + \rho \nabla \cdot \mathbf{u}) \, d\mathbf{x}, \\ &= \int_{\Omega(t)} \left(\frac{D\rho}{Dt} + \rho \nabla \cdot \mathbf{u} \right) d\mathbf{x}. \end{aligned}$$

The substantial derivative

$$\frac{D\rho}{Dt} = \frac{\partial \rho}{\partial t} + \mathbf{u} \cdot \nabla \rho$$

is defined by (1.1). In contrast to the local time derivative $\frac{\partial}{\partial t}$, which represents the rate of change at a fixed point in space, the physical meaning of $\frac{D}{Dt}$ is the time rate of change following a fluid element that moves through space with velocity $\mathbf{u}$.

Since Ω is an arbitrary control volume, we obtain the differential conservation law

$$\frac{D\rho}{Dt} + \rho \nabla \cdot \mathbf{u} = 0. \tag{3.1}$$

By the above definition of the substantial derivative and the vector identity

$$\nabla \cdot (\rho \mathbf{u}) = \mathbf{u} \cdot \nabla \rho + \rho \nabla \cdot \mathbf{u},$$

the Lagrangian form (3.1) of the mass balance is equivalent to the Eulerian form

$$\frac{\partial \rho}{\partial t} + \nabla \cdot (\rho \mathbf{u}) = 0. \tag{3.2}$$

This first-order hyperbolic PDE is known as the *continuity equation* of fluid dynamics. It can also be derived directly from the mass balance for a fixed infinitesimal control volume. The dimensions of the box Ω depicted in Fig. 3.1 are given by Δx_1, Δx_2, and Δx_3. The box is fixed and has the constant volume $V = \Delta x_1 \Delta x_2 \Delta x_3$. The control surface $\partial\Omega$ consists of six rectangular faces. Since the box is infinitesimally small, all properties of the fluid (including density and velocity) are assumed to have constant values on each face.

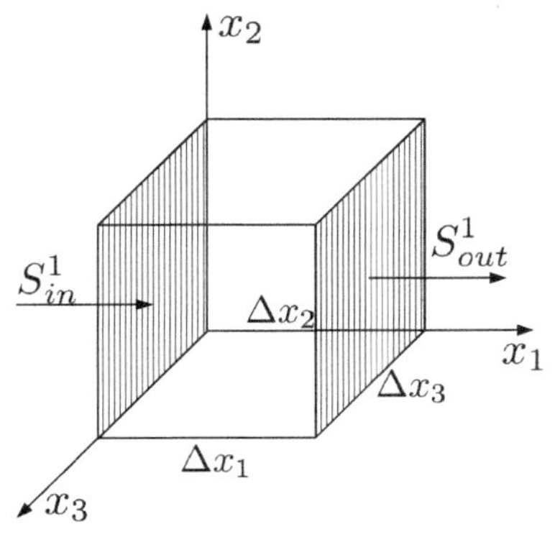

Figure 3.1. *A fixed infinitesimal volume of fluid.*

The values of a fluid property f on two opposite faces are related by the Taylor formula

$$f(x+\Delta x)=f(x)+\frac{\partial f}{\partial x}(x)\Delta x+\frac{1}{2!}\frac{\partial^2 f}{\partial x^2}(x)(\Delta x)^2+\cdots$$
$$\approx f(x)+\frac{\partial f}{\partial x}(x)\Delta x. \tag{3.3}$$

The mass flux across a surface element is defined as the product of density, velocity, and surface area. Consider the left shaded side S^1_{in} of the box depicted in Fig. 3.1. Since the density ρ and velocity component u_1 are constant on this face, the corresponding mass flux, i.e., the amount of mass flowing into the box across S^1_{in} per unit time, is

$$\rho u_1 \Delta x_2 \Delta x_3.$$

Using the above Taylor formula to determine the constant value of $f=\rho u_1$ on the opposite face S^1_{out}, we find that the amount of mass flowing out of the box across this face is

$$\left(\rho u_1+\frac{\partial(\rho u_1)}{\partial x_1}\Delta x_1\right)\Delta x_2\Delta x_3.$$

Hence, the difference between the outgoing and incoming mass fluxes is given by

$$\frac{\partial(\rho u_1)}{\partial x_1}\Delta x_1\Delta x_2\Delta x_3=\frac{\partial(\rho u_1)}{\partial x_1}V.$$

Similarly, the time rate of change due to mass fluxes across the other four faces is

$$\frac{\partial(\rho u_2)}{\partial x_2}V+\frac{\partial(\rho u_3)}{\partial x_3}V.$$

Since the control volume is fixed, the variations of mass ρV and density ρ are related by

$$\frac{\partial}{\partial t}(\rho V)=\frac{\partial \rho}{\partial t}V.$$

This time rate of change should be equal to the difference between the incoming and outgoing fluxes. Summation of the flux differences for the three pairs of faces yields

$$\left(\frac{\partial(\rho u_1)}{\partial x_1}+\frac{\partial(\rho u_2)}{\partial x_2}+\frac{\partial(\rho u_3)}{\partial x_3}\right)V.$$

It follows that density changes due to mass flow across the surface of an infinitesimal control volume centered a fixed point in space are governed by the PDE

$$\left(\frac{\partial \rho}{\partial t}+\frac{\partial(\rho u_1)}{\partial x_1}+\frac{\partial(\rho u_2)}{\partial x_2}+\frac{\partial(\rho u_3)}{\partial x_3}\right)V=0.$$

Dividing this formula by the constant volume V, we obtain the mass conservation law

$$\frac{\partial \rho}{\partial t}+\nabla\cdot(\rho\mathbf{u})=0, \tag{3.4}$$

which is identical to the continuity equation (3.2) derived using a finite control volume.

3.1.2 ▪ Conservation of Momentum

In continuum mechanics, linear momentum is the product of mass and velocity. The total momentum of the fluid contained in a moving control volume $\Omega(t)$ is the vector

$$\int_{\Omega(t)}\rho\mathbf{u}\,\mathrm{d}\mathbf{x}=\left(\int_{\Omega(t)}\rho u_1\,\mathrm{d}\mathbf{x},\int_{\Omega(t)}\rho u_2\,\mathrm{d}\mathbf{x},\int_{\Omega(t)}\rho u_3\,\mathrm{d}\mathbf{x}\right)^T.$$

The evolution of this quantity is governed by Newton's second law. The time rate of change

$$\frac{\mathrm{d}}{\mathrm{d}t}\int_{\Omega(t)}\rho\mathbf{u}\,\mathrm{d}\mathbf{x}$$

is equal to the sum of all external forces acting on the fluid contained in $\Omega(t)$.

The external forces can be classified into *body forces* (e.g., gravitational and electromagnetic forces) and *surface forces* (e.g., pressure and shear stress). Let $\mathbf{g}=(g_1,g_2,g_3)^T$ denote the body force per unit mass. The total body force acting on the fluid in $\Omega(t)$ is

$$\int_{\Omega(t)}\rho\mathbf{g}\,\mathrm{d}\mathbf{x}.$$

The contribution of surface forces is commonly defined in terms of the *Cauchy stress tensor* $\bar{\bar{\boldsymbol{\sigma}}}$ such that $\bar{\bar{\boldsymbol{\sigma}}}\cdot\mathbf{n}$ is the stress vector parallel to the unit normal vector $\mathbf{n}$. The total surface force exerted on the fluid inside $\Omega(t)$ is given by the integral

$$\int_{\partial\Omega(t)}\bar{\bar{\boldsymbol{\sigma}}}\cdot\mathbf{n}\,\mathrm{d}s,$$

where $\mathbf{n}$ denotes the unit outward normal vector to the evolving control surface $\partial\Omega(t)$.

Thus, conservation of momentum in $\Omega(t)$ is described by the integral equation

$$\frac{\mathrm{d}}{\mathrm{d}t}\int_{\Omega(t)}\rho\mathbf{u}\,\mathrm{d}\mathbf{x}=\int_{\Omega(t)}\rho\mathbf{g}\,\mathrm{d}\mathbf{x}+\int_{\partial\Omega(t)}\bar{\bar{\boldsymbol{\sigma}}}\cdot\mathbf{n}\,\mathrm{d}s.$$

The componentwise form of this formula represents a system of integral conservation laws for the three components of the vector field $\rho\mathbf{u}=(\rho u_1,\rho u_2,\rho u_3)^T$. We have

$$\frac{\mathrm{d}}{\mathrm{d}t}\int_{\Omega(t)}\rho u_k\,\mathrm{d}\mathbf{x}=\int_{\Omega(t)}\rho g_k\,\mathrm{d}\mathbf{x}+\int_{\partial\Omega(t)}\boldsymbol{\sigma}_k\cdot\mathbf{n}\,\mathrm{d}s,\qquad k=1,2,3, \tag{3.5}$$

where $\boldsymbol{\sigma}_k=(\sigma_{1k},\sigma_{2k},\sigma_{3k})^T$ denotes the kth column of the Cauchy stress tensor $\bar{\bar{\boldsymbol{\sigma}}}$.

By the Reynolds transport theorem, the left-hand side of (3.5) can be written as

$$\begin{aligned}\frac{\mathrm{d}}{\mathrm{d}t}\int_{\Omega(t)}\rho u_k\,\mathrm{d}\mathbf{x} &= \int_{\Omega(t)}\left(\frac{D(\rho u_k)}{Dt}+(\rho u_k)(\nabla\cdot\mathbf{u})\right)\mathrm{d}\mathbf{x}\\ &= \int_{\Omega(t)}\left(u_k\left(\frac{D\rho}{Dt}+\rho\nabla\cdot\mathbf{u}\right)+\rho\frac{Du_k}{Dt}\right)\mathrm{d}\mathbf{x}\\ &= \int_{\Omega(t)}\rho\frac{Du_k}{Dt}\,\mathrm{d}\mathbf{x}, \qquad k=1,2,3,\end{aligned}$$

where we used the law of mass conservation (3.1). The divergence theorem yields

$$\begin{aligned}\int_{\partial\Omega(t)}\bar{\bar{\boldsymbol{\sigma}}}\cdot\mathbf{n}\,\mathrm{d}s &= \int_{\partial\Omega(t)}(\boldsymbol{\sigma}_1\cdot\mathbf{n},\boldsymbol{\sigma}_2\cdot\mathbf{n},\boldsymbol{\sigma}_3\cdot\mathbf{n})^T\,\mathrm{d}s\\ &= \int_{\Omega(t)}(\nabla\cdot\boldsymbol{\sigma}_1,\nabla\cdot\boldsymbol{\sigma}_2,\nabla\cdot\boldsymbol{\sigma}_3)^T\,\mathrm{d}\mathbf{x}\\ &= \int_{\Omega(t)}\nabla\cdot\bar{\bar{\boldsymbol{\sigma}}}\,\mathrm{d}\mathbf{x}.\end{aligned}$$

After these transformations, the componentwise form of the momentum balance becomes

$$\int_{\Omega(t)}\left(\rho\frac{Du_k}{Dt}-\rho g_k-\nabla\cdot\boldsymbol{\sigma}_k\right)\mathrm{d}\mathbf{x}=0, \qquad k=1,2,3.$$

Since the choice of the control volume $\Omega(t)$ is arbitrary, the expression in parentheses must vanish pointwise. Thus, the differential form of the momentum balance is given by

$$\rho\frac{Du_k}{Dt}=\rho g_k+\nabla\cdot\boldsymbol{\sigma}_k, \qquad k=1,2,3.$$

The representation of inertia in terms of substantial derivatives corresponds to the Lagrangian form of the equations of motion. For brevity, we will frequently use the vector notation

$$\rho\frac{D\mathbf{u}}{Dt}=\rho\mathbf{g}+\nabla\cdot\bar{\bar{\boldsymbol{\sigma}}}. \tag{3.6}$$

By definition of the substantial derivative, the Eulerian componentwise form reads

$$\rho\frac{\partial u_k}{\partial t}+\rho\left(u_1\frac{\partial u_k}{\partial x_1}+u_2\frac{\partial u_k}{\partial x_2}+u_3\frac{\partial u_k}{\partial x_3}\right)=\rho g_k+\frac{\partial\sigma_{1k}}{\partial x_1}+\frac{\partial\sigma_{2k}}{\partial x_2}+\frac{\partial\sigma_{3k}}{\partial x_3}$$

for $k=1,2,3$. The Eulerian vector form of the momentum conservation law reads

$$\rho\frac{\partial\mathbf{u}}{\partial t}+\rho\mathbf{u}\cdot\nabla\mathbf{u}=\rho\mathbf{g}+\nabla\cdot\bar{\bar{\boldsymbol{\sigma}}}. \tag{3.7}$$

Let us now apply Newton's second law to the fixed infinitesimal control volume shown in Fig. 3.1. The incoming flux (value · velocity · area) of the conserved quantity ρu_1 due to fluid flow across the rectangular face S^1_{in} perpendicular to the x_1 axis is

$$\rho u_1\cdot u_1\cdot\Delta x_2\Delta x_3=\rho u_1^2\Delta x_2\Delta x_3.$$

By the Taylor expansion (3.3), the outgoing flux across the face S^1_{out} is

$$\left(\rho u_1^2 + \frac{\partial(\rho u_1^2)}{\partial x_1}\Delta x_1\right)\Delta x_2 \Delta x_3,$$

whence the net loss of momentum (per unit time) equals

$$\frac{\partial(\rho u_1^2)}{\partial x_1} V.$$

By similar arguments, the differences between the outgoing and incoming fluxes across the pairs of faces perpendicular to the x_2 and x_3 axes are given by

$$\left(\rho u_1 u_2 + \frac{\partial(\rho u_1 u_2)}{\partial x_2}\Delta x_2\right)\Delta x_1 \Delta x_3 - \rho u_1 u_2 \Delta x_1 \Delta x_3 = \frac{\partial(\rho u_1 u_2)}{\partial x_2} V$$

and

$$\left(\rho u_1 u_3 + \frac{\partial(\rho u_1 u_3)}{\partial x_3}\Delta x_3\right)\Delta x_1 \Delta x_2 - \rho u_1 u_3 \Delta x_1 \Delta x_2 = \frac{\partial(\rho u_1 u_3)}{\partial x_3} V.$$

The total momentum of the fluid in the box is given by $\rho \mathbf{u} V$. We have

$$\frac{\partial}{\partial t}(\rho u_1 V) = \frac{\partial(\rho u_1)}{\partial t} V.$$

Thus, the balance of momentum in the x_1 direction will have the form

$$\frac{\partial(\rho u_1)}{\partial t} + \frac{\partial(\rho u_1^2)}{\partial x_1} + \frac{\partial(\rho u_1 u_2)}{\partial x_2} + \frac{\partial(\rho u_1 u_3)}{\partial x_3} = F_1,$$

where F_1 stands for the sum of all body and surface forces in the x_1 direction.

The x_1 component of the total body force acting on the fluid in the box is

$$g_1 \Delta x_1 \Delta x_2 \Delta x_3 = g_1 V.$$

The stress per unit area of a surface element perpendicular to the x_1 axis is given by the first column of the Cauchy stress tensor $\bar{\bar{\boldsymbol{\sigma}}}$. On the face S^1_{in}, we have the normal stress

$$\sigma_{11} \Delta x_2 \Delta x_3,$$

and on the opposite face S^1_{out} we have

$$\left(\sigma_{11} + \frac{\partial \sigma_{11}}{\partial x_1}\Delta x_1\right)\Delta x_2 \Delta x_3.$$

The out-of-balance force induced by these normal stresses is

$$\frac{\partial \sigma_{11}}{\partial x_1} V.$$

The variations of momentum in the x_1 direction are also affected by shear forces (viscous friction). On the face S^2_{in} which lies in the plane $x_2 \equiv 0$, we have the tangential stress

$$\sigma_{21} \Delta x_1 \Delta x_3,$$

and on the face S^2_{out} which lies in the plane $x_2 \equiv \Delta x_2$, we have

$$\left(\sigma_{21} + \frac{\partial \sigma_{21}}{\partial x_2}\Delta x_2\right)\Delta x_1 \Delta x_3.$$

The corresponding variation of the shear force component σ_{21} is

$$\frac{\partial \sigma_{21}}{\partial x_2} V.$$

Similarly, the variation of the shear force component σ_{31} is

$$\frac{\partial \sigma_{31}}{\partial x_3} V.$$

Adding the contributions of the body and surface forces together, we obtain

$$F_1 = \rho g_1 + \frac{\partial \sigma_{11}}{\partial x_1} + \frac{\partial \sigma_{21}}{\partial x_2} + \frac{\partial \sigma_{31}}{\partial x_3}.$$

Thus, conservation of linear momentum in the x_1 direction is described by

$$\frac{\partial(\rho u_1)}{\partial t} + \frac{\partial(\rho u_1^2)}{\partial x_1} + \frac{\partial(\rho u_1 u_2)}{\partial x_2} + \frac{\partial(\rho u_1 u_3)}{\partial x_3} = \rho g_1 + \frac{\partial \sigma_{11}}{\partial x_1} + \frac{\partial \sigma_{21}}{\partial x_2} + \frac{\partial \sigma_{31}}{\partial x_3}.$$

The compact vector form of this PDE reads

$$\frac{\partial(\rho u_1)}{\partial t} + \nabla\cdot(\rho u_1 \mathbf{u}) = \rho g_1 + \nabla\cdot\boldsymbol{\sigma}_1.$$

Using the product rule of calculus and the mass conservation law (3.2), one obtains the following equivalent representation of the momentum conservation law:

$$\rho\frac{\partial u_1}{\partial t} + \rho\mathbf{u}\cdot\nabla u_1 = \rho g_1 + \nabla\cdot\boldsymbol{\sigma}_1. \tag{3.8}$$

In a similar vein, the conservation laws for ρu_2 and ρu_3 are given by

$$\rho\frac{\partial u_2}{\partial t} + \rho\mathbf{u}\cdot\nabla u_2 = \rho g_2 + \nabla\cdot\boldsymbol{\sigma}_2,$$

$$\rho\frac{\partial u_3}{\partial t} + \rho\mathbf{u}\cdot\nabla u_3 = \rho g_3 + \nabla\cdot\boldsymbol{\sigma}_3. \tag{3.9}$$

The system of conservation laws (3.8)–(3.9) can be written in the form

$$\rho\frac{\partial \mathbf{u}}{\partial t} + \rho\mathbf{u}\cdot\nabla\mathbf{u} = \rho\mathbf{g} + \nabla\cdot\bar{\bar{\boldsymbol{\sigma}}}. \tag{3.10}$$

For the angular momentum to be conserved, the stress tensor $\bar{\bar{\boldsymbol{\sigma}}}$ must be symmetric, i.e.,

$$\bar{\bar{\boldsymbol{\sigma}}} = \bar{\bar{\boldsymbol{\sigma}}}^T \quad \Leftrightarrow \quad \sigma_{ij} = \sigma_{ji}, \quad i,j = 1,2,3. \tag{3.11}$$

Hence, only six components of $\bar{\bar{\boldsymbol{\sigma}}}$ are independent. The stresses are related to the pressure and velocity gradients via a constitutive law (to be defined below).

3.1.3 ▪ Conservation of Energy

The total energy of the fluid occupying a fixed or moving control volume Ω is given by

$$\int_\Omega \rho\left(e+\frac{1}{2}\mathbf{u}\cdot\mathbf{u}\right)\mathrm{d}\mathbf{x},$$

where e is the specific internal energy and $\frac{1}{2}\mathbf{u}\cdot\mathbf{u}$ is the kinetic energy per unit mass unit.

By the law of energy conservation, the time rate of change of total energy

$$\frac{\mathrm{d}}{\mathrm{d}t}\int_{\Omega(t)} \rho\left(e+\frac{1}{2}\mathbf{u}\cdot\mathbf{u}\right)\mathrm{d}\mathbf{x}$$

is balanced by the energy received from internal heat sources, heat fluxes across the boundary $\partial\Omega(t)$, and the work done by all body and surface forces acting on the fluid in $\Omega(t)$.

Let h denote the rate of volumetric heating per unit mass. The amount of energy produced by heat sources and body forces per unit time is given by

$$\int_{\Omega(t)} \rho(h+\mathbf{g}\cdot\mathbf{u})\,\mathrm{d}\mathbf{x}.$$

The heat flux $\mathbf{q}$ is a vector quantity such that $\mathbf{q}\cdot\mathbf{n}$ is the rate of heat transfer in the direction parallel to $\mathbf{n}$. The rate of energy production by heat fluxes and surface forces is

$$\int_{\partial\Omega(t)} (-\mathbf{q}+\bar{\bar{\sigma}}\cdot\mathbf{u})\cdot\mathbf{n}\,\mathrm{d}s.$$

In light of the above, the integral form of the energy balance equation reads

$$\begin{aligned}\frac{\mathrm{d}}{\mathrm{d}t}\int_{\Omega(t)} \rho\left(e+\frac{1}{2}\mathbf{u}\cdot\mathbf{u}\right)\mathrm{d}\mathbf{x} &= \int_{\Omega(t)} \rho(h+\mathbf{g}\cdot\mathbf{u})\,\mathrm{d}\mathbf{x}\\ &+\int_{\partial\Omega(t)} (-\mathbf{q}+\bar{\bar{\sigma}}\cdot\mathbf{u})\cdot\mathbf{n}\,\mathrm{d}s.\end{aligned}$$

Using the Reynolds transport theorem and the divergence theorem, we obtain

$$\begin{aligned}&\int_{\Omega(t)}\left(\frac{D}{Dt}\left(\rho e+\rho\frac{1}{2}\mathbf{u}\cdot\mathbf{u}\right)+\left(\rho e+\rho\frac{1}{2}\mathbf{u}\cdot\mathbf{u}\right)\nabla\cdot\mathbf{u}\right)\mathrm{d}\mathbf{x}\\ &\quad=\int_{\Omega(t)}\left(\rho h+\rho\mathbf{g}\cdot\mathbf{u}+\nabla\cdot(-\mathbf{q}+\bar{\bar{\sigma}}\cdot\mathbf{u})\right)\mathrm{d}\mathbf{x}.\end{aligned}$$

It follows that the evolution of energy is described by the transport equation

$$\begin{aligned}&\frac{D}{Dt}\left(\rho e+\rho\frac{1}{2}\mathbf{u}\cdot\mathbf{u}\right)+\left(\rho e+\rho\frac{1}{2}\mathbf{u}\cdot\mathbf{u}\right)\nabla\cdot\mathbf{u}\\ &\quad=\rho h+\rho\mathbf{f}\cdot\mathbf{u}-\nabla\cdot\mathbf{q}+\nabla\cdot(\bar{\bar{\sigma}}\cdot\mathbf{u}).\end{aligned}\tag{3.12}$$

The left-hand side of this PDE can be written as

$$\begin{aligned}&\rho\frac{De}{Dt}+\frac{1}{2}\rho\frac{D(\mathbf{u}\cdot\mathbf{u})}{Dt}+\left(e+\frac{1}{2}\mathbf{u}\cdot\mathbf{u}\right)\underbrace{\left(\frac{D\rho}{Dt}+\rho\nabla\cdot\mathbf{u}\right)}_{=0\text{ by }(3.1)}\\ &\quad=\rho\frac{De}{Dt}+\frac{1}{2}\rho\frac{D(\mathbf{u}\cdot\mathbf{u})}{Dt},\end{aligned}\tag{3.13}$$

and the last term on the right-hand side of (3.12) can be transformed as follows:

$$\begin{aligned}
\nabla\cdot(\bar{\bar{\boldsymbol{\sigma}}}\cdot\mathbf{u}) &= \nabla\cdot(\boldsymbol{\sigma}_1\cdot\mathbf{u},\boldsymbol{\sigma}_2\cdot\mathbf{u},\boldsymbol{\sigma}_3\cdot\mathbf{u})^T\\
&= \sum_{k=1}^{3}\frac{\partial}{\partial x_k}(\boldsymbol{\sigma}_k\cdot\mathbf{u}) = \sum_{k=1}^{3}\frac{\partial}{\partial x_k}\left(\sum_{l=1}^{3}\sigma_{kl}u_l\right)\\
&= \sum_{k=1}^{3}\left(\sum_{l=1}^{3}\sigma_{kl}\frac{\partial u_l}{\partial x_k}+\sum_{l=1}^{3}u_l\frac{\partial\sigma_{kl}}{\partial x_k}\right)\\
&= \sum_{k=1}^{3}\sum_{l=1}^{3}\sigma_{kl}\frac{\partial u_l}{\partial x_k}+\sum_{l=1}^{3}u_l\left(\sum_{k=1}^{3}\frac{\partial\sigma_{kl}}{\partial x_k}\right)\\
&= \bar{\bar{\boldsymbol{\sigma}}}:\nabla\mathbf{u}+\sum_{l=1}^{3}u_l(\nabla\cdot\boldsymbol{\sigma}_l)\\
&= \bar{\bar{\boldsymbol{\sigma}}}:\nabla\mathbf{u}+\mathbf{u}\cdot(\nabla\cdot\bar{\bar{\boldsymbol{\sigma}}}).
\end{aligned}$$

This leads us to the following representation of the energy equation:

$$\rho\frac{De}{Dt}+\left(\frac{1}{2}\rho\frac{D(\mathbf{u}\cdot\mathbf{u})}{Dt}-\mathbf{u}\cdot(\nabla\cdot\bar{\bar{\boldsymbol{\sigma}}})-\rho\mathbf{g}\cdot\mathbf{u}\right)=\rho h-\nabla\cdot\mathbf{q}+\bar{\bar{\boldsymbol{\sigma}}}:\nabla\mathbf{u},$$

where

$$\begin{aligned}
\frac{1}{2}\rho\frac{D(\mathbf{u}\cdot\mathbf{u})}{Dt}-\mathbf{u}\cdot(\nabla\cdot\bar{\bar{\boldsymbol{\sigma}}})-\rho\mathbf{g}\cdot\mathbf{u} &= \frac{1}{2}\rho\frac{D}{Dt}\left(\sum_{k=1}^{3}u_k^2\right)-\mathbf{u}\cdot(\nabla\cdot\bar{\bar{\boldsymbol{\sigma}}})-\rho\mathbf{g}\cdot\mathbf{u}\\
&= \rho\mathbf{u}\cdot\frac{D\mathbf{u}}{Dt}-\mathbf{u}\cdot(\nabla\cdot\bar{\bar{\boldsymbol{\sigma}}})-\rho\mathbf{g}\cdot\mathbf{u}=\mathbf{u}\cdot\underbrace{\left(\rho\frac{D\mathbf{u}}{Dt}-\nabla\cdot\bar{\bar{\boldsymbol{\sigma}}}-\rho\mathbf{g}\right)}_{=0\text{ by (3.6)}}=0.
\end{aligned}$$

Thus, the energy conservation law (3.12) can be written in the Lagrangian form

$$\rho\frac{De}{Dt}=\rho h-\nabla\cdot\mathbf{q}+\bar{\bar{\boldsymbol{\sigma}}}:\nabla\mathbf{u}. \tag{3.14}$$

By the mass conservation law (3.1) and the product rule of differentiation, we have

$$\begin{aligned}
\rho\frac{De}{Dt} &= \rho\frac{De}{Dt}+e\underbrace{\left(\frac{D\rho}{Dt}+\rho\nabla\cdot\mathbf{u}\right)}_{=0}=\frac{D(\rho e)}{Dt}+(\rho e)\nabla\cdot\mathbf{u}\\
&= \frac{\partial(\rho e)}{\partial t}+\mathbf{u}\cdot\nabla(\rho e)+(\rho e)\nabla\cdot\mathbf{u}=\frac{\partial(\rho e)}{\partial t}+\nabla\cdot(\rho e\mathbf{u}).
\end{aligned}$$

This transformation leads to the Eulerian form of the energy balance

$$\frac{\partial(\rho e)}{\partial t}+\nabla\cdot(\rho e\mathbf{u})=\rho h-\nabla\cdot\mathbf{q}+\bar{\bar{\boldsymbol{\sigma}}}:\nabla\mathbf{u}. \tag{3.15}$$

The same differential forms of the energy conservation law can be obtained using the fixed infinitesimal control volume model. The total energy flux due to inflow through the face S^1_{in} of the box depicted in Fig. 3.1 is given by

$$\rho u_1 \left(e + \frac{1}{2} \mathbf{u} \cdot \mathbf{u} \right) \Delta x_2 \Delta x_3.$$

The total energy leaving the box through the face S^1_{out} per unit time is

$$\left(\rho u_1 \left(e + \frac{1}{2} \mathbf{u} \cdot \mathbf{u} \right) + \frac{\partial}{\partial x_1} \left(\rho u_1 \left(e + \frac{1}{2} \mathbf{u} \cdot \mathbf{u} \right) \right) \Delta x_1 \right) \Delta x_2 \Delta x_3.$$

The rate of change of total energy due to these convective fluxes is

$$\frac{\partial}{\partial x_1} \left(\rho u_1 \left(e + \frac{1}{2} \mathbf{u} \cdot \mathbf{u} \right) \right) V.$$

Similarly, the rate of change due to flow in the other coordinate directions is

$$\frac{\partial}{\partial x_2} \left(\rho u_2 \left(e + \frac{1}{2} \mathbf{u} \cdot \mathbf{u} \right) \right) V,$$
$$\frac{\partial}{\partial x_3} \left(\rho u_3 \left(e + \frac{1}{2} \mathbf{u} \cdot \mathbf{u} \right) \right) V.$$

The sum of the above convective fluxes and the volumetric rate of change

$$\frac{\partial}{\partial t} \left(\rho \left(e + \frac{1}{2} \mathbf{u} \cdot \mathbf{u} \right) \right) V$$

should be balanced by the rate of internal heating, heat fluxes across the faces of the box, and the work done by all body and surface forces per unit time.

By the usual argument, the contribution of heat fluxes and surface forces is

$$\sum_{k=1}^{3} \left(\frac{\partial q_k}{\partial x_k} V + \frac{\partial}{\partial x_k} (\boldsymbol{\sigma}_k \cdot \mathbf{u}) \right) V.$$

The rate of energy production by internal heat sources and body forces is given by

$$\rho h V + \rho (\mathbf{g} \cdot \mathbf{u}) V.$$

Hence, the law of energy conservation for the fluid in the box is described by

$$\frac{\partial}{\partial t} \left(\rho \left(e + \frac{1}{2} \mathbf{u} \cdot \mathbf{u} \right) \right) + \nabla \cdot \left(\rho \mathbf{u} \left(e + \frac{1}{2} \mathbf{u} \cdot \mathbf{u} \right) \right) = \rho h + \rho \mathbf{g} \cdot \mathbf{u}$$
$$- \nabla \cdot \mathbf{q} + \nabla \cdot (\bar{\bar{\sigma}} \cdot \mathbf{u}),$$

where the left-hand side can be written as follows:

$$\begin{aligned}
&\frac{\partial}{\partial t}\left(\rho\left(e+\frac{1}{2}\mathbf{u}\cdot\mathbf{u}\right)\right)+\sum_{k=1}^{3}\frac{\partial}{\partial x_k}\left(\rho u_k\left(e+\frac{1}{2}\mathbf{u}\cdot\mathbf{u}\right)\right)\\
&=\rho\frac{\partial}{\partial t}\left(e+\frac{1}{2}\mathbf{u}\cdot\mathbf{u}\right)+\left(e+\frac{1}{2}\mathbf{u}\cdot\mathbf{u}\right)\frac{\partial\rho}{\partial t}\\
&+\sum_{k=1}^{3}\left(\frac{\partial\rho}{\partial x_k}u_k\left(e+\frac{1}{2}\mathbf{u}\cdot\mathbf{u}\right)+\rho\frac{\partial u_k}{\partial x_k}\left(e+\frac{1}{2}\mathbf{u}\cdot\mathbf{u}\right)+\rho u_k\frac{\partial}{\partial x_k}\left(e+\frac{1}{2}\mathbf{u}\cdot\mathbf{u}\right)\right)\\
&=\rho\frac{\partial}{\partial t}\left(e+\frac{1}{2}\mathbf{u}\cdot\mathbf{u}\right)+\rho\mathbf{u}\cdot\nabla\left(e+\frac{1}{2}\mathbf{u}\cdot\mathbf{u}\right)\\
&+\left(e+\frac{1}{2}\mathbf{u}\cdot\mathbf{u}\right)\underbrace{\left(\frac{\partial\rho}{\partial t}+\mathbf{u}\cdot\nabla\rho+\rho\nabla\cdot\mathbf{u}\right)}_{=0\text{ by }(3.1)}\\
&=\rho\left(\frac{\partial e}{\partial t}+\mathbf{u}\cdot\nabla e+\frac{1}{2}\frac{\partial(\mathbf{u}\cdot\mathbf{u})}{\partial t}+\frac{1}{2}\mathbf{u}\cdot\nabla(\mathbf{u}\cdot\mathbf{u})\right)\\
&=\rho\frac{De}{Dt}+\frac{1}{2}\rho\frac{D(\mathbf{u}\cdot\mathbf{u})}{Dt}.
\end{aligned}$$

After these transformations, the differential law of energy conservation becomes

$$\rho\frac{De}{Dt}+\frac{1}{2}\rho\frac{D(\mathbf{u}\cdot\mathbf{u})}{Dt}=\rho h+\rho\mathbf{g}\cdot\mathbf{u}-\nabla\cdot\mathbf{q}+\nabla\cdot(\bar{\bar{\sigma}}\cdot\mathbf{u}).$$

This equation is equivalent to (3.12) with substitution (3.13). Due to this equivalence, further transformations can be performed as in the model derived from the integral form.

3.1.4 ▪ Summary of the Governing Equations

In summary, the mathematical model of fluid motion represents a system of four physical conservation laws. The Lagrangian form of the governing equations reads

$$\begin{cases}
\dfrac{D\rho}{Dt}+\rho\nabla\cdot\mathbf{u}=0 & \text{(mass)},\\
\rho\dfrac{D\mathbf{u}}{Dt}-\rho\mathbf{g}-\nabla\cdot\bar{\bar{\sigma}}=0 & \text{(linear momentum)},\\
\bar{\bar{\sigma}}=\bar{\bar{\sigma}}^T & \text{(angular momentum)},\\
\rho\dfrac{De}{Dt}-\rho h+\nabla\cdot\mathbf{q}-\bar{\bar{\sigma}}:\nabla\mathbf{u}=0 & \text{(energy)}.
\end{cases}$$

Recalling the definition of the substantial derivative, one obtains the Eulerian form

$$\begin{cases}
\dfrac{\partial\rho}{\partial t}+\nabla\cdot(\rho\mathbf{u})=0 & \text{(mass)},\\
\rho\dfrac{\partial\mathbf{u}}{\partial t}+\rho\mathbf{u}\cdot\nabla\mathbf{u}-\rho\mathbf{g}-\nabla\cdot\bar{\bar{\sigma}}=0 & \text{(linear momentum)},\\
\bar{\bar{\sigma}}=\bar{\bar{\sigma}}^T & \text{(angular momentum)},\\
\rho\dfrac{\partial e}{\partial t}+\rho\mathbf{u}\cdot\nabla e-\rho h+\nabla\cdot\mathbf{q}-\bar{\bar{\sigma}}:\nabla\mathbf{u}=0 & \text{(energy)}.
\end{cases}$$

The above systems of conservation laws are written in the *nonconservative form*. Using the mass conservation law and the product rule of differentiation, one can show that

$$\rho\frac{\partial \mathbf{u}}{\partial t}+\rho\mathbf{u}\cdot\nabla\mathbf{u}=\frac{\partial(\rho\mathbf{u})}{\partial t}+\nabla\cdot(\rho\mathbf{u}\otimes\mathbf{u}),$$
$$\rho\frac{\partial e}{\partial t}+\rho\mathbf{u}\cdot\nabla e=\frac{\partial(\rho e)}{\partial t}+\nabla\cdot(\rho e\mathbf{u}).$$

The corresponding representation of the momentum and energy equation is referred to as the *conservative form*. At the continuous level, all differential forms of the conservation laws are equivalent, but the choice of a particular form does make a difference when it comes to the design of numerical approximations to the equations of fluid dynamics.

For a system of equations to possess a unique solution, the number of equations must be equal to the number of unknowns. The body force density $\mathbf{g}$ and the heating intensity h can be assumed to be known. Still, we have 14 unknowns (ρ, u_1, u_2, u_3, σ_{11}, σ_{12}, σ_{13}, σ_{22}, σ_{23}, σ_{33}, e, q_1, q_2, q_3) and just $8 = 1 + 3 + 3 + 1$ equations. To obtain a closed system, the general model of fluid motion must be supplemented by a sufficient number of *constitutive laws* which represent additional relations between the unknown variables. Examples of such relations include the ideal gas law, the incompressibility constraint, the stress-strain relationship for Newtonian fluids, and the Fourier law of heat conduction (see below).

In order to formulate a well-posed initial-boundary value problem for the equations of fluid dynamics to be solved in a bounded domain $\Omega \subset \mathbb{R}^3$ over a finite time interval $[0, T]$, the evolution equations must be endowed with problem-dependent initial and boundary conditions. The initial values may be given in terms of the conserved quantities or other variables. The choice of boundary conditions depends on the equation, on the boundary type (inlet, outlet, symmetry axis, solid wall), and on the available information.

3.2 ▪ Navier–Stokes Equations

In many applications of practical interest, the density of the fluid may be assumed to be constant. In this case, the fluid is said to be *incompressible*. Interestingly enough, even compressible gases and liquids behave as incompressible fluids at low Mach numbers, i.e., if the flow velocity is small compared to the speed of sound. The system of conservation laws for the mass and momentum of an incompressible fluid is known as the *incompressible Navier–Stokes equations*. This mathematical model represents one of the seven Millennium Prize Problems [175] and describes an amazing variety of flows. The derivation of the incompressible Navier–Stokes equations in this section is based on [9] and [148].

3.2.1 ▪ Three-Dimensional Form

The stress-strain relationship of a Newtonian fluid is given by the constitutive equation

$$\bar{\bar{\sigma}}=-p\mathbf{I}+2\mu\bar{\bar{\varepsilon}}-\frac{2}{3}\mu(\nabla\cdot\mathbf{u})\mathbf{I}, \tag{3.16}$$

where p is the static pressure, $\mathbf{I}$ is the identity tensor, μ is the dynamic viscosity, and $\bar{\bar{\varepsilon}}$ is the *strain rate* tensor representing the symmetric part of the velocity gradient

$$\bar{\bar{\varepsilon}}=\frac{1}{2}(\nabla\mathbf{u}+\nabla\mathbf{u}^T).$$

The conservation law for angular momentum is satisfied by symmetry of $\bar{\bar{\varepsilon}}$ and $\bar{\bar{\sigma}}$.

The componentwise form of the Newtonian constitutive equation reads

$$\sigma_{ij} = -p\delta_{ij} + 2\mu\varepsilon_{ij} - \frac{2}{3}\mu\left(\frac{\partial u_1}{\partial x_1} + \frac{\partial u_2}{\partial x_2} + \frac{\partial u_3}{\partial x_3}\right)\delta_{ij},$$

$$\varepsilon_{ij} = \frac{1}{2}\left(\frac{\partial u_i}{\partial x_j} + \frac{\partial u_j}{\partial x_i}\right), \qquad i,j = 1,2,3,$$

where the Kronecker delta δ_{ij} represents the corresponding component of $\mathbf{I}$.

The divergence of the Newtonian stress tensor (3.16) is given by

$$\begin{aligned}\nabla\cdot\bar{\bar{\sigma}} &= \nabla\cdot\left(-p\mathbf{I} + 2\mu\bar{\bar{\varepsilon}} - \frac{2}{3}\mu(\nabla\cdot\mathbf{u})\mathbf{I}\right)\\ &= \nabla\cdot(2\mu\bar{\bar{\varepsilon}}) - \nabla p - \nabla\left(\frac{2}{3}\mu(\nabla\cdot\mathbf{u})\right).\end{aligned}$$

Substituting it into the Eulerian momentum conservation law (3.7), we obtain

$$\rho\frac{\partial\mathbf{u}}{\partial t} + \rho\mathbf{u}\cdot\nabla\mathbf{u} - \nabla\cdot(2\mu\bar{\bar{\varepsilon}}) + \nabla p + \nabla\left(\frac{2}{3}\mu(\nabla\cdot\mathbf{u})\right) = \rho\mathbf{g}. \tag{3.17}$$

If the density ρ is constant, the law of mass conservation (3.1) reduces to

$$\nabla\cdot\mathbf{u} = 0,$$

and the last term on the left-hand side of the momentum equation (3.17) vanishes. The simplified conservation laws constitute the incompressible Navier–Stokes equations

$$\begin{cases}\rho\dfrac{\partial\mathbf{u}}{\partial t} + \rho\mathbf{u}\cdot\nabla\mathbf{u} - \nabla\cdot(2\mu\bar{\bar{\varepsilon}}) + \nabla p = \rho\mathbf{g},\\ \nabla\cdot\mathbf{u} = 0.\end{cases} \tag{3.18}$$

This vector form of the momentum equation is just a shorthand notation for a coupled system of three evolution equations, one for each velocity component:

$$\rho\frac{\partial u_k}{\partial t} + \rho\mathbf{u}\cdot\nabla u_k - \nabla\cdot(2\mu\boldsymbol{\varepsilon}_k) + \frac{\partial p}{\partial x_k} = \rho g_k, \qquad k = 1,2,3,$$

where $\boldsymbol{\varepsilon}_k := (\varepsilon_{1k}, \varepsilon_{2k}, \varepsilon_{3k})^T$. The pressure p is implicitly defined by the constraint

$$\frac{\partial u_1}{\partial x_1} + \frac{\partial u_2}{\partial x_2} + \frac{\partial u_3}{\partial x_3} = 0$$

and is generally nonunique. If $(\mathbf{u}, p)$ is a solution to the Navier–Stokes equations, then so is $(\mathbf{u}, p + c)$ for any constant c. This is a direct consequence of $\nabla(p + c) = \nabla p$. Hence, the pressure of an incompressible fluid is defined up to an additive constant. The value of this constant can be fixed, e.g., by using a boundary condition that depends on p.

If the viscosity μ is constant, a further simplification is possible. We have

$$\nabla\cdot\bar{\bar{\varepsilon}} = \frac{1}{2}\nabla\cdot\nabla\mathbf{u} + \frac{1}{2}\nabla\cdot\nabla\mathbf{u}^T = \frac{1}{2}\Delta\mathbf{u} + \frac{1}{2}\nabla\underbrace{\nabla\cdot\mathbf{u}}_{=0} = \frac{1}{2}\Delta\mathbf{u}.$$

It follows that $\nabla\cdot(2\mu\bar{\bar{\varepsilon}}) = \mu\Delta\mathbf{u}$ in the case $\mu = const$. This transformation leads to the following representation of the incompressible Navier–Stokes equations:

$$\begin{cases} \rho\dfrac{\partial\mathbf{u}}{\partial t} + \rho\mathbf{u}\cdot\nabla\mathbf{u} - \mu\Delta\mathbf{u} + \nabla p = \rho\mathbf{g}, \\ \nabla\cdot\mathbf{u} = 0. \end{cases} \tag{3.19}$$

If the body force density $\mathbf{g}$ is parallel to a unit vector $\mathbf{e}_k$, it can be written in the form

$$\mathbf{g} = -g\mathbf{e}_k = -g\nabla x_k, \qquad k\in\{1,2,3\}.$$

It follows that $\nabla p - \rho\mathbf{g} = \nabla p_g$, where $p_g = p + \rho g x_k$. The momentum equation becomes

$$\rho\frac{\partial\mathbf{u}}{\partial t} + \rho\mathbf{u}\cdot\nabla\mathbf{u} - \mu\Delta\mathbf{u} + \nabla p_g = 0. \tag{3.20}$$

That is, potential body forces can be "hidden" in the gradient of the modified pressure.

In many texts, the momentum equation is divided by the constant density, which gives

$$\frac{\partial\mathbf{u}}{\partial t} + \mathbf{u}\cdot\nabla\mathbf{u} - \nu\Delta\mathbf{u} + \nabla P = 0, \tag{3.21}$$

where $\nu = \frac{\mu}{\rho}$ is the kinematic viscosity of the fluid and $P = \frac{p_g}{\rho} = \frac{p}{\rho} - g x_k$.

In general, the viscosity μ may depend on the temperature T. In the presence of temperature gradients, the general form (3.18) of the momentum equation must be employed. Also, the energy equation needs to be solved to determine the temperature distribution.

The temperature gradients may also change the density of the fluid. In many cases, the dependence of the density on the temperature is negligible. However, in buoyancy-driven flows the variations of density represent the main cause of fluid motion and must be taken into account. A popular simplified model of this *natural convection* phenomenon is based on the *Boussinesq approximation*, which boils down to using a temperature-dependent density ρ in the gravity force and a constant reference density ρ_0 elsewhere. The resulting modification of the general incompressible flow model (3.18) is as follows:

$$\begin{cases} \rho_0\dfrac{\partial\mathbf{u}}{\partial t} + \rho_0\mathbf{u}\cdot\nabla\mathbf{u} - \nabla\cdot(2\mu\bar{\bar{\varepsilon}}) + \nabla p = \rho\mathbf{g}, \\ \nabla\cdot\mathbf{u} = 0, \end{cases} \tag{3.22}$$

where the density $\rho = \rho_0[1 - \beta(T - T_0)]$ depends on the temperature field T, on the reference temperature T_0, and on the *thermal expansion coefficient*

$$\beta \approx -\frac{1}{\rho_0}\left(\frac{\partial\rho}{\partial T}\right)_{T=T_0}.$$

The general form $\nabla\cdot(2\mu\bar{\bar{\varepsilon}})$ of the viscous term must be employed whenever the viscosity μ exhibits strong variations. In particular, the approximation by $\mu\Delta\mathbf{u}$ is unacceptable if an eddy viscosity turbulence model or a non-Newtonian constitutive relation is employed. Moreover, some types of boundary conditions may require that the general form (3.18) of the Navier–Stokes equations be used even if the density and viscosity are constant.

3.2.2 ▪ Planar and Axisymmetric Forms

Many mathematical and numerical studies of incompressible flow problems are performed using the 2D (planar) and axisymmetric forms of the Navier–Stokes equations, which are easier to analyze and implement than the general 3D model.

The 2D form of the Navier-Stokes equations (3.18) is obtained by omitting the momentum equation for u_3 and setting $u_3 = 0$ in the other equations. This gives

$$\rho\frac{\partial u_1}{\partial t} + \rho u_1\frac{\partial u_1}{\partial x_1} + \rho u_2\frac{\partial u_1}{\partial x_2} - \frac{\partial}{\partial x_1}\left(2\mu\frac{\partial u_1}{\partial x_1}\right) - \frac{\partial}{\partial x_2}\left(\mu\left(\frac{\partial u_1}{\partial x_2}+\frac{\partial u_2}{\partial x_1}\right)\right) + \frac{\partial p}{\partial x_1} = \rho g_1,$$

$$\rho\frac{\partial u_2}{\partial t} + \rho u_1\frac{\partial u_2}{\partial x_1} + \rho u_2\frac{\partial u_2}{\partial x_2} - \frac{\partial}{\partial x_1}\left(\mu\left(\frac{\partial u_1}{\partial x_2}+\frac{\partial u_2}{\partial x_1}\right)\right) - \frac{\partial}{\partial x_2}\left(2\mu\frac{\partial u_2}{\partial x_2}\right) + \frac{\partial p}{\partial x_2} = \rho g_2.$$

The continuity equation (i.e., the law of mass conservation) becomes

$$\frac{\partial u_1}{\partial x_1} + \frac{\partial u_2}{\partial x_2} = 0. \tag{3.23}$$

If the viscosity μ is constant, the above 2D model reduces to

$$\begin{cases} \rho\frac{\partial u_1}{\partial t} + \rho\left(u_1\frac{\partial u_1}{\partial x_1} + u_2\frac{\partial u_1}{\partial x_2}\right) - \mu\left(\frac{\partial^2 u_1}{\partial x_1^2}+\frac{\partial^2 u_1}{\partial x_2^2}\right) + \frac{\partial p}{\partial x_1} = \rho g_1, \\ \rho\frac{\partial u_2}{\partial t} + \rho\left(u_1\frac{\partial u_2}{\partial x_1} + u_2\frac{\partial u_2}{\partial x_2}\right) - \mu\left(\frac{\partial^2 u_2}{\partial x_1^2}+\frac{\partial^2 u_2}{\partial x_2^2}\right) + \frac{\partial p}{\partial x_2} = \rho g_2, \\ \frac{\partial u_1}{\partial x_1} + \frac{\partial u_2}{\partial x_2} = 0. \end{cases}$$

The axisymmetric form of the incompressible Navier–Stokes equations is derived using a transformation to the cylindrical coordinates (r, z, θ). We have

$$\begin{cases} \rho\frac{\partial u_z}{\partial t} + \rho\mathbf{u}\cdot\nabla u_z - \frac{\partial \sigma_{zz}}{\partial z} - \frac{1}{r}\frac{\partial}{\partial r}(r\sigma_{rz}) = \rho g_z, \\ \rho\frac{\partial u_r}{\partial t} + \rho\mathbf{u}\cdot\nabla u_r - \frac{\partial \sigma_{rz}}{\partial z} - \frac{1}{r}\frac{\partial}{\partial r}(r\sigma_{rr}) + \frac{1}{r}\sigma_{\theta\theta} = \rho g_r, \\ \frac{\partial u_z}{\partial z} + \frac{1}{r}\frac{\partial}{\partial r}(r u_r) = 0, \end{cases}$$

where $\mathbf{u} = (u_z, u_r)$. The four components of the stress tensor are given by

$$\sigma_{zz} = -p + 2\mu\frac{\partial u_z}{\partial z},$$

$$\sigma_{rz} = \mu\left(\frac{\partial u_z}{\partial r} + \frac{\partial u_r}{\partial z}\right),$$

$$\sigma_{rr} = -p + 2\mu\frac{\partial u_r}{\partial r},$$

$$\sigma_{\theta\theta} = -p + 2\mu\frac{u_r}{r}.$$

Substituting the constitutive equations for $\sigma_{zz}, \sigma_{rz}, \sigma_{rr}, \sigma_{\theta\theta}$ into the axisymmetric form of the momentum conservation laws, we obtain the following flow model:

$$\left\{\begin{aligned} \rho\frac{\partial u_z}{\partial t} + \rho\mathbf{u}\cdot\nabla u_z - \frac{\partial}{\partial z}\left(2\mu\frac{\partial u_z}{\partial z}\right) - \frac{1}{r}\frac{\partial}{\partial r}\left(r\mu\left(\frac{\partial u_z}{\partial r} + \frac{\partial u_r}{\partial z}\right)\right) + \frac{\partial p}{\partial z} &= \rho g_z, \\ \rho\frac{\partial u_r}{\partial t} + \rho\mathbf{u}\cdot\nabla u_r - \frac{\partial}{\partial z}\left(\mu\left(\frac{\partial u_z}{\partial r} + \frac{\partial u_r}{\partial z}\right)\right) - \frac{1}{r}\frac{\partial}{\partial r}\left(2r\mu\frac{\partial u_r}{\partial r}\right) + 2\mu\frac{u_r}{r^2} + \frac{\partial p}{\partial r} &= \rho g_r, \\ \frac{\partial u_z}{\partial z} + \frac{1}{r}\frac{\partial}{\partial r}(r u_r) &= 0. \end{aligned}\right.$$

In the case of a constant viscosity, the axisymmetric equations reduce to

$$\left\{\begin{aligned} \rho\frac{\partial u_z}{\partial t} + \rho\mathbf{u}\cdot\nabla u_z - \mu\frac{\partial^2 u_z}{\partial z^2} - \mu\frac{1}{r}\frac{\partial}{\partial r}\left(r\frac{\partial u_z}{\partial r}\right) + \frac{\partial p}{\partial z} &= \rho g_z, \\ \rho\frac{\partial u_r}{\partial t} + \rho\mathbf{u}\cdot\nabla u_r - \mu\frac{\partial^2 u_r}{\partial z^2} - \mu\frac{1}{r}\frac{\partial}{\partial r}\left(r\frac{\partial u_r}{\partial r}\right) + \mu\frac{u_r}{r^2} + \frac{\partial p}{\partial r} &= \rho g_r, \\ \frac{\partial u_z}{\partial z} + \frac{1}{r}\frac{\partial}{\partial r}(r u_r) &= 0. \end{aligned}\right.$$

The axisymmetric flow model will not be handled further in this book, but it forms a natural basis, e.g., for some analytical studies of fully developed pipe flows. For this reason, the axisymmetric Navier–Stokes equations are also presented in this chapter.

3.2.3 ▪ Dimensionless Form

In the equations presented so far, the physical dimensions of all variables are defined using the International System of Units (SI). For example, the SI unit of density is $\frac{\text{kg}}{\text{m}^3}$, the SI unit of velocity is $\frac{\text{m}}{\text{s}}$, and the SI unit of energy is $\frac{\text{kg}\cdot\text{m}^2}{\text{s}^2}$. The SI units of other variables and model parameters can be determined using *dimensional analysis*. The equations of incompressible fluid dynamics can also be written in terms of the dimensionless variables

$$t^* := \frac{tU}{L}, \quad x_k^* := \frac{x_k}{L}, \quad u_k^* := \frac{u_k}{U}, \quad p^* := \frac{p}{\rho U^2}, \quad g_k^* := \frac{g_k L}{U^2},$$

where U is a reference velocity, L is a reference length, and ρ is the constant density.

Dividing equation (3.19) by the density ρ and substituting $\mathbf{u} = \mathbf{u}^* U$, we obtain

$$U\frac{\partial \mathbf{u}^*}{\partial t} + U^2\mathbf{u}^*\cdot\nabla\mathbf{u}^* - \frac{\mu U}{\rho}\Delta\mathbf{u}^* + \nabla\left(\frac{p}{\rho}\right) = \mathbf{g}.$$

The partial derivatives w.r.t. the dimensionless variables are defined by

$$\frac{\partial}{\partial t} = \frac{\partial}{\partial t^*}\frac{\partial t^*}{\partial t} = \frac{\partial}{\partial t^*}\frac{U}{L},$$

$$\frac{\partial}{\partial x_k} = \frac{\partial}{\partial x_k^*}\frac{\partial x_k^*}{\partial x_k} = \frac{\partial}{\partial x_k^*}\frac{1}{L}.$$

This yields

$$\frac{U^2}{L}\frac{\partial \mathbf{u}^*}{\partial t^*} + \frac{U^2}{L}\mathbf{u}^*\cdot\nabla^*\mathbf{u}^* - \frac{\mu U}{\rho L^2}\Delta^*\mathbf{u}^* + \frac{1}{L}\nabla^*\left(\frac{p}{\rho}\right) = \mathbf{g}.$$

Multiplying by L/U^2, we obtain the dimensionless Navier–Stokes equations

$$\begin{cases} \dfrac{\partial \mathbf{u}^*}{\partial t^*} + \mathbf{u}^* \cdot \nabla^* \mathbf{u}^* - \dfrac{1}{\mathrm{Re}} \Delta^* \mathbf{u}^* + \nabla^* p^* = \mathbf{g}^*, \\ \nabla^* \cdot \mathbf{u}^* = 0, \end{cases}$$

where

$$\mathrm{Re} := \frac{\rho U L}{\mu}$$

is the *Reynolds number* which is commonly used to describe the flow regime in experimental and computational fluid dynamics. At low Reynolds numbers, the momentum equation is dominated by the viscous term. At high Reynolds numbers, the nonlinear convective term becomes dominant, and a suitable turbulence model may need to be employed.

Incompressible flows characterized by the same Reynolds number behave in a similar way even if the actual physical dimensions differ by orders of magnitude. In experimental fluid dynamics, the concept of *flow similarity* makes it possible to scale up the measurement results obtained for a laboratory-scale model. In CFD, the dimensionless form of the governing equations may be employed, e.g., if the use of SI units would require error-prone arithmetic operations with very large or very small numbers.

3.2.4 ▪ Boundary Conditions

As a rule of thumb, the number of boundary conditions to be imposed on the velocity of a viscous fluid is equal to the number of space dimensions (two in 2D, three in 3D). In addition, an initial condition for the velocity field must be provided if the problem at hand is time-dependent. In this section, we will discuss the imposition of boundary conditions for a typical flow configuration with an inlet Γ_{in}, an outlet Γ_{out}, and a solid wall Γ_{wall}.

Dirichlet Boundary Conditions

In viscous incompressible flow models, Dirichlet boundary conditions may be imposed to specify the values of individual velocity components. For example, an experimentally measured velocity profile may be used to define the *inflow boundary condition*

$$\mathbf{u} = \mathbf{u}_{\mathrm{in}} \qquad \text{on } \Gamma_{\mathrm{in}}.$$

On a solid wall, the velocity of a viscous fluid is equal to the velocity of the wall. If the wall is fixed, the corresponding *no-slip boundary condition* is given by

$$\mathbf{u} = 0 \qquad \text{on } \Gamma_{\mathrm{wall}}.$$

The validity of this boundary condition has been confirmed in experimental studies [105].

Neumann Boundary Conditions

Instead of specifying the boundary value of a velocity component, one may prescribe the corresponding component of the *traction vector*, i.e., of the surface force

$$\bar{\bar{\sigma}} \cdot \mathbf{n} = (\boldsymbol{\sigma}_1 \cdot \mathbf{n}, \ldots, \boldsymbol{\sigma}_d \cdot \mathbf{n})^T. \tag{3.24}$$

Such Neumann boundary conditions are usually imposed at the outlet Γ_{out} of a bounded domain or on artificial open boundaries in external flow problems.

If the stress tensor is given by the Newtonian constitutive law $\bar{\bar{\boldsymbol{\sigma}}} = -p\mathbf{I} + 2\mu\bar{\bar{\varepsilon}}$, the Neumann boundary conditions for the components of the traction can be written as

$$2\mu\boldsymbol{\varepsilon}_k \cdot \mathbf{n} - pn_k = s_k, \qquad k = 1,2,3. \tag{3.25}$$

For example, the 2D Neumann boundary conditions are given by

$$\begin{cases} \sigma_{11}n_1 + \sigma_{12}n_2 = 2\mu(\varepsilon_{11}n_1 + \varepsilon_{12}n_2) - pn_1 = s_1, \\ \sigma_{21}n_1 + \sigma_{22}n_2 = 2\mu(\varepsilon_{21}n_1 + \varepsilon_{22}n_2) - pn_2 = s_2. \end{cases}$$

Setting both components of the traction vector equal to zero, one obtains the *do-nothing boundary condition*, which is frequently used by default on outflow boundaries.

Mixed Boundary Conditions

The homogeneous Dirichlet boundary condition for the normal velocity component

$$\mathbf{u} \cdot \mathbf{n} = 0$$

is known as the *no-penetration condition* and may be combined with Neumann boundary conditions for the tangential components of the traction vector $\bar{\bar{\boldsymbol{\sigma}}} \cdot \mathbf{n}$.

If the problem at hand is symmetric, it can be solved in a reduced domain using the above no-penetration condition and the homogeneous Neumann boundary condition for the tangential traction on the axis of symmetry. In some turbulent flow models, a logarithmic *wall law* for the tangential stress is derived using the boundary layer theory.

In other applications, some components of the velocity and traction vectors are assumed to be related by a Newton (or Robin) boundary condition of the form

$$\alpha_k u_k + \boldsymbol{\sigma}_k \cdot \mathbf{n} = s_k, \tag{3.26}$$

where α_k is a given parameter. The Dirichlet boundary condition $u_k = 0$ is recovered in the limit $\alpha_k \to \infty$, whereas the Neumann boundary condition $\boldsymbol{\sigma}_k \cdot \mathbf{n} = s_k$ corresponds to the choice $\alpha_k = 0$. Examples of Newton boundary conditions include the *Navier slip condition* and porous wall boundary conditions (see Example 3.2 below).

Natural Boundary Conditions

Multiplying the residual of the momentum equation (3.18) by an admissible test function $\mathbf{w}$ and integrating over the domain Ω, one obtains a weighted residual formulation which forms the basis for the construction of finite element approximations (see Chapter 2 and the following chapters for details). Integration by parts produces the boundary integral

$$\int_\Gamma \mathbf{w} \cdot \left(\bar{\bar{\boldsymbol{\sigma}}} \cdot \mathbf{n}\right) \mathrm{d}s = \int_\Gamma \mathbf{w} \cdot (2\mu\bar{\bar{\varepsilon}} - p\mathbf{I})\,\mathrm{d}s. \tag{3.27}$$

Using test functions of the form $\mathbf{w} = w_k\mathbf{e}_k$, where $\mathbf{e}_k$ is the kth unit vector, we obtain

$$\int_\Gamma w_k(\boldsymbol{\sigma}_k \cdot \mathbf{n})\,\mathrm{d}s = \int_\Gamma w_k(2\mu\boldsymbol{\varepsilon}_k \cdot \mathbf{n} - pn_k)\,\mathrm{d}s. \tag{3.28}$$

If the Neumann condition (3.25) is imposed on a boundary component $\Gamma_2 \subseteq \Gamma$, we have

$$\int_{\Gamma_2} w_k(2\mu\boldsymbol{\varepsilon}_k \cdot \mathbf{n} - pn_k)\,\mathrm{d}s = \int_{\Gamma_2} w_k s_k\,\mathrm{d}s.$$

Similarly, the imposition of the Newton condition (3.26) on $\Gamma_3 \subseteq \Gamma$ implies that

$$\int_{\Gamma_2} w_k(2\mu\varepsilon_k \cdot \mathbf{n} - pn_k)\,ds = \int_{\Gamma_2} w_k s_k\,ds - \alpha_k \int_{\Gamma_2} w_k u_k\,ds.$$

Since Neumann and Newton boundary conditions can be incorporated into the variational formulation in this natural way, they are called *natural boundary conditions.*

If the general form (3.18) of the Navier–Stokes equations is replaced by the simplified form (3.19), the boundary integral in the weak formulation becomes

$$\int_{\Gamma} \mathbf{w} \cdot (\mu\nabla\mathbf{u} - p\mathbf{I}) \cdot \mathbf{n}\,ds. \tag{3.29}$$

Substituting the test function $\mathbf{w}_k = w_k\mathbf{e}_k$, we end up with the boundary integral

$$\int_{\Gamma} w_k(\mu\nabla u_k \cdot \mathbf{n} - pn_k)\,ds, \tag{3.30}$$

which is generally *not* equal to (3.28). Hence, Neumann and Newton boundary conditions can no longer be taken into account in a natural way. A notable exception to this rule is the typical case of an outflow boundary condition under the assumption that the flow is fully developed and the normal traction is equal to zero (see the following example).

Example 3.1. Steady incompressible flow over a backward facing step.

Let us exemplify the imposition of boundary conditions using the steady laminar 2D flow over the backward facing step sketched in Fig. 3.2 as a model problem. The density and viscosity of the fluid are assumed to be constant. We prescribe the Dirichlet boundary condition $\mathbf{u} = \mathbf{u}_{\text{in}}$ at the inlet Γ_{in} and the standard no-slip condition $\mathbf{u} = 0$ on the wall boundary Γ_{wall}. If the outflow boundary Γ_{out} is located far from the step, it is reasonable to assume that the flow is fully developed and parallel to the walls of the channel. Thus, the homogeneous Dirichlet boundary condition $u_2 = 0$ may be used for the tangential velocity component on Γ_{out}. However, the boundary values of the normal velocity component u_1 are generally unknown and difficult to guess. The use of wrong Dirichlet boundary values for u_1 on Γ_{out} may disturb the flow pattern and result in a violation of the mass conservation law. Hence, outflow boundary conditions of Neumann or Newton type are preferred.

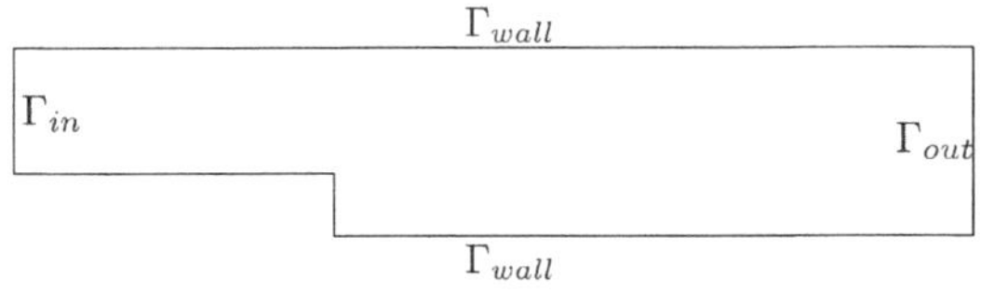

Figure 3.2. *2D flow over a backward facing step.*

Since the tangential velocity component u_2 is set to zero at the outlet, we have

$$\frac{\partial u_2}{\partial x_2} = 0.$$

Invoking the 2D continuity equation (3.23), we conclude that

$$\frac{\partial u_1}{\partial x_1} = 0.$$

It follows that the normal component of the Newtonian stress tensor on Γ_{out} is given by

$$\sigma_{11} = \mu \frac{\partial u_1}{\partial x_1} - p = -p.$$

Since the actual value of the pressure p has no influence on the velocity, we set the exit pressure equal to zero and prescribe the homogeneous Neumann boundary condition

$$\sigma_{11} n_1 + \sigma_{12} n_2 = \sigma_{11} = 0 \qquad \text{on } \Gamma_{out},$$

where $(n_1, n_2) = \mathbf{n} = (1, 0)$ is the unit outward normal to the outflow boundary.

In summary, a usable set of boundary conditions for our model problem is as follows:

$$\begin{cases} \text{inlet: } u_1 = u_{in}^{(1)}, \\ \qquad u_2 = u_{in}^{(2)} = 0, \\ \text{walls: } u_1 = 0, \\ \qquad u_2 = 0, \\ \text{outlet: } \sigma_{11} n_1 + \sigma_{12} n_2 = \sigma_{11} = 0, \\ \qquad u_2 = 0. \end{cases}$$

Note that exactly two boundary conditions are prescribed on each boundary component. ∎

Remark 3.1. For a fully developed outflow, the normal traction is given by

$$\bar{\bar{\boldsymbol{\sigma}}}_1 \cdot \mathbf{n} = \sigma_{11} = \mu \nabla u_1 \cdot \mathbf{n} - p n_1.$$

Thus, the normal components of the boundary integrals (3.28) and (3.30) are both equal to zero. The tangential components of these integrals are also zero on Γ_{out}, where the trace of the test function w vanishes by virtue of the Dirichlet boundary condition imposed on u_2. It follows that the weak forms of the Navier–Stokes systems (3.18) and (3.19) are equivalent in the case of constant viscosity for the above choice of the boundary conditions.

Remark 3.2. If the simplified form (3.30) of the Navier–Stokes equations is employed, the Dirichlet condition $u_2 = 0$ may be replaced by the homogeneous Neumann condition

$$\frac{\partial u_2}{\partial x_1} = 0 \qquad \text{on } \Gamma_{out}.$$

However, the homogeneous Neumann condition for the tangential traction

$$\bar{\bar{\boldsymbol{\sigma}}}_2 \cdot \mathbf{n} = \mu \left(\frac{\partial u_1}{\partial x_2} + \frac{\partial u_2}{\partial x_1} \right)$$

is physically incorrect because $\frac{\partial u_2}{\partial x_1} \approx 0$ near the outlet, whereas $\frac{\partial u_1}{\partial x_2} \neq 0$ for a fully developed (parabolic) velocity profile. Therefore, the general form (3.18) of the Navier–Stokes equations should be used in conjunction with the Dirichlet condition $u_2 = 0$.

Example 3.2. Steady incompressible flow in the headbox of a paper machine.

To illustrate the use of Newton boundary conditions, we consider the flow domain sketched in Fig. 3.3. This configuration models the headbox of a paper machine [94].

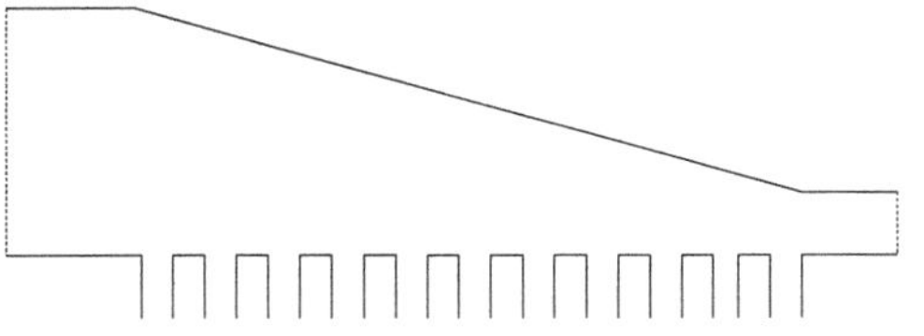

Figure 3.3. *2D flow in the headbox of a paper machine.*

The outflow boundary consists of many small tubes built into the perforated wall of the tapered header. In principle, a Neumann boundary condition could be prescribed at the entrance of each tube. In real-life applications, the number of tubes is usually so large, and their radius so small, that a direct numerical simulation would require the use of extremely fine computational meshes and become prohibitively expensive. For this reason, we will approximate the perforated wall by a straight line and use a homogenized boundary condition.

The traction in the direction of the outward normal vector $\mathbf{n} = (0, -1)$ is given by

$$\sigma_{21} n_1 + \sigma_{22} n_2 = -\sigma_{22} = p - 2\mu \frac{\partial u_2}{\partial x_2} \approx p$$

because the pressure is typically an order of magnitude larger than the viscous force.

The pressure p on the outflow boundary of the header is defined by

$$p = \Delta p(\bar{u}) + p_{\text{ext}},$$

where p_{ext} is the external pressure and Δp is the major head loss which is assumed to be a known function of the average velocity $\bar{u}$. Using the model

$$\Delta p(\bar{u}) = -\alpha_2(\bar{u}) u_2,$$

which depends on a given function α_2, we obtain the outflow boundary condition

$$\sigma_{21} n_1 + \sigma_{22} n_2 = p = -\alpha_2 u_2 + p_{\text{ext}},$$

which is equivalent to

$$\sigma_{21} n_1 + \sigma_{22} n_2 + \alpha_2 u_2 = p_{\text{ext}}.$$

This is a Newton boundary condition of the form (3.26), where $s_2 = p_{\text{ext}}$ and $\alpha_2 u_2$ represents the head losses in the tube bank as a function of the average velocity. ■

3.3 ▪ General Heat Equation

In the presence of temperature gradients and nonisothermal flow effects, the incompressible Navier–Stokes equations (3.18) must be supplemented by an evolution equation for the temperature. The general heat equation is derived from the law of energy conservation,

$$\rho \frac{De}{Dt} - \rho h + \nabla \cdot \mathbf{q} - \bar{\bar{\boldsymbol{\sigma}}} : \nabla \mathbf{u} = 0,$$

where $\bar{\bar{\boldsymbol{\sigma}}}$ is the Newtonian stress tensor. Neglecting the viscous dissipation rate $2\mu \bar{\bar{\boldsymbol{\varepsilon}}} : \nabla \mathbf{u}$, we approximate the energy production term $\bar{\bar{\boldsymbol{\sigma}}} : \nabla \mathbf{u}$ by $-p \nabla \cdot \mathbf{u}$. This gives

$$\rho \frac{De}{Dt} - \rho h + \nabla \cdot \mathbf{q} + p \nabla \cdot \mathbf{u} = 0.$$

By the law of mass conservation (3.1), we have

$$\nabla\cdot\mathbf{u}=-\frac{1}{\rho}\frac{D\rho}{Dt}.$$

Using the formula

$$\frac{D(\frac{1}{\rho})}{Dt}=\frac{D(\rho^{-1})}{Dt}=-\frac{1}{\rho^2}\frac{D\rho}{Dt},$$

we obtain

$$p\nabla\cdot\mathbf{u}=-p\frac{1}{\rho}\frac{D\rho}{Dt}=p\rho\left(-\frac{1}{\rho^2}\frac{D\rho}{Dt}\right)=p\rho\frac{D(\frac{1}{\rho})}{Dt},$$

which yields

$$\rho\frac{De}{Dt}+p\rho\frac{D(\frac{1}{\rho})}{Dt}-\rho h+\nabla\cdot\mathbf{q}=0.$$

If the internal energy e and density ρ are related to the temperature T and pressure p by constitutive laws of the form $e=e(T,p)$ and $\rho=\rho(T,p)$, the chain rule gives

$$\frac{De}{Dt}=\frac{\partial e}{\partial T}\frac{DT}{Dt}+\frac{\partial e}{\partial p}\frac{Dp}{Dt}$$

and

$$\frac{D(\frac{1}{\rho})}{Dt}=\frac{\partial(\frac{1}{\rho})}{\partial T}\frac{DT}{Dt}+\frac{\partial(\frac{1}{\rho})}{\partial p}\frac{Dp}{Dt}.$$

The contribution of the substantial derivative of pressure may be neglected. Thus

$$\rho\left(\frac{De}{Dt}+p\frac{D(\frac{1}{\rho})}{Dt}\right)=\rho\left(\frac{\partial e}{\partial T}+p\frac{\partial(\frac{1}{\rho})}{\partial T}\right)\frac{DT}{Dt}=\rho h-\nabla\cdot\mathbf{q}.$$

The specific heat capacity at constant pressure c_p is defined as follows:

$$c_p=\frac{\partial e}{\partial T}+p\frac{\partial(\frac{1}{\rho})}{\partial T},$$

whence the law of energy conservation becomes

$$\rho c_p\frac{DT}{Dt}=\rho h-\nabla\cdot\mathbf{q}. \tag{3.31}$$

The temperature T cannot be determined from this equation, because the heat flux $\mathbf{q}$ is still unknown. Therefore, a constitutive law is needed to express $\mathbf{q}$ as a function of the temperature T. By the *Fourier law* of heat conduction, the heat flux is parallel to the local temperature gradient and points in the opposite direction. That is,

$$\mathbf{q}=-k\nabla T, \tag{3.32}$$

where k is the *thermal conductivity*. Substituting (3.32) into (3.31), one obtains

$$\rho c_p\frac{DT}{Dt}=\rho h+\nabla\cdot(k\nabla T).$$

The equivalent Eulerian form of this general heat equation is given by

$$\rho c_p \frac{\partial T}{\partial t} + \rho c_p \mathbf{u} \cdot \nabla T - \nabla \cdot (k \nabla T) = \rho h. \tag{3.33}$$

In the case $\mathbf{u} = 0$, $h = 0$, $k = const$, it reduces to the classical heat equation

$$\rho c_p \frac{\partial T}{\partial t} - k \Delta T = 0.$$

The axisymmetric form of the above heat transfer model reads

$$\rho c_p \frac{\partial T}{\partial t} + \rho c_p \left(u_z \frac{\partial T}{\partial z} + u_r \frac{\partial T}{\partial r} \right) - \frac{\partial}{\partial z}\left(k \frac{\partial T}{\partial z} \right) - \frac{1}{r}\frac{\partial}{\partial r}\left(rk \frac{\partial T}{\partial r} \right) = \rho h.$$

The selection of boundary conditions for the temperature field depends on whether the temperature or the heat flux is known on the given boundary component. If the temperature distribution T_D is given, it is natural to use the Dirichlet boundary condition

$$T = T_D.$$

If the boundary represents an insulated wall, then the heat flux $\mathbf{q} \cdot \mathbf{n}$ is zero. By (3.32), this assumption corresponds to the homogeneous Neumann boundary condition

$$-k \nabla T \cdot \mathbf{n} = 0.$$

If there is a heat flux to the wall, one may use the Newton boundary condition

$$-k \nabla T \cdot \mathbf{n} = h_c (T - T_{\text{ext}}),$$

where h_c is the heat transfer coefficient and T_{ext} is the external temperature. The rate of radiative heat transfer can be modeled using the nonlinear relation

$$-k \nabla T \cdot \mathbf{n} = \varepsilon \sigma (T^4 - T_{\text{ext}}^4),$$

where ε is the emissivity and σ is the Stefan–Boltzmann constant. The right-hand side of this Newton boundary condition can be written in the form

$$\varepsilon \sigma (T^4 - T_{\text{ext}}^4) = \underbrace{\varepsilon \sigma (T^3 + T^2 T_{\text{ext}} + T T_{\text{ext}}^2 + T_{\text{ext}}^3)}_{=:h_r(T)} (T - T_{\text{ext}}).$$

This leads to the following representation of the radiative boundary condition

$$-k \nabla T \cdot \mathbf{n} = h_r(T)(T - T_{\text{ext}}),$$

where

$$h_r(T) = \varepsilon \sigma (T^3 + T^2 T_{\text{ext}} + T T_{\text{ext}}^2 + T_{\text{ext}}^3)$$

is an effective heat transfer coefficient which depends on T in a nonlinear way.

3.4 ▪ Turbulence Modeling

In fluid dynamics, the *turbulent flow* regime is characterized by a highly irregular unsteady motion of a great number of vortices (so-called *eddies*) which vary greatly in their size and interact with one another in a random manner. Turbulence is not a physical property of the fluid but a phenomenon caused by the multiscale nature of fluid motion at high Reynolds numbers. A famous statement by Hinze [103] defines turbulent flows as follows: "Turbulent fluid motion is an irregular condition of the flow in which the various quantities show a random variation with time and space coordinates, so that statistically distinct average values can be discerned." For an in-depth introduction to turbulence, we refer the reader to the books by Pope [189], Tennekes and Lumley [223], and Wilcox [243].

In principle, the transient 3D Navier–Stokes equations are valid for both laminar and turbulent flows. However, direct numerical simulation (DNS) of turbulence requires that the mesh size be fine enough to resolve the smallest scales of turbulent flow motion. Even today, the DNS approach is only feasible for moderately high Reynolds numbers since the mesh size h scales with the Reynolds number as $\mathrm{Re}^{9/4}$. In addition, the time step Δt depends on Reynolds number as $\Delta t \approx (\mathrm{Re})^{-1/2}$. As a consequence, the cost of DNS tends to become prohibitively high as the Reynolds number increases.

If the presence of microscopic turbulent eddies makes DNS too expensive, the unresolvable fine-scale effects may be eliminated by filtering (averaging) the Navier–Stokes equations. The most popular turbulence models of this kind are based on the Reynolds-averaged Navier–Stokes (RANS) equations. The influence of the unresolvable fine scales on the resolvable coarse scales is taken into account via a subgrid stress tensor which is modeled using additional algebraic relations and/or PDEs. In this section, we present the so-called k–ε model [152], which is widely used in engineering applications. In this model, the turbulent stress tensor is assumed to depend on the mean strain rate in the same way as the laminar viscous stress, and the turbulent eddy viscosity is determined by solving a pair of extra transport equations for the turbulent kinetic energy k and its dissipation rate ε.

3.4.1 ▪ The RANS Equations

In engineering practice, the long-time behavior of fluid flows is described by time-averaged equations with empirical closures. The statistical average of a generic variable ϕ is denoted by $\bar{\phi}$. The instantaneous deviation from this average state is denoted by $\phi' = \phi - \bar{\phi}$.

By definition, the time average $\bar{\phi}'$ of a fluctuation ϕ' is equal to zero. A time-averaged partial derivative equals the partial derivative of the time average. We have

$$\overline{\frac{\partial \phi}{\partial t}} = \frac{\partial \bar{\phi}}{\partial t}, \qquad \overline{\frac{\partial \phi}{\partial x_k}} = \frac{\partial \bar{\phi}}{\partial x_k}, \qquad k = 1,2,3. \tag{3.34}$$

In the context of turbulence modeling, the generic function ϕ may represent, e.g., the pressure field, a component of the velocity vector, or a product of velocity components.

The general form of the incompressible Navier–Stokes equations is given by

$$\rho\frac{\partial \mathbf{u}}{\partial t} + \rho\mathbf{u}\cdot\nabla\mathbf{u} - \nabla\cdot(2\mu\bar{\bar{\varepsilon}}) + \nabla p_g = 0,$$

$$\nabla\cdot\mathbf{u} = 0, \qquad \bar{\bar{\varepsilon}} = \frac{1}{2}(\nabla\mathbf{u} + \nabla\mathbf{u}^T).$$

For simplicity, we assume that the fluid has constant density ρ and viscosity μ. The potential gravity force $\rho\mathbf{g}$ has been absorbed into the modified pressure p_g.

By virtue of (3.34), the time-averaged divergence of the instantaneous velocity

$$\mathbf{u}=\bar{\mathbf{u}}+\mathbf{u}'$$

is given by $\nabla\cdot\bar{\mathbf{u}}$. Therefore, the averaged velocity field $\bar{\mathbf{u}}$ is divergence-free:

$$\nabla\cdot\bar{\mathbf{u}}=0.$$

The time derivative and gradient operators also commute with averaging

$$\overline{\frac{\partial\mathbf{u}}{\partial t}}=\frac{\partial\bar{\mathbf{u}}}{\partial t},\qquad \overline{\nabla\mathbf{u}}=\nabla\bar{\mathbf{u}},\qquad \overline{\nabla p_g}=\nabla\bar{p}_g,$$

$$\overline{\nabla\cdot(2\mu\bar{\bar{\varepsilon}})}=\nabla\cdot[\mu(\nabla\bar{\mathbf{u}}+\nabla\bar{\mathbf{u}}^T)].$$

The only term which changes its appearance in the process of averaging is $\rho\mathbf{u}\cdot\nabla\mathbf{u}$. Substituting $\mathbf{u}=\bar{\mathbf{u}}+\mathbf{u}'$ into the equivalent conservative form $\nabla\cdot(\rho\mathbf{u}\otimes\mathbf{u})$, one obtains

$$\overline{\rho\mathbf{u}\cdot\nabla\mathbf{u}}=\rho\bar{\mathbf{u}}\cdot\nabla\bar{\mathbf{u}}-\nabla\cdot\bar{\bar{\tau}},$$

where $\bar{\bar{\tau}}$ stands for the *Reynolds stress tensor* defined by

$$\bar{\bar{\tau}}=-\overline{\rho\mathbf{u}'\otimes\mathbf{u}'}.$$

The resulting system of RANS equations reads

$$\begin{aligned}\rho\frac{\partial\bar{\mathbf{u}}}{\partial t}+\rho\bar{\mathbf{u}}\cdot\nabla\bar{\mathbf{u}}-\nabla\cdot[\mu(\nabla\bar{\mathbf{u}}+\nabla\bar{\mathbf{u}}^T)+\bar{\bar{\tau}}]+\nabla\bar{p}_g=0,\\ \nabla\cdot\bar{\mathbf{u}}=0.\end{aligned}\tag{3.35}$$

The averaged equations of motion still provide an exact description of the mean scales of turbulent flow, but constitutive equations must be provided for the turbulent stresses:

$$\tau_{ij}=-\overline{\rho u_i'u_j'}.$$

By definition, the Reynolds stress tensor $\bar{\bar{\tau}}=\{\tau_{ij}\}$ is symmetric. There are three independent components in two dimensions and six independent components in three dimensions. These components must be expressed in terms of time-averaged quantities in order to obtain a closed system.

3.4.2 ▪ Boussinesq Hypothesis

The simplest closures for the Reynolds stress tensor $\bar{\bar{\tau}}$ are based on the observation that unresolvable turbulent eddies increase the effective mixing rate and have the same effect as shear stresses proportional to the mean velocity gradients. The *Boussinesq eddy viscosity approximation* defines $\bar{\bar{\tau}}$ in terms of the mean strain rate as follows:

$$\bar{\bar{\tau}}=\mu_T(\nabla\bar{\mathbf{u}}+\nabla\bar{\mathbf{u}}^T)-\frac{2}{3}\rho k\mathbf{I},$$

where μ_T is the turbulent eddy viscosity and k is the turbulent kinetic energy

$$k = \frac{\overline{(u_1')^2} + \overline{(u_2')^2} + \overline{(u_3')^2}}{2}.$$

Substituting $\bar{\bar{\tau}}$ into the averaged momentum equation, one obtains

$$\rho\frac{\partial \bar{\mathbf{u}}}{\partial t} + \rho\bar{\mathbf{u}}\cdot\nabla\bar{\mathbf{u}} - \nabla\cdot[(\mu+\mu_T)(\nabla\bar{\mathbf{u}}+\nabla\bar{\mathbf{u}}^T)] + \nabla\bar{p}_* = 0,$$

$$\nabla\cdot\bar{\mathbf{u}} = 0,$$

where $\bar{p}_* = \bar{p}_g + \frac{2}{3}\rho k$ is a new modified pressure. Since the mean static pressure is typically much greater than the term $\frac{2}{3}\rho k$, the difference between $\bar{p}_*$ and $\bar{p}_g$ is negligible.

The above system of RANS equations has the same structure as the standard Navier–Stokes system for an incompressible fluid with the effective viscosity

$$\mu_{\text{eff}} = \mu + \mu_T.$$

The turbulent eddy viscosity μ_T represents an additional unknown which must be determined using a constitutive relation. Dimensional analysis indicates that

$$\mu_T = \rho\hat{V}\hat{L},$$

where ρ is the constant density, $\hat{V}$ is a *turbulence velocity scale* describing the intensity of fluctuations, and $\hat{L}$ is a *turbulence length scale* related to the size of the large energy-containing eddies. The local values of $\hat{V}$ and $\hat{L}$ must be determined using algebraic models or PDEs. A closure based on $m \in \{0,1,2\}$ transport equations is called an m-equation turbulence model. The fully algebraic approach corresponds to a zero-equation model. The k–ε model belongs to the family of two-equation models.

3.4.3 ▪ The Standard k–ε Model

Let the turbulence velocity scale be defined as $\hat{V} = \sqrt{k}$, where k is the turbulent kinetic energy. The rate at which k is dissipated into internal energy is given by

$$\varepsilon = \frac{\nu}{2}\sum_{i=1}^{3}\sum_{j=1}^{3}\left(\frac{\partial u_i}{\partial x_j} + \frac{\partial u_j}{\partial x_i}\right)^2 = \frac{\nu}{2}\overline{|\nabla\mathbf{u}' + (\nabla\mathbf{u}')^T|^2},$$

where $\nu = \mu/\rho$ is the kinematic viscosity of the fluid. The dissipation rate ε represents the work done by the fluctuating viscous stresses against the fluctuating strain rate.

Using the turbulence length scale $\hat{L} = \frac{k^{3/2}}{\varepsilon}$, one obtains the constitutive equation

$$\mu_T = \rho C_\mu \frac{k^2}{\varepsilon}, \tag{3.36}$$

where C_μ is a constant. The exact evolution equation for k can be derived by averaging the product of the fluctuation velocity $\mathbf{u}'$ with the momentum equation for $\mathbf{u}'$. The exact transport equation for ε is obtained by taking the curl of the equation for $\mathbf{u}'$ [176].

As in the case of the RANS equations for the averaged velocity and pressure, the exact equations for k and ε cannot be used in practical computations because they contain correlations of unresolvable fluctuating components. In order to obtain a closed system, the noncomputable terms are approximated by functions of averaged variables. The classical k–ε model [152] is based on the following pair of approximate transport equations:

$$\begin{cases} \rho\dfrac{\partial k}{\partial t}+\rho\mathbf{u}\cdot\nabla k-\nabla\cdot\left(\left(\mu+\dfrac{\mu_T}{\sigma_k}\right)\nabla k\right)=P_k-\rho\varepsilon, \\ \rho\dfrac{\partial \varepsilon}{\partial t}+\rho\mathbf{u}\cdot\nabla \varepsilon-\nabla\cdot\left(\left(\mu+\dfrac{\mu_T}{\sigma_\varepsilon}\right)\nabla \varepsilon\right)=C_1\dfrac{\varepsilon}{k}P_k-\rho C_2\dfrac{\varepsilon^2}{k}, \end{cases} \tag{3.37}$$

where the rate of turbulent energy production is modeled by

$$P_k=\mu_T\sum_{i=1}^{3}\sum_{j=1}^{3}\left(\frac{\partial u_i}{\partial x_j}+\frac{\partial u_j}{\partial x_i}\right)\frac{\partial u_i}{\partial x_j}=\frac{\mu_T}{2}|\nabla\mathbf{u}+\nabla\mathbf{u}^T|^2. \tag{3.38}$$

In the 2D case, the production term reduces to

$$P_k=\mu_T\left[2\left(\frac{\partial u_1}{\partial x_1}\right)^2+\left(\frac{\partial u_1}{\partial x_2}+\frac{\partial u_2}{\partial x_1}\right)^2+2\left(\frac{\partial u_2}{\partial x_2}\right)^2\right].$$

The axisymmetric form of P_k is given by

$$P_k=\mu_T\left[2\left(\frac{\partial u_z}{\partial z}\right)^2+\left(\frac{\partial u_z}{\partial r}+\frac{\partial u_r}{\partial z}\right)^2+2\left(\frac{\partial u_r}{\partial r}\right)^2+2\left(\frac{u_r}{r}\right)^2\right].$$

The RANS equations (3.35), the eddy viscosity formula (3.36), and the two coupled transport equations (3.37) form a closed system of equations for $\bar{\mathbf{u}}, \bar{p}_*, \mu_T, k$, and ε.

The k–ε model contains five empirical constants which have the default values

$$C_\mu=0.09,\qquad C_1=1.44,\qquad C_2=1.92,\qquad \sigma_k=1.0,\qquad \sigma_\varepsilon=1.3.$$

These universal constants are suitable for all types of turbulent flows. Fine-tuning of the model may be performed using experimental measurements for a specific problem. For the derivation and analysis of the k–ε model, we refer the reader to Mohammadi and Pironneau [176].

3.4.4 ▪ Boundary Conditions for k and ε

Due to the presence of second-order terms in the transport equations for k and ε, a boundary condition for the transported quantity must be prescribed at each boundary point.

At the inlet Γ_{in}, the profiles of k and ε are given by the Dirichlet boundary conditions

$$\begin{cases} k=k_{\text{in}}, \\ \varepsilon=\varepsilon_{\text{in}}. \end{cases}$$

The problem-dependent values of k_{in} and ε_{in} can be measured experimentally or estimated, e.g., from the turbulence intensity and length scale of the inlet geometry.

At the outlet Γ_{out}, the homogeneous Neumann boundary conditions

$$\begin{cases} \mathbf{n}\cdot\nabla k=0, \\ \mathbf{n}\cdot\nabla\varepsilon=0 \end{cases}$$

are commonly prescribed under the assumption that the outflow is fully developed.

The imposition of boundary conditions on a solid wall Γ_w is a matter of paramount importance in turbulence models. For physical reasons, it is appropriate to set $k = 0$ and $\mathbf{n}\cdot\nabla\varepsilon = 0$ on Γ_w. However, the standard k–ε model is only valid for fully turbulent (high Reynolds number) flows. Due to the existence of a thin boundary layer in which the flow is laminar, the distribution of k and ε in this layer must be determined using a modified (low Reynolds number) k–ε model (see Section 8.3.5). Moreover, the computational mesh must be fine enough to capture the steep velocity gradients in the boundary layer.

The need for integration up to the wall can be circumvented using empirical *wall laws* based on analytical solutions to the boundary layer equations. The wall functions for the turbulent kinetic energy and its dissipation rate are defined as follows [153]:

$$\begin{cases} k_w = \dfrac{u_\tau^2}{\sqrt{C_\mu}}, \\ \varepsilon_w = \dfrac{u_\tau^3}{\varkappa y}, \end{cases} \tag{3.39}$$

where $\varkappa$ is the *von Karman constant*, y is the dimensional distance from the wall, and

$$u_\tau = \sqrt{\frac{|\tau_w|}{\rho}} \tag{3.40}$$

is the dimensionless *friction velocity*. The *wall shear stress* τ_w is defined by

$$\tau_w = 2\mu[\bar{\bar{\varepsilon}}\cdot\mathbf{n} - \mathbf{n}(\bar{\bar{\varepsilon}}\cdot\mathbf{n})\cdot\mathbf{n}].$$

In the 2D case, it reduces to

$$\tau_w = \mu\left(\frac{\partial u}{\partial y}\right)_{y=0},$$

where u denotes the flow velocity parallel to the wall and y is the wall distance.

The wall laws for k and ε are valid outside the *viscous sublayer*. They must be applied at a positive distance y from the wall, where the normal velocity is still zero but the tangential velocity is nonvanishing. Hence, the homogeneous Dirichlet boundary condition for the mean tangential velocity $\bar{\mathbf{u}} - \mathbf{n}(\bar{\mathbf{u}}\cdot\mathbf{n})$ must be replaced by a law of the wall.

The tangential velocity profile depends on the dimensionless wall distance

$$y^+ = \frac{\rho u_\tau y}{\mu},$$

where y is the dimensional distance from the wall. In the viscous sublayer, we have

$$|\bar{\mathbf{u}}| = u_\tau y^+,$$

i.e., the magnitude of the velocity increases linearly with the dimensionless distance y^+.

In the *logarithmic layer*, the profile of the mean velocity is described by the wall law

$$|\bar{\mathbf{u}}| = \frac{u_\tau}{\varkappa}\ln(Ey^+),$$

where E is a roughness coefficient. The continuous velocity profile is defined by

$$|\bar{\mathbf{u}}(y^+)| = \begin{cases} u_\tau y^+, & y^+ \le y_*^+, \\ \frac{u_\tau}{\varkappa}\ln(Ey^+), & y^+ > y_*^+. \end{cases} \tag{3.41}$$

The typical values of model parameters are $\varkappa = 0.41$ and $E = 9.0$. Substituting these values into the nonlinear equation $y^+ = \frac{1}{\varkappa}\ln(Ey^+)$, one obtains $y_*^+ \approx 11.3$.

Many additional issues arise when it comes to practical implementation of wall laws for the standard k–ε model. We will address these issues in Section 8.3.5.

3.5 ▪ Summary

In this chapter, we derived the equations of fluid dynamics from the conservation principles for mass, momentum, and total energy. Using the incompressible Navier–Stokes equations as the model problem, we discussed the imposition of boundary conditions and derived a transport equation for the temperature field. We also gave a brief introduction to turbulence modeling for incompressible flows and described the k–ε model in some detail.

The incompressible Navier–Stokes equations form the basis for construction of more advanced CFD models. For example, a system of transport equations for the concentrations of chemical species may be included. These equations can also be derived from conservation principles applied to a fixed or moving control volume. The standard k–ε model has certain limitations and may need to be replaced by a more sophisticated turbulence model. The coarse scales of turbulent flow motion may be resolved within the framework of large eddy simulation with explicit or implicit subgrid scale modeling [69, 88, 119, 204]. In some applications, one has to deal with moving boundaries or interfaces separating immiscible fluids with different physical properties. Many other extensions and improvements may be envisaged, but the main focus of our guidebook is on numerical methods for flow problems, and the basic models presented in this chapter will suffice for illustration purposes.

Chapter 4

The Basics of the Finite Element Method for the One-Dimensional Heat Equation

4.1 ▪ General Concepts and Definitions

In this chapter, we give a general introduction to the finite element method (FEM) and apply the presented concepts to the 1D heat equation. The construction of basic finite element approximations to the 2D heat equation and to the incompressible Navier–Stokes equations for laminar flows is addressed in the next chapter.

The design of numerical methods for CFD begins with the choice of a mathematical model that describes the *physical phenomenon* to be studied, e.g., fluid flow or heat transfer. Next, a well-posed (initial-)boundary value problem is formulated for the *governing equations* to be solved in a bounded domain during a finite interval of time. If no analytical solution can be obtained, which is usually the case in applications of practical interest, an *approximate solution* is calculated using numerical approximations to convert PDEs into algebraic systems (see the general introduction to CFD in Chapter 2). In this guidebook, we will focus on the finite element method.

As shown in Chapter 3, mathematical models of fluid flow and heat transfer are derived from conservation laws. The geometry of the *computational domain* Ω and the length of the time interval $(0,\tau)$ depend on the nature of the problem to be solved. The following example illustrates the process of mathematical modeling for a typical flow problem.

Example 4.1. Mathematical modeling of incompressible fluid flows.

physical phenomenon	steady water flow in a tube of diameter D and length L
governing equations	$\rho\mathbf{u}\cdot\nabla\mathbf{u}-\mu\Delta\mathbf{u}+\nabla p=\rho\mathbf{g}, \quad \nabla\cdot\mathbf{u}=0$
computational domain	$\Omega:=\{\mathbf{x}\in\mathbb{R}^3\,\vert\,0<x_1<L,\ \sqrt{x_2^2+x_3^2}<\frac{1}{2}D\}$
boundary conditions	$\mathbf{u}=\mathbf{u}_{\text{in}}$ at the inlet, $\mathbf{u}=0$ on the walls, fully developed outflow profile ■

The discretization process begins with a subdivision of the computational domain into a finite number of small subdomains which form the computational mesh. The mesh cells are also called control volumes or *elements*, whence the name *the finite element method*.

In the 1D case, each element is an interval delimited by two neighboring grid points (see Fig. 4.1). A linear *shape function* can be defined by specifying its values at two *nodes* which are commonly located at the endpoints of the 1D element. Higher-order (quadratic, cubic) shape functions can be constructed by introducing additional nodal points inside the element. The typical 2D element shapes are triangles and quadrilaterals, whereas tetrahedral and hexahedral elements are typically employed in three dimensions. The nodal points (black bullets in Figs. 4.1 and 4.2) are usually located at the corners of the polygonal/polyhedral elements. In addition, there can be nodes on the edges or faces of 2D and 3D elements.

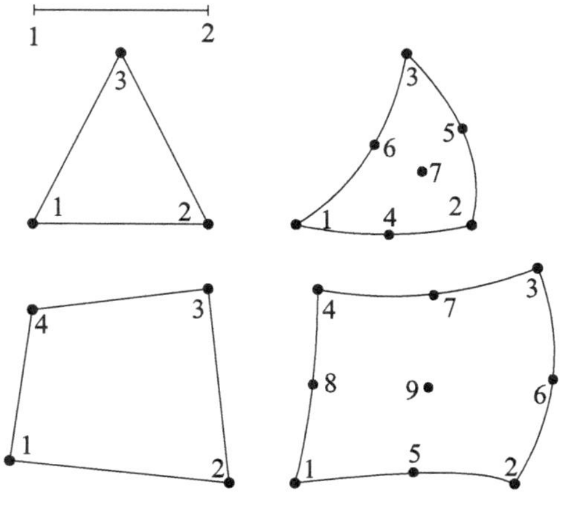

Figure 4.1. *Typical finite elements in one and two dimensions.*

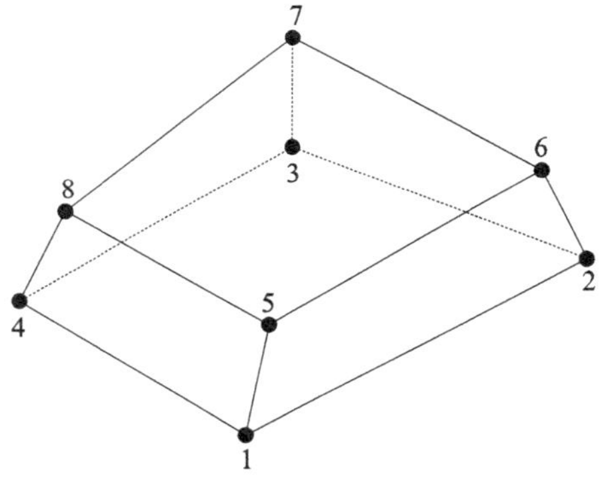

Figure 4.2. *A general hexahedral finite element in three dimensions.*

The finite element mesh is often called a *triangulation* even if it consists of rectangles, tetrahedra, or hexahedra. The elements may not overlap, and their union must cover the whole computational domain (up to approximation errors at curved boundaries). The mesh is called *conforming* if two elements can intersect only at a common vertex, edge, or face. A vertex of an element is called a *hanging node* if it lies in the interior of an edge or face of another element. Such nodes may arise, e.g., in the process of local mesh refinement and require special treatment to maintain continuity across interelement boundaries. A mesh with hanging nodes is *nonconforming*. In this text, we will only use conforming meshes.

As already mentioned in Chapter 2, finite element approximations are based on the method of weighted residuals, which leads to a variational form of the governing equation. Approximating the exact weak solution by a linear combination of piecewise-polynomial basis functions, the FEM transforms a stationary PDE into an algebraic system. A time-dependent PDE is transformed into a system of differential-algebraic equations (DAEs) which must be discretized in time to obtain an algebraic system. The corresponding matrices and right-hand sides are assembled element by element in a fully automatic manner. The size of the global matrix depends on the number of nodal points, whereas its sparsity pattern depends on the connectivity of the underlying mesh and on the node numbering.

The finite element analysis of flow problems is based on the following road map:

- *Application data*
  - quantities of interest
  - domain geometry
  - time interval
- *Mathematical model*
  - governing equations
  - boundary conditions
  - initial conditions
  - weak formulation
- *Discretization*
  - mesh generation
  - basis functions
  - matrix assembly
  - time integration
- *Simulation run*
  - solution of linear systems
  - postprocessing (visualization)
  - model validation and analysis

The quality of numerical solutions and the cost of the simulation run depend on the accuracy of the finite element approximation and efficiency of the employed linear algebra tools. In this book, we will focus on discretization aspects. For a presentation of fast iterative solvers with applications to incompressible fluid dynamics, we refer the reader to Elman, Silvester, and Wathen [61].

4.2 ▪ An Example of the FEM Usage

To give a typical example that illustrates the usage of the FEM, we apply the procedure outlined above to the backward facing step sketched in Fig. 3.2. A detailed description of this popular benchmark problem can be found in [179]. The step geometry is formed by a 10 cm high channel that suddenly expands to a 20 cm high channel. The flow enters from left through the narrow channel, which is assumed to be long enough for the incoming flow to be fully developed. The outflow boundary of the wide channel is located far enough downstream to have no influence on the flow phenomena close to the step.

In the absence of gravity effects, i.e., body forces, the steady laminar isothermal flow of an incompressible fluid is governed by the system of Navier–Stokes equations

$$\begin{cases} \rho \mathbf{u} \cdot \nabla \mathbf{u} - \nabla \cdot (2\mu \bar{\bar{\varepsilon}}) + \nabla p = 0, \\ \nabla \cdot \mathbf{u} = 0. \end{cases} \tag{4.1}$$

If the inflow velocity is known, we can prescribe the following boundary conditions:

$$\begin{cases} \text{inlet:} & u_1 = u_{\text{in}}^{(1)}, \\ & u_2 = u_{\text{in}}^{(2)} = 0, \\ \text{walls:} & u_1 = 0, \\ & u_2 = 0, \\ \text{outlet:} & \mu \dfrac{\partial u_1}{\partial x_1} - p = 0, \\ & u_2 = 0. \end{cases}$$

In accordance with the outlined discretization procedure, the 2D polygonal domain is discretized using the block-structured quadrilateral mesh displayed in Fig. 4.3. Each mesh element consists of four vertices connected by four edges. Note that the mesh is denser in the near-wall region to capture the larger velocity gradients in boundary layers, as well as flow separation at the corner of the step and reattachment at the stagnation point.

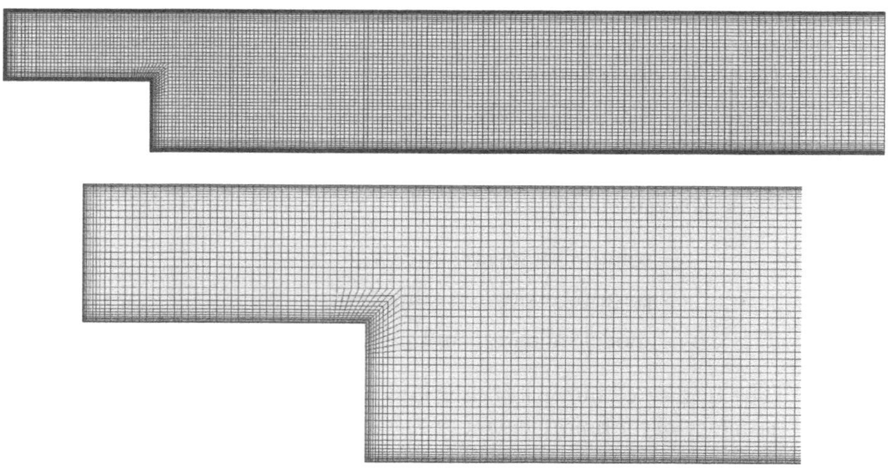

Figure 4.3. *Backward facing step: a finite element mesh and a zoom of the corner.*

The objective of the numerical study is to calculate the values of the approximate solution at the vertices of the mesh. At each node, we have three degrees of freedom which represent the unknown values of u_1, u_2, and p. In fact, the static pressure can be eliminated, leaving two unknowns per node and eight unknowns per element. Then the total number of unknowns is twice the number of mesh vertices N. The vector of nodal values

$$u = (u_{1,1}, u_{2,1}, \ldots, u_{1,N}, u_{2,N})^T$$

satisfies an algebraic system of the form $Au = f$. In the finite element literature, A is called a *stiffness matrix* and f is called a *load vector*. These names have their origins in applications of the FEM to solid mechanics. Due to the nonlinearity of the convective term $\rho\mathbf{u}\cdot\nabla\mathbf{u}$, the entries of the stiffness matrix A arising from the discretization of the Navier-Stokes equations depend on the unknown solution, i.e., the algebraic system is nonlinear. The solution vector u should provide an accurate approximation to the exact solution of the governing equations, and it should also satisfy the imposed boundary conditions.

In finite element codes, the global matrix A and the right-hand side b are assembled from element contributions. The element stiffness matrix $A^{(e)}$, the vector of local degrees of freedom $u^{(e)}$, and the element load vector $f^{(e)}$ of our model problem can be written as

$$A^{(e)} = \begin{pmatrix} a_{11}^{(e)} & a_{12}^{(e)} & \dots & a_{18}^{(e)} \\ a_{21}^{(e)} & a_{22}^{(e)} & \dots & a_{28}^{(e)} \\ \vdots & \vdots & \ddots & \vdots \\ a_{81}^{(e)} & a_{82}^{(e)} & \dots & a_{88}^{(e)} \end{pmatrix}, \qquad u^{(e)} = \begin{pmatrix} u_{1,1}^{(e)} \\ u_{2,1}^{(e)} \\ \vdots \\ u_{1,4}^{(e)} \\ u_{2,4}^{(e)} \end{pmatrix}, \qquad f^{(e)} = \begin{pmatrix} f_{1,1}^{(e)} \\ f_{2,1}^{(e)} \\ \vdots \\ f_{1,4}^{(e)} \\ f_{2,4}^{(e)} \end{pmatrix},$$

where $u_{k,i}^{(e)}$ denotes the value of the kth velocity component at the ith node of the element.

The FEM is based on the weighted residual formulation of the governing equations, so the coefficients of $A^{(e)}$ and $f^{(e)}$ represent volume and/or boundary integrals to be evaluated exactly or using numerical quadrature rules. All terms that depend on the unknown solution are represented by the matrix-vector product $A^{(e)}u^{(e)}$, whereas the data of the local problem are stored in the load vector $f^{(e)}$. Since we have neglected the body forces, all components of $f^{(e)}$ are equal to zero in our case study.

In essence, the components of $u^{(e)}$ represent pointers to some entries of the global solution vector u. For example, if the first node of the given element lies at a mesh vertex with the global number $k \in \{1,\dots,N\}$, we have $u_{1,1}^{(e)} = u_{1,k}$ and $u_{2,1}^{(e)} = u_{2,k}$. In general, each of the element nodes has a unique global number, which makes it possible to assemble the matrix A and the right-hand side f of the global system element by element.

The mesh shown in Fig. 4.3 has 9,696 vertices, i.e., $N = 9{,}696$, which corresponds to $2N = 19{,}392$ unknowns in the global system. For an optimal numbering of mesh nodes, the bandwidth of the global stiffness matrix A equals 178. Since the band also contains some zero entries, more than 99% of the matrix entries are equal to zero (there are $178 \times 19{,}392$ entries in the band and $19{,}392 \times 19{,}392$ entries in the whole matrix). Due to the large size of the algebraic system, the storage of zero entries and arithmetic operations with zeros would result in a waste of computational resources. For this reason, a sparse matrix representation is used in finite element codes. In Section 7.6.11, we describe the *compressed row storage* (CRS) format, which is suitable for general sparse matrices. The diagonals of a banded matrix with a small bandwidth can be stored compactly in a 2D array.

The numerical solution of the global system yields the approximate values of the two velocity components at each mesh point. The results of a simulation run can be visualized using postprocessing software. As an example, we present the plots of the velocity magnitude and pressure distribution for the 2D backward facing step in Fig. 4.4. It should be borne in mind that the results produced by the FEM and other numerical methods are just an approximation to the exact solution. The magnitude of numerical errors depends on the type of finite elements, on the geometric properties of the underlying mesh, and on the mesh size. The use of higher-order shape functions or denser meshes typically

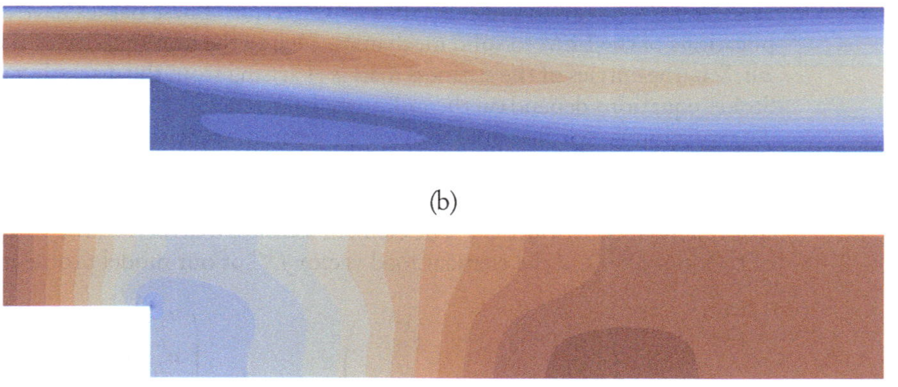

Figure 4.4. *Backward facing step:* (a) *velocity magnitude and* (b) *pressure distribution.*

leads to more accurate solutions but also results in larger algebraic systems and longer simulation times.

4.3 ▪ One-Dimensional Heat Equation

Real-life fluid flows and heat transfer processes are mostly 3D and transient in nature. However, this introductory chapter is aimed at readers with little or no knowledge of the FEM. For this reason, we begin with 1D "toy" problems. Another reason for the frequent use of 1D models in this book is the availability of analytical solutions which are more difficult or impossible to obtain for 2D and 3D problems. Since a comparison to the exact solution is the most reliable way of quantifying numerical errors, the development, analysis, and preliminary evaluation of numerical methods are commonly performed in one dimension. The extension of resulting schemes to structured grid approximations in two and three dimensions is straightforward, while the design of unstructured grid methods for multidimensional problems requires a more careful generalization of 1D concepts and algorithms.

Let us begin with simple 1D heat transfer models. The 1D counterpart of the general heat equation (3.33) derived from the energy conservation law is given by

$$\begin{aligned}\rho(x,t)c_p(x,t)\frac{\partial T(x,t)}{\partial t}&+\rho(x,t)c_p(x,t)u(x,t)\frac{\partial T(x,t)}{\partial x}\\-\frac{\partial}{\partial x}\left(k(x,t)\frac{\partial T}{\partial x}(x,t)\right)&=f(x,t),\end{aligned}$$

where $f(x,t)=\rho(x,t)h(x,t)$ denotes the rate of internal heating per unit length.

The domain of interest is an open interval $\Omega=(0,c)$, as shown in Fig. 4.5. The temperature $T(x,t)$ is a function of $x\in[0,c]$ and $t\in[0,\tau]$, where $\tau>0$ is the final time. Note that a general domain (a,b) can always be transformed into $(0,b-a)$ by a simple shift. Likewise, the assumption that the simulation begins at the time $t_0=0$ is not essential.

Assuming that the temperature T is known on the boundaries, i.e., at the endpoints of the interval $(0,c)$, we prescribe the Dirichlet boundary conditions

$$\begin{aligned}T(0)&=T_0,\\T(c)&=T_c.\end{aligned}$$

Figure 4.5. *1D computational domain.*

The temperature distribution at the time $t = 0$ is given by the initial condition

$$T(x,0) = T_0(x), \qquad x \in (0,c).$$

As a further simplification, we will neglect the time derivative and model steady heat transfer in the 1D domain $\Omega = (0,c)$ using the ODE

$$\rho(x)c_p(x)u(x)\frac{\partial T(x)}{\partial x} - \frac{\partial}{\partial x}\left(k(x)\frac{\partial T}{\partial x}(x)\right) = f(x) \tag{4.2}$$

endowed with the above Dirichlet boundary conditions. This two-point boundary value problem will serve as a starting point for our presentation, which is based on [28, 95].

4.4 ▪ Finite Element Approximations

The approximate solution to the steady 1D heat equation is denoted by $T_h(x)$, where the subscript h indicates dependence on the mesh size. In contrast to the finite difference method, T_h is defined not only at the grid points but also in the interior of mesh elements.

The main ingredients of finite element approximation techniques are

- a representation of T_h as a linear combination of basis functions;
- criteria for calculating the coefficients of the best approximation;
- estimates of the approximation error and proofs of convergence.

The most general form of a 1D finite element approximation reads

$$T_h(x) = \varphi_0(x) + T_1\varphi_1(x) + T_2\varphi_2(x) + \cdots + T_N\varphi_N(x), \tag{4.3}$$

where $T_1,\ldots,T_N$ are the unknown degrees of freedom and $\varphi_0,\ldots,\varphi_N$ are given functions constructed so as to satisfy the prescribed Dirichlet boundary conditions. In finite element methods, the basis functions φ_j have small support; i.e., they vanish outside a small patch of adjacent mesh elements. The restriction of T_h to a single element is a polynomial of certain degree which is constructed by interpolating T_h at the nodal points.

Before using the above representation of T_h to discretize the 1D heat equation, we consider a simplified problem. Suppose that the function $T : [0,c] \to \mathbb{R}$ is given and we can readily evaluate it at any point $x \in [0,c]$. How do we find the coefficients $T_1,\ldots,T_N$ such that the resulting finite-dimensional approximation T_h is as close to T as possible?

In order to determine the optimal values of the N unknown coefficients, we need N equations. The best approximation of the form (4.3) is defined by the L^2 projection

$$\int_0^c \varphi_i(x)(T_h(x) - T(x))\mathrm{d}x = 0, \qquad i = 1,\ldots,N.$$

Substituting the definition of T_h in terms of basis functions, we obtain the linear system

$$\sum_{j=1}^{N}\left(\int_0^c \varphi_i(x)\varphi_j(x)\mathrm{d}x\right)T_j = \int_0^c \varphi_i T(x)\mathrm{d}x - \int_0^c \varphi_i\varphi_0\mathrm{d}x, \qquad i = 1,\ldots,N.$$

The matrix of this linear system is regular, so it has a unique solution $(T_1,\ldots,T_N)^T$.

The same approach can be used to determine the approximate nodal values of an unknown function $T(x)$ satisfying a PDE of the form

$$\mathscr{L}T(x) = f(x),$$

where $\mathscr{L}$ is a differential operator and f is a given function. In this case, the nodal values of the best approximation T_h are determined by solving the following system:

$$\int_0^c \varphi_i(x)[\mathscr{L}T_h(x) - f(x)]\mathrm{d}x = 0, \qquad i = 1,\dots,N.$$

Second-order partial derivatives (if any) can be eliminated using integration by parts. Each test function φ_i provides one algebraic equation for the calculation of the coefficients $T_1,\dots,T_N$. The resulting approximation T_h is optimal in the sense that the residual

$$R(x) = \mathscr{L}T_h(x) - f(x)$$

is orthogonal (w.r.t. the L^2 scalar product defined in Chapter 1) to any function that belongs to the N-dimensional space spanned by the test functions $\varphi_1,\dots,\varphi_N$.

The above approach to numerical solution of PDEs is known as the *Galerkin method.* In the next section, we apply it to a simplified 1D heat equation. We remark that finite element approximations can also be constructed by solving a discrete minimization problem for a suitably defined functional. Examples of such methods include the *Ritz method* (which is only applicable to some elliptic problems and leads to the same algebraic systems as the Galerkin method) and the *least squares method.* In this text, we will focus on the Galerkin method. For a presentation of other methods, we refer the reader to [28, 248].

4.5 ▪ Galerkin Method in One Dimension

Let us consider the stationary 1D heat equation (4.2) with the constant coefficients

$$\rho = c_p = u = k = 1.$$

In the absence of heat sources ($f = 0$), our model reduces to

$$\frac{\partial T(x)}{\partial x} - \frac{\partial^2 T(x)}{\partial x^2} = 0. \tag{4.4}$$

We will solve this equation in the domain $\Omega = (0,2)$ subject to the boundary conditions

$$T(0) = 2, \tag{4.5}$$
$$T(2) = 4. \tag{4.6}$$

The exact solution of the above two-point boundary value problem is given by

$$T(x) = \frac{2(e^2-2)+2e^x}{e^2-1}.$$

The Galerkin approximation to the solution of our problem is a function of the form

$$T_h(x) = \varphi_0(x) + T_1\varphi_1(x) + T_2\varphi_2(x) + \cdots + T_N\varphi_N(x),$$

which is optimal in the sense of weighted residuals and satisfies the boundary conditions

$$T_h(0) = \varphi_0(0) + T_1\varphi_1(0) + T_2\varphi_2(0) + \cdots + T_N\varphi_N(0) = 2, \tag{4.7}$$

$$T_h(2) = \varphi_0(2) + T_1\varphi_1(2) + T_2\varphi_2(2) + \cdots + T_N\varphi_N(2) = 4 \tag{4.8}$$

for any values of the unknown coefficients $T_1, \ldots, T_N$. Setting all degrees of freedom equal to zero, we find that φ_0 must satisfy the same boundary conditions as T_h, namely,

$$\varphi_0(0) = 2, \tag{4.9}$$

$$\varphi_0(2) = 4. \tag{4.10}$$

For example, a linear function φ_0 with desired properties is given by

$$\varphi_0(x) = 2 + x.$$

In the case $T_i \neq 0$ and $T_j = 0$ for all $j \neq i$, equations (4.7) and (4.8) reduce to

$$T_h(0) = \varphi_0(0) + T_i\varphi_i(0) = 2, \tag{4.11}$$

$$T_h(2) = \varphi_0(2) + T_i\varphi_i(2) = 4, \tag{4.12}$$

where the boundary values of φ_0 are defined by (4.9) and (4.10). It follows that

$$\varphi_i(0) = \varphi_i(2) = 0, \qquad i = 1, \ldots, N. \tag{4.13}$$

Thus, all basis functions $\varphi_1, \ldots, \varphi_N$ must vanish at the endpoints of the interval $(0, 2)$.

To illustrate the construction of finite element basis functions, let us approximate the exact solution T of the 1D heat equation by the third order polynomial

$$T_h(x) = a + bx + cx^2 + dx^3. \tag{4.14}$$

By virtue of the boundary condition (4.5), we have

$$T_h(0) = a = 2.$$

Invoking the boundary condition (4.6), we obtain

$$T_h(2) = a + 2b + 4c + 8d = 2 + 2b + 4c + 8d = 4,$$

whence

$$b = 1 - 2c - 4d.$$

Substituting the values of a and b into equation (4.14), we write it as

$$T_h(x) = 2 + x + c(x^2 - 2x) + d(x^3 - 4x).$$

This result leads us to the desired representation

$$T_h(x) = \varphi_0(x) + T_1\varphi_1(x) + T_2\varphi_2(x), \tag{4.15}$$

where $T_1 = c$ and $T_2 = d$ are the degrees of freedom and

$$\varphi_0(x) = 2 + x,$$

$$\varphi_1(x) = x^2 - 2x,$$

$$\varphi_2(x) = x^3 - 4x$$

are the basis functions satisfying the boundary conditions (4.9), (4.10), and (4.13).

So far we have generated the basis functions for approximating T by a cubic polynomial T_h. Two out of four unknown coefficients have been eliminated by using the boundary conditions. The remaining two coefficients, T_1 and T_2, will be determined such that the approximation is as close to the exact solution T as possible. Here we will use the Galerkin method based on the concept of weighted residuals to find optimal coefficients.

The residual $R(x)$ of the simplified 1D heat equation (4.4) is defined by

$$R(x) = \frac{\partial T_h(x)}{\partial x} - \frac{\partial^2 T_h(x)}{\partial x^2}. \tag{4.16}$$

Differentiating the approximate solution T_h, as defined by (4.15), we find that

$$\begin{aligned} R(x) &= \frac{\partial}{\partial x}(\varphi_0(x) + T_1\varphi_1(x) + T_2\varphi_2(x)) \\ &\quad - \frac{\partial^2}{\partial x^2}(\varphi_0(x) + T_1\varphi_1(x) + T_2\varphi_2(x)) \\ &= 1 + T_1(2x-4) + T_2(3x^2-6x-4). \end{aligned}$$

The value of $R(x)$ depends on the position x and measures the accuracy of the finite element approximation at the given point. It is generally impossible to find the coefficients T_1 and T_2 such that the residual becomes zero everywhere. Therefore, the Galerkin method employs a weighted residual formulation of the governing equation. Multiplying $R(x)$ by a test function $w(x)$ and integrating over $(0,c)$, we obtain the Galerkin condition

$$\int_0^c w(x)R(x)\,\mathrm{d}x = 0. \tag{4.17}$$

For the discrete problem to have a unique solution, the number of such conditions must be equal to the number of basis functions associated with the unknown degrees of freedom.

In our example with two unknown coefficients, the Galerkin conditions are given by

$$\begin{aligned} \int_0^c \varphi_1(x)R(x)\,\mathrm{d}x &= 0, \\ \int_0^c \varphi_2(x)R(x)\,\mathrm{d}x &= 0. \end{aligned}$$

Using the above definitions of $R(x)$, $\varphi_1(x)$, and $\varphi_2(x)$ for our problem, we obtain

$$\begin{aligned} \int_0^2 (x^2-2x)\big(1+T_1(2x-4)+T_2(3x^2-6x-4)\big)\,\mathrm{d}x &= 0, \\ \int_0^2 (x^3-4x)\big(1+T_1(2x-4)+T_2(3x^2-6x-4)\big)\,\mathrm{d}x &= 0. \end{aligned}$$

Calculating the integrals on the left-hand side, we end up with the linear system

$$\begin{aligned} \frac{8}{3}T_1 + \frac{128}{15}T_2 &= \frac{4}{3}, \\ \frac{112}{15}T_1 + \frac{128}{5}T_2 &= 4, \end{aligned}$$

which can easily be solved for T_1 and T_2. The unique solution is given by

$$T_1 = 0,$$
$$T_2 = \frac{5}{32}.$$

Substituting these values of T_1 and T_2 into (4.15), we obtain the Galerkin solution

$$T_h(x) = 2 + x + \frac{5}{32}(x^3 - 4x).$$

The graphs of the approximate and exact solutions are presented in Fig. 4.6. We can hardly see the difference between T and T_h, although we used a very simple approximation.

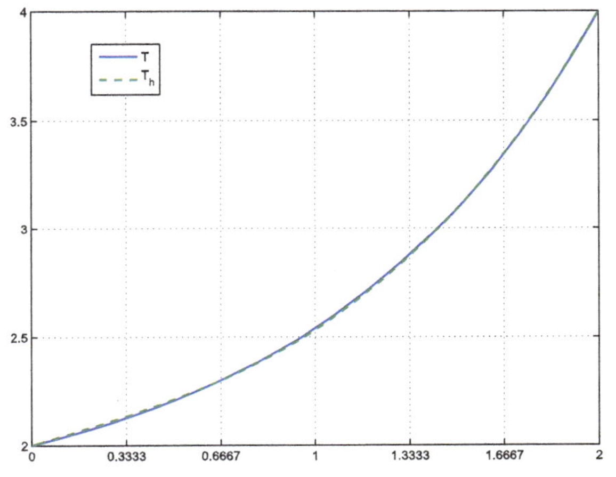

Figure 4.6. *The approximate solution T_h vs. the exact solution T.*

This example shows clearly how the Galerkin method transforms a linear ODE into a system of linear algebraic equations for the coefficients of the best approximation $T_h(x)$ to the exact solution $T(x)$ of the problem at hand. Even if it is impossible to find the analytical solution, the Galerkin approximation to a well-posed boundary value problem will lead to an algebraic system with a regular matrix.

The approximation properties of the Galerkin method depend on the type and number of basis functions, as well as on the structure of the exact solution. In the above example, we obtained a reasonable numerical solution by looking for the best approximation in the space of globally defined cubic polynomials. Of course, not every function can be approximated accurately by a cubic polynomial. A major strength of the Galerkin method is the great flexibility in the choice of basis functions for constructing approximate solutions of the form (4.3) and their multidimensional counterparts. The degree of polynomials or, more generally, the mathematical form of the basis functions may be selected more or less arbitrarily. Instead of cubic polynomials, we could use higher-order polynomials or piecewise-polynomial approximations defined in terms of local basis functions on mesh elements. Increasing the number of basis functions, one usually obtains more accurate approximations, but the computations become more laborious. The discrete form of our toy problem was easy to solve analytically, but algebraic systems for a large number of unknown degrees of freedom are solved numerically using direct or iterative methods.

In our 1D example, the inhomogeneous boundary conditions were taken into account using φ_0. This auxiliary function may also be expressed as a linear combination of suitably defined basis functions, and the equations that define the coefficients of this linear combination may be added to the algebraic system for the unknown degrees of freedom. We will use this approach to numerical implementation of Dirichlet boundary conditions later in this chapter. It is very general and readily applicable to multidimensional problems.

4.6 ▪ Generic FEM Algorithm

The numerical solution of fluid dynamics problems by finite element methods begins with the formulation of a well-posed mathematical problem and ends with postprocessing of the numerical results. Before going into detail, we present the main cornerstones of numerical simulation and a road map that we will follow in the remainder of this chapter.

The finite element analysis process is based on mathematical analysis and technical calculations. The theoretical part of this multistage procedure is needed to set up the general mathematical framework and consists of tasks to be accomplished using paper and pencil. The computational part includes all operations to be carried out by a computer during the simulation run. While the theoretical work needs to be done just once and leads to general-purpose algorithms, the computational steps may depend on problem-specific data such as the geometry of the computational domain and the prescribed boundary conditions.

The following ten-step algorithm summarizes the finite element analysis procedure.

Theoretical Part

(i) Problem statement: derivation of the governing equations (with appropriate modifications or simplifications); possible types of initial/boundary conditions.

(ii) Variational formulation: multiplication of the governing equations by the weighting functions and integration over the domain; integration by parts using Green's formula, substitution of the boundary conditions into the boundary integrals.

(iii) Discretization: Galerkin approximation to the continuous integral-form equations.

(iv) Localization: decomposition of global matrices and vectors into element contributions.

Computational Part

(v) Generation of a finite element mesh for the given computational domain.

(vi) Assembly of the global algebraic system from element matrices and vectors.

(vii) Incorporating problem-specific boundary conditions into the algebraic system.

(viii) Numerical solution of the algebraic system using direct or iterative methods.

(ix) Repeating the appropriate computational steps in applications to time-dependent problems, domains with moving boundaries, and nonlinear problems.

(x) Postprocessing: computation of derived quantities and visualization.

Some of the above steps require further explanation, which we give in the following sections.

4.7 ▪ Steady One-Dimensional Heat Equation

Let us now apply the Galerkin method to the steady 1D heat transfer model

$$\rho c_p u \frac{\partial T}{\partial x} - \frac{\partial}{\partial x}\left(k \frac{\partial T}{\partial x}\right) = f, \tag{4.18}$$

where the parameters ρ, c_p, u, and k are assumed to be constant. At the endpoints of the general 1D domain $\Omega = (0, c)$, we prescribe the Dirichlet boundary conditions

$$T(0) = T_0,$$
$$T(c) = T_c.$$

We will discretize this two-point boundary value problem following the general procedure outlined in the previous section. The below presentation is mainly based on [129].

(i) Problem Statement

The steady 1D heat equation (4.18) represents a simplification of the unsteady multidimensional heat transfer model derived in Chapter 3 from the law of energy conservation. The equation at hand is linear since the coefficients are constant and independent of T.

Since the time derivative is neglected, no initial condition is required. A boundary condition for T or $\frac{\partial T}{\partial x}$ is required at each of the two endpoints of the interval $(0, c)$.

(ii) Variational Formulation

The residual of the second-order ODE (4.18) is defined by

$$R(x) = \rho c_p u \frac{\partial T}{\partial x} - k \frac{\partial^2 T}{\partial x^2} - f$$

and equals zero if T is the exact solution. Multiplying $R(x)$ by a test function w from a suitable space V, we obtain the weighted residual form of the governing equation

$$\int_0^c w(x) R(x)\, \mathrm{d}x = 0. \tag{4.19}$$

By definition of $R(x)$, each test function $w \in V$ yields an integral equation of the form

$$\int_0^c w \left(\rho c_p u \frac{\partial T}{\partial x} - k \frac{\partial^2 T}{\partial x^2} - f \right) \mathrm{d}x = 0. \tag{4.20}$$

For this problem to be well-defined, the solution T must have square integrable second derivatives. Using integration by parts (Green's formula), we obtain

$$\rho c_p u \int_0^c w \frac{\partial T}{\partial x}\, \mathrm{d}x + k \int_0^c \frac{\partial w}{\partial x} \frac{\partial T}{\partial x}\, \mathrm{d}x - k w \frac{\partial T}{\partial x}\Big|_0^c = \int_0^c w f\, \mathrm{d}x. \tag{4.21}$$

This variational formulation is more general because it is well-defined for $T, w \in H^1((0, c))$, where H^1 is the Sobolev space of weakly differentiable functions (see Chapter 1).

The trace of the test function w must vanish on the Dirichlet boundaries. Thus, the space of the test functions is given by $H_0^1((0, c)) = \{v \in H^1((0, c)) \,|\, v(0) = v(c) = 0\}$ and the boundary terms drop out. This leads us to the following problem statement:

Find a function $T \in H^1((0,c))$ such that $T(0) = T_0$, $T(c) = T_c$, and

$$\rho c_p u \int_0^c w \frac{\partial T}{\partial x}\, dx + k \int_0^c \frac{\partial w}{\partial x} \frac{\partial T}{\partial x}\, dx = \int_0^c w f \, dx \qquad \forall w \in H_0^1((0,c)). \tag{4.22}$$

The solution to this *continuous variational problem* is a generalized solution to (4.18).

(iii) Discretization

Since the Sobolev spaces $H^1((0,c))$ and $H_0^1((0,c))$ contain infinitely many functions, we will look for approximations in a finite-dimensional subspace $V_h \subset H^1((0,c))$ and use test functions from the same subspace. Given a set of basis functions $\{\varphi_j\}$ spanning V_h, we define the approximate solution T_h as a linear combination of these basis functions:

$$T_h(x) = \sum_{j=1}^{N} T_j \varphi_j(x). \tag{4.23}$$

Since we have two boundary conditions, we could eliminate two degrees of freedom as we did in the previous example. However, we will treat all coefficients as unknown for the time being and insert the Dirichlet boundary conditions into the algebraic system later.

Replacing the temperature T by its approximation T_h in the weighted residual formulation (4.22), we obtain the associated *discrete variational problem*:

$$\rho c_p u \int_0^c w_h \frac{\partial T_h}{\partial x}\, dx + k \int_0^c \frac{\partial w_h}{\partial x} \frac{\partial T_h}{\partial x}\, dx = \int_0^c w_h f \, dx, \qquad w_h \in V_h, \tag{4.24}$$

where T_h is the approximate solution defined by (4.23). The optimal values of the coefficients $T_1, \ldots, T_N$ can be determined using the two boundary conditions and $N-2$ Galerkin conditions of the form (4.24). In practice, it is more convenient to assemble the system of N Galerkin conditions and overwrite two equations by the Dirichlet boundary conditions $T_h(0) = T_0$ and $T_h(c) = T_c$. Using $w_h = \varphi_i$ as a test function, we obtain

$$\rho c_p u \int_0^c \varphi_i \frac{\partial}{\partial x}\left(\sum_{j=1}^{N} T_j \varphi_j\right) dx + k \int_0^c \frac{\partial \varphi_i}{\partial x} \frac{\partial}{\partial x}\left(\sum_{j=1}^{N} T_j \varphi_j\right) dx = \int_0^c \varphi_i f \, dx$$

or, equivalently,

$$\sum_{j=1}^{N}\left(\rho c_p u \int_0^c \varphi_i \frac{\partial \varphi_j}{\partial x}\, dx\right) T_j + \sum_{j=1}^{N}\left(k \int_0^c \frac{\partial \varphi_i}{\partial x} \frac{\partial \varphi_j}{\partial x}\, dx\right) T_j = \int_0^c \varphi_i f \, dx. \tag{4.25}$$

The system of equations for $i = 1, \ldots, N$ can be represented in the matrix form

$$(C + K)T = F, \tag{4.26}$$

where C is the discrete convection operator, K is the discrete diffusion operator, and F is the discretized source term. The coefficients of the algebraic system are given by

$$\begin{aligned} c_{ij} &= \rho c_p u \int_0^c \varphi_i \frac{\partial \varphi_j}{\partial x}\, dx, && i,j = 1, \ldots, N, \\ k_{ij} &= k \int_0^c \frac{\partial \varphi_i}{\partial x} \frac{\partial \varphi_j}{\partial x}\, dx, && i,j = 1, \ldots, N, \\ F_i &= \int_0^c \varphi_i f \, dx, && i = 1, \ldots, N. \end{aligned}$$

The stiffness matrix $A = C + K$ models the transport of T by convective fluxes and heat conduction. The load vector F models the generation of heat by internal sources.

Remark 4.1. To make the $N \times N$ system matrix A nonsingular, at least one equation must be replaced by a Dirichlet or Newton boundary condition. If an inhomogeneous Neumann or Newton boundary condition is employed, the boundary term that results from integration by parts must be included in the variational formulation (see Section 4.8).

(iv) Localization

In Chapter 4.5, we approximated the exact solution T of the simplified 1D heat equation by a cubic polynomial T_h defined on the whole domain $(0,2)$. This strategy is not to be recommended in general because it leads to linear systems with full matrices, and the entries of these matrices are expensive to calculate. Moreover, the procedure that we used to construct the basis functions was not general enough. The FEM is based on the idea of using local basis functions to construct interpolation polynomials defined on small subdomains which represent the elements of a given computational mesh. The globally defined approximate solution T_h is obtained by piecing together the resulting shape functions. The corresponding piecewise-polynomial basis functions have small support, leading to sparse matrices. When it comes to matrix assembly, the integrals are decomposed into element contributions which may be calculated using numerical integration.

Let us discretize the interval $(0,c)$ using N grid points $x_1 < x_2 < \cdots < x_N$ such that $x_1 = 0$ and $x_N = c$. The 1D mesh consists of $N-1$ elements (see Fig. 4.7),

$$E_e = [x_e, x_{e+1}], \qquad e = 1, \ldots, N-1.$$

For simplicity, we assume that all mesh elements have the same length,

$$h = x_{e+1} - x_e = \frac{c}{N-1}, \qquad e = 1, \ldots, N-1.$$

In this case, the coordinate of the eth grid point is given by

$$x_e = (e-1)h, \qquad e = 1, \ldots, N.$$

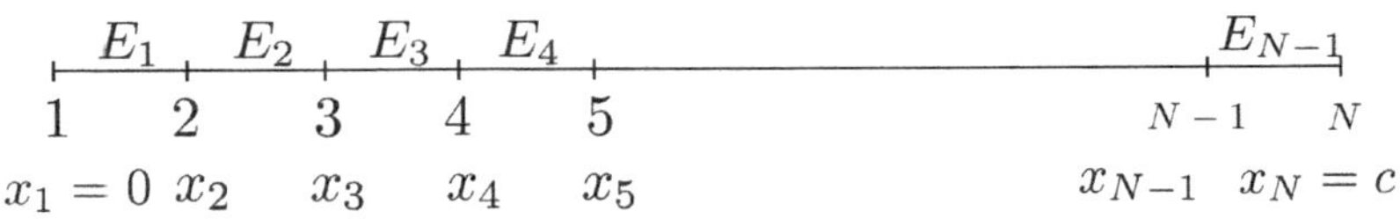

Figure 4.7. *Elements and nodes of a uniform mesh in one dimension.*

The construction of the approximate solution T_h begins with the definition of the shape functions $T_h^{(e)} = T_h|_{E_e}$ for $e = 1, \ldots, N-1$. For example, a piecewise-linear approximation T_h can be defined by gluing together the first-order polynomials:

$$T_h^{(e)}(x) = \alpha_1^{(e)} + \alpha_2^{(e)} x, \qquad x \in [x_e, x_{e+1}]. \tag{4.27}$$

A linear shape function $T_h^{(e)}$ is depicted in Fig. 4.8. The values of the coefficients $\alpha_1^{(e)}$ and $\alpha_2^{(e)}$ may be determined by fitting the linear polynomial $T_h^{(e)}$ to the solution values at two nodal points, i.e., by solving a local linear interpolation problem (see Fig. 4.9).

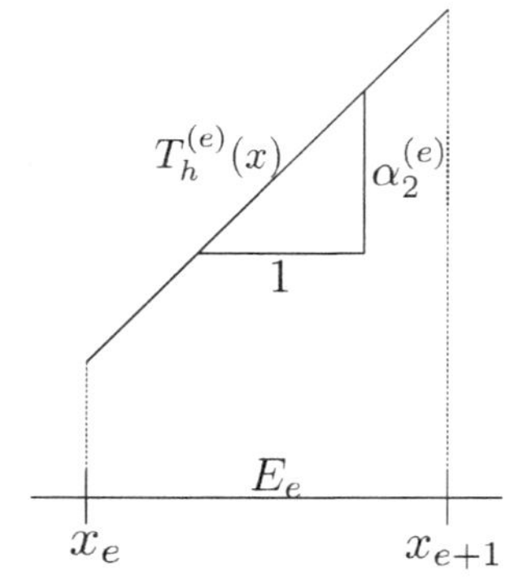

Figure 4.8. *Linear shape functions $T_h^{(e)}$.*

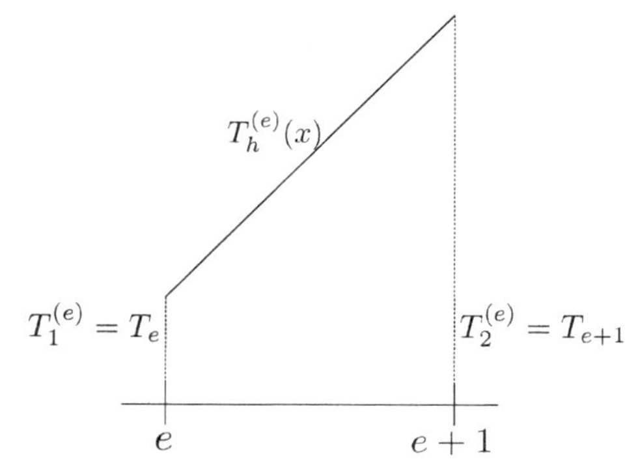

Figure 4.9. *Local interpolation problem.*

To define a continuous piecewise-linear finite element approximation T_h in terms of its values at the nodes of the 1D mesh, we will use the interpolation conditions

$$T_h^{(e)}(x_e) = T_1^{(e)} := T_h(x_e), \tag{4.28}$$

$$T_h^{(e)}(x_{e+1}) = T_2^{(e)} := T_h(x_{e+1}). \tag{4.29}$$

The evaluation of the linear polynomial (4.27) at the nodal points yields

$$\alpha_1^{(e)} + \alpha_2^{(e)}(e-1)h = T_1^{(e)},$$
$$\alpha_1^{(e)} + \alpha_2^{(e)}eh = T_2^{(e)}.$$

This linear system can be solved for $\alpha_1^{(e)}$ and $\alpha_2^{(e)}$. The solution is given by

$$\alpha_1^{(e)} = eT_1^{(e)} - (e-1)T_2^{(e)},$$
$$\alpha_2^{(e)} = \frac{1}{h}(T_2^{(e)} - T_1^{(e)}).$$

Substituting the coefficients $\alpha_1^{(e)}$ and $\alpha_2^{(e)}$ into equation (4.27), we obtain

$$T_h(x) = T_1^{(e)}\left(\frac{-x}{h} + e\right) + T_2^{(e)}\left(\frac{x}{h} - e + 1\right).$$

The so-defined linear interpolation polynomial T_h can be written in the form

$$T_h(x) = T_1^{(e)}\varphi_1^{(e)}(x) + T_2^{(e)}\varphi_2^{(e)}(x), \tag{4.30}$$

where $T_1^{(e)}$ and $T_2^{(e)}$ represent the local degrees of freedom and

$$\begin{aligned} \varphi_1^{(e)}(x) &= \frac{-x}{h} + e = \frac{x_{e+1} - x}{h}, \\ \varphi_2^{(e)}(x) &= \frac{x}{h} - e + 1 = \frac{x - x_e}{h} \end{aligned} \tag{4.31}$$

are the local basis functions associated with the element E_e (see Fig. 4.10).

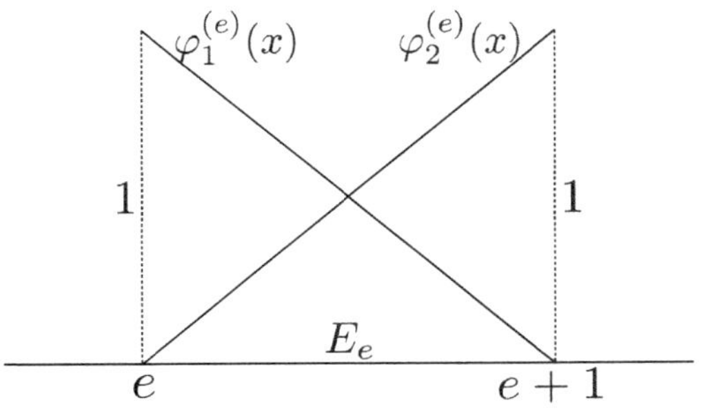

Figure 4.10. *Local basis functions for the element E_e.*

The shape function $T_h^{(e)}$ defined by (4.30) is linear and so are the local basis functions $\varphi_1^{(e)}(x)$ and $\varphi_2^{(e)}(x)$. The interpolation conditions (4.28), (4.29) are satisfied because

$$\begin{aligned} \varphi_1^{(e)}(x_e) &= 1, \\ \varphi_1^{(e)}(x_{e+1}) &= 0 \end{aligned}$$

and

$$\begin{aligned} \varphi_2^{(e)}(x_e) &= 0, \\ \varphi_2^{(e)}(x_{e+1}) &= 1. \end{aligned}$$

Each element provides a pair of local basis functions $\varphi_1^{(e)}(x)$ and $\varphi_2^{(e)}(x)$ which represent the *Lagrange basis polynomials* for the local interpolation problem:

$$E_1: \quad \begin{cases} \varphi_1^{(1)}(x) = \dfrac{-x}{h} + 1, \\ \varphi_2^{(1)}(x) = \dfrac{x}{h}, \end{cases}$$

$$E_2: \quad \begin{cases} \varphi_1^{(2)}(x) = \dfrac{-x}{h} + 2, \\ \varphi_2^{(2)}(x) = \dfrac{x}{h} - 1, \end{cases}$$

and so on for all elements. The last pair of Lagrange polynomials is

$$E_{N-1}: \quad \begin{cases} \varphi_1^{(N-1)}(x) = \dfrac{-x}{h} + N - 1, \\ \varphi_2^{(N-1)}(x) = \dfrac{x}{h} - N + 2. \end{cases}$$

The corresponding element-based shape functions $T_h^{(e)} = T_h|_{E_e}$ are defined by

$$\begin{aligned} T_h^{(1)}(x) &= T_1^{(1)}\varphi_1^{(1)}(x) + T_2^{(1)}\varphi_2^{(1)}(x), \qquad x \in E_1, \\ T_h^{(2)}(x) &= T_1^{(2)}\varphi_1^{(2)}(x) + T_2^{(2)}\varphi_2^{(2)}(x), \qquad x \in E_2, \\ &\dots \\ T_h^{(N-1)}(x) &= T_1^{(N-1)}\varphi_1^{(N-1)}(x) + T_2^{(N-1)}\varphi_2^{(N-1)}(x), \qquad x \in E_{N-1}. \end{aligned} \tag{4.32}$$

The above pairs of local basis functions and the resulting piecewise-linear approximation T_h on the interval $(0, c)$ are sketched in Figs. 4.11 and 4.12, respectively.

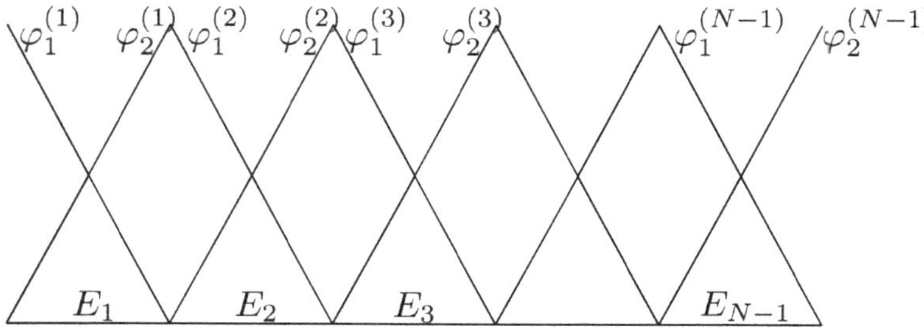

Figure 4.11. *Local basis functions on the elements* $E_1, \dots, E_{N-1}$.

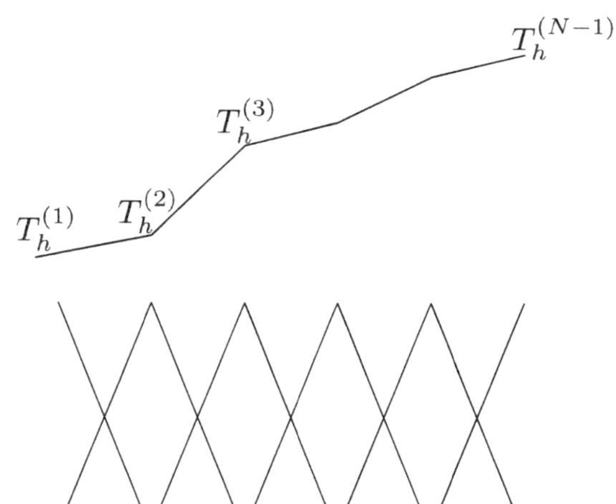

Figure 4.12. *Piecewise-linear approximation* T_h *on the interval* $(0, c)$.

So far we have derived the local basis functions and used them to construct a local approximation $T_h^{(e)}$ on each element. The corresponding global approximation T_h can be written as a linear combination of *global basis functions* $\varphi_1, \varphi_2, \dots, \varphi_N$ such that

$$T_h(x) = \sum_{j=1}^{N} T_j \varphi_j(x).$$

The unknown coefficients T_j represent the values of T_h at the grid points. We have

$$T_j = T_h(x_j).$$

The restriction of the basis function φ_j to each element is a linear polynomial and

$$\varphi_j(x_i)=\delta_{ij}, \qquad i,j=1,\dots,N.$$

It follows that φ_j is a piecewise-linear continuous function which is equal to 1 at the grid point $x_j=jh$ and equal to zero at any other grid point (see Fig. 4.13).

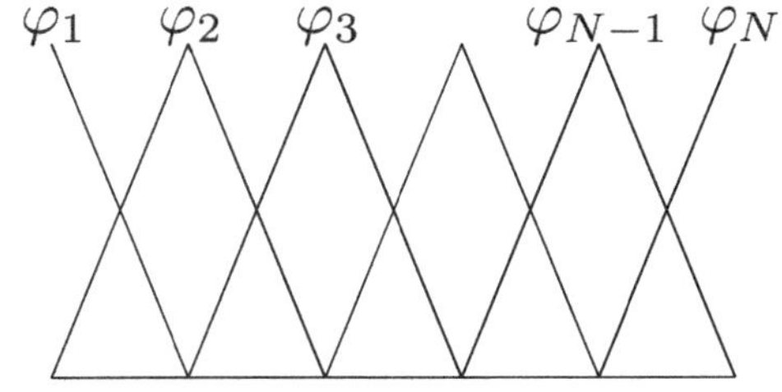

Figure 4.13. *Global piecewise-linear basis functions* $\varphi_1,\dots,\varphi_N$.

The global basis functions $\varphi_1,\dots,\varphi_N$ are associated with the grid points $x_1,\dots,x_N$ and can be represented in terms of the local basis functions as follows:

$$\varphi_1(x)=\begin{cases}\varphi_1^{(1)}(x), & x\in E_1=(0,h),\\ 0 & \text{elsewhere},\end{cases}$$

$$\varphi_2(x)=\begin{cases}\varphi_2^{(1)}(x), & x\in E_1=(0,h),\\ \varphi_1^{(2)}(x), & x\in E_2=(h,2h),\\ 0 & \text{elsewhere},\end{cases}$$

$$\varphi_{N-1}(x)=\begin{cases}\varphi_2^{(N-2)}(x), & x\in E_{N-2}=((N-3)h,(N-2)h),\\ \varphi_1^{(N-1)}(x), & x\in E_{N-1}=((N-2)h,(N-1)h),\\ 0 & \text{elsewhere},\end{cases}$$

$$\varphi_N(x)=\begin{cases}\varphi_2^{(N-1)}(x), & x\in E_{N-1}=((N-2)h,(N-1)h),\\ 0 & \text{elsewhere}.\end{cases}$$

The basis function φ_e is associated with the grid point x_e. Its support, i.e., the set of points where φ_e is not equal to zero, consists of at most two elements. If x_e is not a boundary point, the shape of the "hat" function φ_e, $e\in\{2,\dots,N-1\}$, is defined by

$$\varphi_e(x)=\begin{cases}\dfrac{x}{h}-e+2, & x\in((e-2)h,(e-1)h),\\ \dfrac{-x}{h}+e, & x\in((e-1)h,eh),\\ 0 & \text{elsewhere}.\end{cases} \tag{4.33}$$

Remark 4.2. Due to the property $\varphi_j(x_i)=\delta_{ij}$, the basis functions $\varphi_2,\dots,\varphi_{N-1}$ must vanish at $x_0=0$ and $x_N=c$. The approximate solution T_h can be written as

$$T_h(x)=\varphi_0(x)+\sum_{j=2}^{N-1}T_j\varphi_j(x),$$

where

$$\varphi_0(x) = T_1\varphi_1(x) + T_N\varphi_N(x)$$

is a piecewise-linear function satisfying the Dirichlet boundary conditions

$$T_1 = \varphi_0(x_0) = T_h(0) = T_0,$$
$$T_N = \varphi_0(x_N) = T_h(c) = T_c.$$

Now that we have defined the local and global basis/test functions for the Galerkin finite element approximation, we proceed to the assembly of the algebraic system for the degrees of freedom $T_1, \ldots, T_N$. Recall that entries of the global matrices C and K and the components of the load vector F are integrals over $(0, c)$. Since the approximate solution T_h and the test functions $w_h \in \{\varphi_1, \ldots, \varphi_N\}$ are defined in terms of their restrictions to individual elements, it makes sense to decompose each integral over $(0, c)$ into a sum of integrals over the elements E_e, $e = 1, \ldots, N-1$. The entries of the corresponding element matrices $C^{(e)}$, $K^{(e)}$ and of the element load vector $F^{(e)}$ are defined as follows:

$$c_{ij}^{(e)} = \rho c_p u \int_{E_e} \varphi_i^{(e)} \frac{\partial \varphi_j^{(e)}}{\partial x}\, \mathrm{d}x, \qquad i,j = 1,2, \tag{4.34}$$

$$k_{ij}^{(e)} = k \int_{E_e} \frac{\partial \varphi_i^{(e)}}{\partial x} \frac{\partial \varphi_j^{(e)}}{\partial x}\, \mathrm{d}x, \qquad i,j = 1,2, \tag{4.35}$$

$$F_i^{(e)} = \int_{E_e} \varphi_i^{(e)} f \, \mathrm{d}x, \qquad i = 1,2. \tag{4.36}$$

The element matrices and vectors represent restrictions of their global counterparts to a single element. Integration over $(0, c)$ is replaced by integration over E_e, and global basis functions are replaced by local basis functions. The size of element matrices/vectors is determined by the number of local degrees of freedom ($N^{(e)} = 2$ for linear elements in one dimension).

When it comes to global matrix assembly, the element contributions stored in

$$C^{(e)} = \begin{pmatrix} c_{11}^{(e)} & c_{12}^{(e)} \\ c_{21}^{(e)} & c_{22}^{(e)} \end{pmatrix}, \qquad K^{(e)} = \begin{pmatrix} k_{11}^{(e)} & k_{12}^{(e)} \\ k_{21}^{(e)} & k_{22}^{(e)} \end{pmatrix}, \qquad F^{(e)} = \begin{pmatrix} F_1^{(e)} \\ F_2^{(e)} \end{pmatrix} \tag{4.37}$$

are used to update the coefficients of the global algebraic system in a loop over mesh elements. We will explain the assembly process below in step (vi) of the general procedure.

Remark 4.3. In this section, we defined the linear shape functions by interpolating T_h at the endpoints of mesh elements. In principle, the local interpolation polynomials could have been defined using any other pair of points. However, the use of nodes located in the interior of mesh elements would increase the number of global degrees of freedom to $2N-2$ and make it impossible to ensure the continuity of global basis functions at the common boundary of two elements (leading to a *discontinuous Galerkin* approximation).

Remark 4.4. The accuracy of a finite element approximation could be improved by using shape functions defined by higher-order (quadratic, cubic) local interpolation polynomials. In the 1D case, one needs $k+1$ points to construct a polynomial of degree k. To maintain global continuity and benefit from the sharing of degrees of freedom, two nodal points should be placed at the endpoints of the element. In the case $k \geq 2$, additional interpolation points can be defined, e.g., as nodes of the uniform local grid that corresponds to a subdivision of the given element into k intervals of equal length. For example, a quadratic shape function can be constructed by interpolating the solution values at the center and at the endpoints of the element. Since the number of local degrees of freedom N_{loc} is equal to the number of nodes per element, quadratic finite element approximations lead to 3×3 element matrices and 3×1 element vectors in one dimension. The number of global degrees of freedom becomes equal to $N+(N-1)=2N-1$. For details, we refer the reader to the book by Cuvelier, Segal, and van Steenhoven [44].

(v) Mesh Generation

To generate the computational mesh for simulating steady heat transfer in the 1D domain $\Omega=(0,2)$, we subdivide this interval into three elements of equal length $h=\frac{2}{3}$. In this case, the number of grid points is $N=4$, and their coordinates are given by

$$x_1=0, \qquad x_2=\frac{2}{3}, \qquad x_3=\frac{4}{3}, \qquad x_4=2.$$

The number of global degrees of freedom for a piecewise-linear approximation on this uniform mesh is also equal to 4, but two degrees of freedom are actually known from the Dirichlet boundary conditions. In our 1D simulation, we will use the boundary values

$$T_1=2, \qquad T_4=4.$$

For a continuous 1D finite element approximation of degree $k>1$, the numerical solution T_h is defined by its values at the N grid points and $(N-1)(k-1)$ nodes located in the interior of mesh elements. In our example with $N=4$ grid points and $N-1=3$ elements, the number of global degrees of freedom is given by $N_{\text{DOF}}=4+3(k-1)=3k+1$. The use of quadratic elements ($k=2$) would lead to $N_{\text{DOF}}=7$.

(vi) Assembly of the Global System

The global algebraic system for our toy problem is assembled by updating the coefficients of the 4×4 matrices C and K and of the 4×1 load vector F step by step in a loop over the three elements of the 1D mesh generated at step (v). Since the global basis functions and their derivatives are nonvanishing on at most two elements, the integrals containing their products with other functions will consist of at most two element contributions.

The coefficients of the local matrices $C^{(e)}$ and $K^{(e)}$ and of the local vector $F^{(e)}$ are defined by equations (4.34)–(4.36). The corresponding integrals over E_e can be calculated using paper and pencil or numerical integration. Since our toy problem has constant coefficients, all element matrices are the same. We present them below in Example 4.2.

In the process of global assembly, the entries of local matrices and vectors are added to the corresponding entries of their global counterparts. For example, we have

$$c_{11}=\rho c_p u\int_\Omega \varphi_1\frac{\partial \varphi_1}{\partial x}\,\mathrm{d}x=\rho c_p u\int_{E_1}\varphi_1^{(1)}\frac{\partial \varphi_1^{(1)}}{\partial x}\,\mathrm{d}x=c_{11}^{(1)},$$

$$c_{12}=\rho c_p u\int_\Omega \varphi_1\frac{\partial \varphi_2}{\partial x}\,\mathrm{d}x=\rho c_p u\int_{E_1}\varphi_1^{(1)}\frac{\partial \varphi_2^{(1)}}{\partial x}\,\mathrm{d}x=c_{12}^{(1)}$$

and

$$k_{11}=k\int_\Omega \frac{\partial \varphi_1}{\partial x}\frac{\partial \varphi_1}{\partial x}\,\mathrm{d}x=k\int_{E_1}\frac{\partial \varphi_1^{(1)}}{\partial x}\frac{\partial \varphi_1^{(1)}}{\partial x}\,\mathrm{d}x=k_{11}^{(1)},$$

$$k_{12}=k\int_\Omega \frac{\partial \varphi_1}{\partial x}\frac{\partial \varphi_2}{\partial x}\,\mathrm{d}x=k\int_{E_1}\frac{\partial \varphi_1^{(1)}}{\partial x}\frac{\partial \varphi_2^{(1)}}{\partial x}\,\mathrm{d}x=k_{12}^{(1)}.$$

Since the global basis functions φ_3 and φ_4 are equal to zero in element E_1, while the basis function φ_1 is equal to zero in all other mesh elements, it follows that

$$c_{13} = c_{14} = \rho c_p u \int_\Omega 0 \, dx = 0,$$

$$k_{13} = k_{14} = k \int_\Omega 0 \, dx = 0.$$

In general, the global matrix will have a zero entry in row i and column j if the supports of the basis functions φ_i and φ_j do not overlap, i.e., if $\varphi_i \varphi_j = 0$ in Ω.

So far we have determined the coefficients of the first equation in our linear algebraic system. The right-hand side of this algebraic equation is given by the integral

$$F_1 = \int_\Omega \varphi_1 f \, dx = \int_{E_1} \varphi_1^{(1)} f \, dx = F_1^{(1)}.$$

The coefficients of the second equation are given by the integrals in which φ_2 is used as a test function. The support of φ_2 overlaps the supports of φ_1 and φ_3 on the elements $E_1 = [x_1, x_2]$ and $E_2 = [x_2, x_3]$, respectively. The supports of φ_2 and φ_4 are nonoverlapping. Hence, the second rows of the global matrices C and K are given by

$$\begin{aligned}
c_{21} &= \rho c_p u \int_\Omega \varphi_2 \frac{\partial \varphi_1}{\partial x} \, dx = \rho c_p u \int_{E_1} \varphi_2^{(1)} \frac{\partial \varphi_1^{(1)}}{\partial x} \, dx = c_{21}^{(1)}, \\
c_{22} &= \rho c_p u \int_\Omega \varphi_2 \frac{\partial \varphi_2}{\partial x} \, dx \\
&= \rho c_p u \int_{E_1} \varphi_2^{(1)} \frac{\partial \varphi_2^{(1)}}{\partial x} \, dx + \rho c_p u \int_{E_2} \varphi_1^{(2)} \frac{\partial \varphi_1^{(2)}}{\partial x} \, dx \\
&= c_{22}^{(1)} + c_{11}^{(2)}, \\
c_{23} &= \rho c_p u \int_\Omega \varphi_2 \frac{\partial \varphi_3}{\partial x} \, dx = \rho c_p u \int_{E_2} \varphi_1^{(2)} \frac{\partial \varphi_2^{(2)}}{\partial x} \, dx = c_{12}^{(2)}, \\
c_{24} &= \rho c_p u \int_\Omega 0 \, dx = 0
\end{aligned}$$

and

$$\begin{aligned}
k_{21} &= k \int_\Omega \frac{\partial \varphi_2}{\partial x} \frac{\partial \varphi_1}{\partial x} \, dx = k \int_{E_1} \frac{\partial \varphi_2^{(1)}}{\partial x} \frac{\partial \varphi_1^{(1)}}{\partial x} \, dx = k_{21}^{(1)}, \\
k_{22} &= k \int_\Omega \frac{\partial \varphi_2}{\partial x} \frac{\partial \varphi_2}{\partial x} \, dx \\
&= k \int_{E_1} \frac{\partial \varphi_2^{(1)}}{\partial x} \frac{\partial \varphi_2^{(1)}}{\partial x} \, dx + k \int_{E_2} \frac{\partial \varphi_1^{(2)}}{\partial x} \frac{\partial \varphi_1^{(2)}}{\partial x} \, dx \\
&= k_{22}^{(1)} + k_{11}^{(2)}, \\
k_{23} &= k \int_\Omega \frac{\partial \varphi_2}{\partial x} \frac{\partial \varphi_3}{\partial x} \, dx = k \int_{E_2} \frac{\partial \varphi_1^{(2)}}{\partial x} \frac{\partial \varphi_2^{(2)}}{\partial x} \, dx = k_{12}^{(2)}, \\
k_{24} &= k \int_\Omega 0 \, dx = 0.
\end{aligned}$$

The right-hand side of the second equation is a sum of two element contributions,

$$F_2 = \int_\Omega \varphi_2 f \, \mathrm{d}x = \int_{E_1} \varphi_2^{(1)} f \, \mathrm{d}x + \int_{E_2} \varphi_1^{(2)} f = F_2^{(1)} + F_1^{(2)}.$$

In the third equation of our algebraic system, φ_3 serves as a test function. The support of φ_3 overlaps the supports of φ_2 and φ_4 on the elements $E_2 = [x_2, x_3]$ and $E_3 = [x_3, x_4]$, respectively. The supports of φ_3 and φ_1 are nonoverlapping. This gives

$$\begin{aligned}
c_{31} &= \rho c_p u \int_\Omega 0 \, \mathrm{d}x = 0, \\
c_{32} &= \rho c_p u \int_\Omega \varphi_3 \frac{\partial \varphi_2}{\partial x} \, \mathrm{d}x = \rho c_p u \int_{E_2} \varphi_2^{(2)} \frac{\partial \varphi_1^{(2)}}{\partial x} \, \mathrm{d}x = c_{21}^{(2)}, \\
c_{33} &= \rho c_p u \int_\Omega \varphi_3 \frac{\partial \varphi_3}{\partial x} \, \mathrm{d}x \\
&= \rho c_p u \int_{E_2} \varphi_2^{(2)} \frac{\partial \varphi_2^{(2)}}{\partial x} \, \mathrm{d}x + \rho c_p u \int_{E_3} \varphi_1^{(3)} \frac{\partial \varphi_1^{(3)}}{\partial x} \, \mathrm{d}x \\
&= c_{22}^{(2)} + c_{11}^{(3)}, \\
c_{34} &= \cdots = c_{12}^{(3)}
\end{aligned}$$

and

$$\begin{aligned}
k_{31} &= k \int_\Omega 0 \, \mathrm{d}x = 0, \\
k_{32} &= k \int_\Omega \frac{\partial \varphi_3}{\partial x} \frac{\partial \varphi_2}{\partial x} \, \mathrm{d}x = k \int_{E_2} \frac{\partial \varphi_2^{(2)}}{\partial x} \frac{\partial \varphi_1^{(2)}}{\partial x} \, \mathrm{d}x = k_{21}^{(2)}, \\
k_{33} &= k \int_\Omega \frac{\partial \varphi_3}{\partial x} \frac{\partial \varphi_3}{\partial x} \, \mathrm{d}x \\
&= k \int_{E_2} \frac{\partial \varphi_2^{(2)}}{\partial x} \frac{\partial \varphi_2^{(2)}}{\partial x} \, \mathrm{d}x + k \int_{E_3} \frac{\partial \varphi_1^{(3)}}{\partial x} \frac{\partial \varphi_1^{(3)}}{\partial x} \, \mathrm{d}x \\
&= k_{22}^{(2)} + k_{11}^{(3)}, \\
k_{34} &= \cdots = k_{12}^{(3)}.
\end{aligned}$$

The right-hand side of the third equation is again a sum of two element contributions,

$$F_3 = \int_\Omega \varphi_3 f \, \mathrm{d}x = \int_{E_2} \varphi_2^{(2)} f \, \mathrm{d}x + \int_{E_3} \varphi_1^{(3)} f = F_2^{(2)} + F_1^{(3)}.$$

The support of φ_4 overlaps the support of φ_3 on the element $E_3 = [x_3, x_4]$. Since there is no overlap with the supports of φ_1 and φ_2, the last equation is defined by

$$\begin{aligned}
c_{41} &= \rho c_p u \int_\Omega 0 \, \mathrm{d}x = 0, \\
c_{42} &= \rho c_p u \int_\Omega 0 \, \mathrm{d}x = 0, \\
c_{43} &= \cdots = c_{21}^{(3)}, \\
c_{44} &= \cdots = c_{22}^{(3)},
\end{aligned}$$

$$
\begin{aligned}
k_{41} &= k\int_\Omega 0 \,\mathrm{d}x = 0,\\
k_{42} &= k\int_\Omega 0 \,\mathrm{d}x = 0,\\
k_{43} &= \cdots = k_{21}^{(3)},\\
k_{44} &= \cdots = k_{22}^{(3)},
\end{aligned}
$$

and

$$F_4 = \int_\Omega \varphi_4 f \,\mathrm{d}x = \int_{E_3} \varphi_2^{(3)} f \,\mathrm{d}x = F_2^{(3)}.$$

In summary, the relationship between the coefficients of the global 4×4 matrices C and K and the local 2×2 element matrices $C^{(e)}$ and $K^{(e)}$ is as follows:

$$C = \begin{pmatrix} c_{11}^{(1)} & c_{12}^{(1)} & 0 & 0 \\ c_{21}^{(1)} & c_{22}^{(1)} + c_{11}^{(2)} & c_{12}^{(2)} & 0 \\ 0 & c_{21}^{(2)} & c_{22}^{(2)} + c_{11}^{(3)} & c_{12}^{(3)} \\ 0 & 0 & c_{21}^{(3)} & c_{22}^{(3)} \end{pmatrix}, \tag{4.38}$$

$$K = \begin{pmatrix} k_{11}^{(1)} & k_{12}^{(1)} & 0 & 0 \\ k_{21}^{(1)} & k_{22}^{(1)} + k_{11}^{(2)} & k_{12}^{(2)} & 0 \\ 0 & k_{21}^{(2)} & k_{22}^{(2)} + k_{11}^{(3)} & k_{12}^{(3)} \\ 0 & 0 & k_{21}^{(3)} & k_{22}^{(3)} \end{pmatrix}. \tag{4.39}$$

The components of the global load vector F and its local counterpart $F^{(e)}$ are related by

$$F = \begin{pmatrix} F_1^{(1)} \\ F_2^{(1)} + F_1^{(2)} \\ F_2^{(2)} + F_1^{(3)} \\ F_2^{(3)} \end{pmatrix}. \tag{4.40}$$

In the above derivation, we used the *node-by-node* approach to global assembly. In practice, it is more convenient to initialize $A = C + K$ and F by zeros and add element contributions $A^{(e)} = C^{(e)} + K^{(e)}$ and $F^{(e)}$ in a loop over elements.

The linear element E_e, as shown in Fig. 4.14, has two nodes. The left node has the global number e and local number (1). The right node has the global number $e+1$ and local number (2). The restriction of the global basis functions φ_e and φ_{e+1} to E_e is given by the local basis functions $\varphi_1^{(e)}$ and $\varphi_2^{(e)}$. The entries of the element matrix $A^{(e)}$ are addressed using the local node numbers $I, J \in \{1,2\}$, whereas the corresponding entries of the global matrix A are addressed using the global node numbers $i, j \in \{e, e+1\}$. If I and J are the

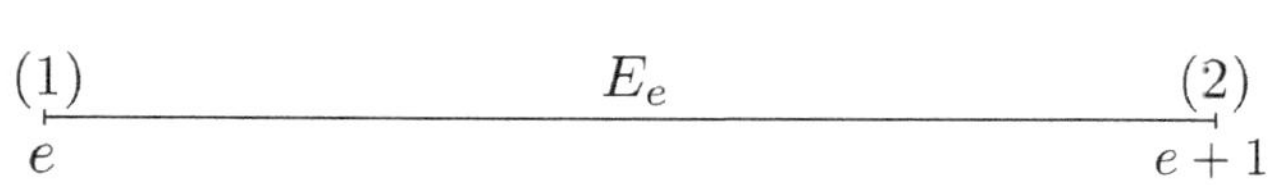

Figure 4.14. *Element E_e and its nodal points.*

local numbers of the nodal points x_i and x_j, the entry $A_{IJ}^{(e)}$ of the element matrix $A^{(e)}$ should be added to A_{ij} and the element contribution $F_I^{(e)}$ should be added to F_i. The following algorithm illustrates the process of *element-by-element* assembly for 1D linear elements.

Algorithm 4.1

For $e = 1,\ldots,N-1$, update the coefficients of the global algebraic system thus:

1. Calculate the four entries of the element stiffness matrix,

$$A_{IJ}^{(e)} = \rho c_p u \int_{E_e} \varphi_I^{(e)} \frac{\partial \varphi_J^{(e)}}{\partial x} \mathrm{d}x + k \int_{E_e} \frac{\partial \varphi_I^{(e)}}{\partial x} \frac{\partial \varphi_J^{(e)}}{\partial x} \mathrm{d}x, \qquad I,J = 1,2,$$

and the two entries of the element load vector,

$$F_I^{(e)} = \int_{E_e} f \varphi_I^{(e)} \mathrm{d}x, \qquad I = 1,2.$$

2. Determine the correspondence between the global and local node numbers:

$$i = \text{node}(e,I), \qquad j = \text{node}(e,J), \qquad I,J = 1,2.$$

3. Update the four entries of the global stiffness matrix,

$$A_{ij} = A_{ij} + A_{IJ}^{(e)}, \qquad I,J = 1,2,$$

and the two entries of the global load vector,

$$F_i = F_i + F_I^{(e)}, \qquad I = 1,2.$$

If the mesh points and elements are numbered consecutively, the above correspondence between the local and global node numbers is given by

$$\text{node}(e,1) = e, \qquad \text{node}(e,2) = e+1.$$

Example 4.2. On a uniform mesh of linear 1D elements, the evaluation of the integrals in equations (4.34) and (4.35) leads to the constant element matrices

$$C^{(e)} = \rho c_p u \begin{pmatrix} -\frac{1}{2} & \frac{1}{2} \\ -\frac{1}{2} & \frac{1}{2} \end{pmatrix}, \qquad K^{(e)} = \frac{k}{h} \begin{pmatrix} 1 & -1 \\ -1 & 1 \end{pmatrix}, \qquad e = 1,2,3.$$

If the domain $\Omega = (0,2)$ is discretized using 3 elements of length $h = \frac{2}{3}$, the element-by-element insertion of $C^{(e)}$ and $K^{(e)}$ into the global matrices C and K leads to

$$C = \rho c_p u \begin{pmatrix} -\frac{1}{2} & \frac{1}{2} & 0 & 0 \\ -\frac{1}{2} & 0 & \frac{1}{2} & 0 \\ 0 & -\frac{1}{2} & 0 & \frac{1}{2} \\ 0 & 0 & -\frac{1}{2} & \frac{1}{2} \end{pmatrix}, \tag{4.41}$$

$$K = \frac{k}{h} \begin{pmatrix} 1 & -1 & 0 & 0 \\ -1 & 2 & -1 & 0 \\ 0 & -1 & 2 & -1 \\ 0 & 0 & -1 & 1 \end{pmatrix}. \tag{4.42}$$

■

Remark 4.5 In many practical applications, it is worthwhile to perform local mesh refinement in local subdomains, where small-scale effects are present. The finite element assembly procedure is readily applicable to nonuniform meshes. In our 1D example, the element matrix $C^{(e)}$ is independent of the local mesh size $h_e = x_{e+1} - x_{\epsilon}$, but

$$K^{(e)} = \frac{k}{h_e} \begin{pmatrix} 1 & -1 \\ -1 & 1 \end{pmatrix}$$

does depend on h_e, $e = 1, \ldots, N-1$, and must be updated in the loop over elements.

(vii) Dirichlet Boundary Conditions

The Galerkin finite element discretization of the governing equation (4.18) on a uniform mesh of 4 linear elements has led us to the global algebraic system

$$(C+K)T = F,$$

where C and K are the global matrices given by (4.41) and (4.42), T is the vector of nodal values, and F is a forcing vector which depends on the heat source f.

Introducing the 4×4 stiffness matrix $A = C + K$, we write our system in the form

$$\begin{pmatrix} a_{11} & a_{12} & a_{13} & a_{14} \\ a_{21} & a_{22} & a_{23} & a_{24} \\ a_{31} & a_{32} & a_{33} & a_{34} \\ a_{41} & a_{42} & a_{43} & a_{44} \end{pmatrix} \begin{pmatrix} T_1 \\ T_2 \\ T_3 \\ T_4 \end{pmatrix} = \begin{pmatrix} F_1 \\ F_2 \\ F_3 \\ F_4 \end{pmatrix}. \tag{4.43}$$

Since the matrices C and K have zero row sums in our example, so does their sum A. It follows that $AT = A(T+C)$ for any constant vector C. Even if a solution to $AT = F$ exists, it is nonunique because adding C to T would give us another solution.

The reason for the ill-posedness of the above discrete problem lies in the fact that we have "forgotten" the Dirichlet boundary conditions

$$T_1 = T_h(2) = T_0,$$
$$T_4 = T_h(4) = T_c.$$

Indeed, the current form of system (4.43) provides the best approximation to equation (4.18) but does not take the boundary conditions into account. Since the values of the coefficients T_1 and T_4 are known, we have assembled a system of four equations for two unknowns (T_2 and T_3). This system is overdetermined and cannot have a unique solution.

There are at least three ways in which the Dirichlet boundary conditions can incorporated into system (4.43). The first option is to reduce the size of the system by eliminating two equations and substituting the known values of T_1 and T_4 into the remaining two equations. In order to see which equations are redundant, we recall that the space of test functions for the continuous variational problem was defined as $H_0^1(\Omega)$ to get rid of the boundary terms. Since we required that the trace of the test function be zero at the continuous level, the discrete test functions should also vanish on the boundary. Otherwise, the space of discrete test functions will not be a subspace of $H_0^1(\Omega)$. Thus, the equations to be eliminated from the algebraic system are the ones associated with the basis functions φ_1 and φ_4. In practice, elimination is performed by removing the corresponding rows/columns of the matrix and two components of the right-hand side (F_1 and F_4). The contribution of the eliminated degrees of freedom T_1 and T_4 is added to the right-hand side of the reduced system

$$\begin{pmatrix} a_{22} & a_{23} \\ a_{32} & a_{33} \end{pmatrix} \begin{pmatrix} T_2 \\ T_3 \end{pmatrix} = \begin{pmatrix} F_2 - a_{21} T_0 - a_{24} T_c \\ F_3 - a_{31} T_0 - a_{34} T_c \end{pmatrix}. \tag{4.44}$$

Another way to implement the Dirichlet boundary conditions is to treat the degrees of freedom T_1 and T_4 as unknowns but overwrite the corresponding algebraic equations by the Dirichlet boundary conditions. To this end, the first and last rows of the matrix A are replaced by rows of the identity matrix, and two components of the right-hand side (F_1 and F_4) are replaced by the given boundary values. The modified system is given by

$$\begin{pmatrix} 1 & 0 & 0 & 0 \\ a_{21} & a_{22} & a_{23} & a_{23} \\ a_{31} & a_{32} & a_{33} & a_{34} \\ 0 & 0 & 0 & 1 \end{pmatrix} \begin{pmatrix} T_1 \\ T_2 \\ T_3 \\ T_4 \end{pmatrix} = \begin{pmatrix} T_0 \\ F_2 \\ F_3 \\ T_c \end{pmatrix}. \tag{4.45}$$

It consists of two Galerkin equations (for φ_2 and φ_3) and two constraints (for T_1 and T_4).

Essentially the same effect can be achieved by adding penalty terms of the form $\alpha(T_1 - T_0)$ and $\alpha(T_4 - T_c)$ to the first and last equations of the linear system. This gives

$$\begin{pmatrix} a_{11}+\alpha & a_{12} & a_{13} & a_{14} \\ a_{21} & a_{22} & a_{23} & a_{24} \\ a_{31} & a_{32} & a_{33} & a_{34} \\ a_{41} & a_{42} & a_{43} & a_{44}+\alpha \end{pmatrix} \begin{pmatrix} T_1 \\ T_2 \\ T_3 \\ T_4 \end{pmatrix} = \begin{pmatrix} F_1 + \alpha T_0 \\ F_2 \\ F_3 \\ F_4 + \alpha T_c \end{pmatrix} \tag{4.46}$$

so that the Dirichlet boundary conditions are satisfied approximately for large values of the penalty parameter α. Since only diagonal entries of the left-hand side matrix are modified, this implementation technique preserves the symmetry of A if the original matrix is symmetric (e.g., in the case of the 1D heat equation with vanishing convective velocity).

(viii) Solving the Algebraic System

In addition to the computational domain and Dirichlet boundary conditions, the user must provide a formula for calculating the right-hand side f of the governing equation (4.18). Setting f equal to zero, we will solve the homogeneous 1D heat equation in the domain $\Omega = (0,2)$ subject to the Dirichlet boundary conditions chosen as in Section 4.5:

$$T(0) = 2,$$
$$T(2) = 4.$$

The coefficients of the governing equation and the constant mesh size are defined by

$$\rho = c_p = u = k = 1, \qquad h = \frac{2}{3}.$$

The overdetermined system of four Galerkin equations is given by (4.43) with

$$F_1 = F_2 = F_3 = F_4 = 0.$$

Using the elimination method to implement the Dirichlet boundary conditions, we obtain

$$\begin{pmatrix} c_{22}+k_{22} & c_{23}+k_{23} \\ c_{32}+k_{32} & c_{33}+k_{33} \end{pmatrix} \begin{pmatrix} T_2 \\ T_3 \end{pmatrix} = -\begin{pmatrix} (c_{21}+k_{21})T_1 \\ (c_{34}+k_{34})T_4 \end{pmatrix},$$

where the entries of matrices C and K are defined by (4.41) and (4.42). Substituting the boundary values $T_1 = 2$ and $T_4 = 4$, we end up with the reduced system

$$\begin{pmatrix} 3 & -1 \\ -2 & 3 \end{pmatrix}\begin{pmatrix} T_2 \\ T_3 \end{pmatrix} = \begin{pmatrix} 4 \\ 4 \end{pmatrix},$$

which gives

$$T_2 = \frac{16}{7},$$
$$T_3 = \frac{20}{7}.$$

Hence, the piecewise-linear Galerkin approximation is given by

$$T_h(x) = 2\varphi_1(x) + \frac{16}{7}\varphi_2(x) + \frac{20}{7}\varphi_3(x) + 4\varphi_4(x),$$

where $\varphi_1, \varphi_2, \varphi_3, \varphi_4$ are the global basis functions defined on the whole interval $(0, 2)$.

(ix) Further Iterations?

Since we are solving a linear stationary toy problem, T_h is the final solution.

(x) Postprocessing

Fig. 4.15 displays the exact solution of our toy problem and the approximate solution obtained at step (viii). It can readily be seen that the quality of the Galerkin solution is quite good, although we used just three elements and linear basis functions. Of course, it is more difficult to assess the accuracy of a numerical solution if the exact solution is unknown, which is usually the case in practical applications. Therefore, it is worthwhile to perform grid convergence studies by solving the problem on successively refined meshes.

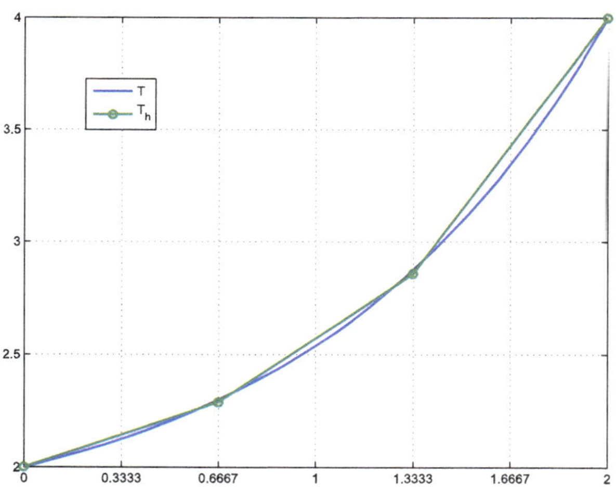

Figure 4.15. *Dirichlet problem: T_h vs. T.*

The extension of the presented discretization procedure to higher-order basis functions and general 1D meshes is relatively straightforward. The size of the element matrices and vectors depends on the degree of local interpolation polynomials. The size of the global system also depends on the number of mesh elements. Not only the number but also the location of mesh points may have a significant impact on the accuracy of numerical solutions. For this reason, adaptive meshes are frequently employed in finite element codes.

4.8 ▪ Natural Boundary Conditions

So far we have only considered the Dirichlet boundary value problem for the 1D heat equation. In situations when the temperature T is unknown at $x = c$, the heat flux can be prescribed instead of $T(c)$. Let us consider a flux boundary condition of the form

$$-k\frac{\partial T}{\partial x}\bigg|_{x=c} = h_c T(c) + g, \tag{4.47}$$

where h_c and g are given constants which determine the rate of heat conduction at $x = c$.

If the heat flux is independent of the temperature, we can set the parameter h_c equal to zero. This simplification leads us to the Neumann boundary condition

$$-k\frac{\partial T}{\partial x}\bigg|_{x=c} = g. \tag{4.48}$$

Using $h_c \neq 0$ and $g = -h_c T_{\text{ext}}$, we obtain the Newton boundary condition

$$-k\frac{\partial T}{\partial x}\bigg|_{x=c} = h_c(T(c) - T_{\text{ext}}), \tag{4.49}$$

where T_{ext} is a given external temperature. The dependence of the heat flux on the temperature may also be nonlinear, as in the radiative boundary condition

$$-k\frac{\partial T}{\partial x}\bigg|_{x=c} = \sigma\varepsilon(T^4(c) - T_{\text{ext}}^4), \tag{4.50}$$

where σ is the Stefan–Boltzmann constant and ε is the emissivity (see Chapter 3). We will discuss the implementation of this boundary condition at the end of this chapter.

(i) Problem Statement

In this section, our model problem will be the steady 1D heat equation

$$\rho c_p u \frac{\partial T(x)}{\partial x} - k\frac{\partial^2 T(x)}{\partial x^2} = f(x), \qquad x \in (0,c), \tag{4.51}$$

endowed with the boundary conditions

$$T(0) = T_0, \tag{4.52}$$

$$-k\frac{\partial T}{\partial x}\bigg|_{x=c} = h_c T(c) + g. \tag{4.53}$$

Remark 4.6. The general form of the Newton boundary condition in $\mathbb{R}^d$, $d = 1,2,3$, is

$$-k\nabla T \cdot \mathbf{n} = h_c T + g, \tag{4.54}$$

where $\mathbf{n} = (n_1, \ldots, n_d)^T$ is a unit outward normal to the boundary. In our 1D example, we have $n_1 = -1$ at the boundary point $x = 0$ and $n_1 = 1$ at the boundary point $x = c$. The application of (4.54) at $x = 0$ leads to the Newton boundary condition

$$k\frac{\partial T}{\partial x}\bigg|_{x=0} = h_c T(0) + g.$$

The replacement of a Dirichlet boundary condition by a flux boundary condition requires just a minor modification of the algorithm that we used in Section 4.7. Therefore, we will only discuss the changes caused by the use of (4.53) instead of $T(c) = T_c$.

(ii) Variational Formulation

The governing equation (4.51) is the same as in Section 4.7. Using the method of weighted residuals and integration by parts, we obtain the variational problem

$$\rho c_p u \int_0^c w\frac{\partial T}{\partial x}\mathrm{d}x + k\int_0^c \frac{\partial w}{\partial x}\frac{\partial T}{\partial x}\,\mathrm{d}x - kw\frac{\partial T}{\partial x}\bigg|_0^c = \int_0^c wf\,\mathrm{d}x \qquad \forall w \in V. \tag{4.55}$$

Since we no longer constrain the value of $T(c)$, the space of test functions V becomes

$$V = \{v \in H^1((0,c)) | v(0) = 0\}.$$

That is, the test function w vanishes at the boundary point $x = 0$, where the Dirichlet boundary condition (4.52) is applied, but may have a nonzero value at $x = c$. Substituting $w(0) = 0$ into (4.55) and invoking the Newton boundary condition (4.53), we obtain

$$\rho c_p u \int_0^c w\frac{\partial T}{\partial x}\,\mathrm{d}x + k\int_0^c \frac{\partial w}{\partial x}\frac{\partial T}{\partial x}\,\mathrm{d}x + w(c)(h_c T(c) + g) = \int_0^c wf\,\mathrm{d}x. \tag{4.56}$$

This integral equation holds for all $w \in V$. In contrast to the Dirichlet boundary condition which must be incorporated into the definition of the space V, the flux boundary condition is imposed weakly via an additional boundary term in the variational formulation.

Remark 4.7. In the calculus of variations for minimization problems, Dirichlet boundary conditions are called *essential boundary conditions* because they modify the definition of functional spaces. Boundary conditions of Neumann and Newton type are built into the definition of the functional to be minimized and are called *natural boundary conditions* [44]. The same terminology is often used in the literature on Galerkin methods for boundary value problems which do necessarily possess an equivalent minimization problem.

Remark 4.8. In applications to multidimensional problems, the generic Newton/Neumann boundary condition (4.53) gives rise to a boundary integral of the form

$$-\int_{\partial\Omega} kw\nabla T \cdot \mathbf{n}\,\mathrm{d}s = \int_{\partial\Omega} w(h_c T + g)\,\mathrm{d}s, \qquad w \in V.$$

(iii) Discretization

Let $\{\varphi_1,\ldots,\varphi_N\}$ be the set of global finite element basis functions spanning a finite-dimensional space $V_h \subset H^1((0,c))$. The approximate solution $T_h \in V_h$ is given by

$$T_h(x) = \sum_{j=1}^{N} T_j \varphi_j(x).$$

The discrete counterpart of the continuous variational problem (4.56) reads

$$\begin{aligned} \rho c_p u \int_0^c w_h \frac{\partial T_h}{\partial x}\,\mathrm{d}x + k\int_0^c \frac{\partial w_h}{\partial x}\frac{\partial T_h}{\partial x}\,\mathrm{d}x + w_h(c)(h_c T_h(c)+g) \\ = \int_0^c w_h f\,\mathrm{d}x \qquad \forall w_h \in V_h. \end{aligned} \tag{4.57}$$

Substituting the above representation of T_h as a linear combination of N basis functions and using $w_h \in \{\varphi_1,\ldots,\varphi_N\}$ as test functions, we obtain the linear system

$$\begin{aligned} \sum_{j=1}^{N}\left(\rho c_p u \int_0^c \varphi_i \frac{\partial \varphi_j}{\partial x}\,\mathrm{d}x\right) T_j + \sum_{j=1}^{N}\left(k\int_0^c \frac{\partial \varphi_i}{\partial x}\frac{\partial \varphi_j}{\partial x}\,\mathrm{d}x\right) T_j + \sum_{j=1}^{N} h_c \varphi_i(c)\varphi_j(c) T_j \\ = \int_0^c f\varphi_i\,\mathrm{d}x - \varphi_i(c) g, \qquad i=1,\ldots,N. \end{aligned} \tag{4.58}$$

The matrix form of the above sparse linear system is given by

$$(C+K+B)T = F - G, \tag{4.59}$$

where C, K, F are defined as in equation (4.26). The contribution of the natural boundary condition is represented by the matrix $B = \{b_{ij}\}$ and vector $G = \{G_i\}$. We have

$$b_{ij} = h_c \varphi_i(c)\varphi_j(c), \qquad i,j = 1,\ldots,N, \tag{4.60}$$

$$G_i = \varphi_i(c) g, \qquad i = 1,\ldots,N, \tag{4.61}$$

where $c = x_N$ is the last grid point. Due to the interpolation property $\varphi_i(x_j) = \delta_{ij}$, the only nonvanishing entry of B is b_{NN} and the only nonvanishing entry of G is G_N.

(iv) Localization

As before, we discretize the domain $(0,c)$ using a uniform mesh of $N-1$ linear finite elements. The corresponding mesh size is $h = c/(N-1)$. The local basis functions

$$\varphi_1^{(e)}(x) = \frac{-x}{h} + e = \frac{x_{e+1}-x}{h},$$

$$\varphi_2^{(e)}(x) = \frac{x}{h} - e + 1 = \frac{x - x_e}{h}$$

and their global counterparts $\varphi_1,\ldots,\varphi_N$ are defined as in Section 4.7. The element matrices $C^{(e)}$, $K^{(e)}$ and element vector $F^{(e)}$ are given by equations (4.34)–(4.36).

The last element, E_{N-1}, also makes a nontrivial contribution to B and G. The entries of the element matrix $B^{(N-1)}$ and element vector $G^{(N-1)}$ are given by

$$b_{ij}^{(N-1)} = h_c \varphi_i^{(N-1)}(c)\varphi_j^{(N-1)}(c), \qquad i,j = 1,2, \tag{4.62}$$

$$G_i^{(N-1)} = \varphi_i^{(N-1)}(c) g, \qquad i = 1,2. \tag{4.63}$$

Since $c = x_N$ is a node, we have $\varphi_1^{(N-1)}(c) = 0$ and $\varphi_2^{(N-1)}(c) = 1$. This gives

$$B^{(N-1)} = h_c \begin{pmatrix} 0 & 0 \\ 0 & 1 \end{pmatrix}, \qquad G^{(N-1)} = g \begin{pmatrix} 0 \\ 1 \end{pmatrix}.$$

The trivial contribution of any other element E_e to B and G is represented by

$$B^{(e)} = 0, \qquad G^{(e)} = 0, \qquad e = 1, \dots, N-2.$$

(vi) Assembly of the Global System

The element-by-element insertion of local matrices $C^{(e)}, K^{(e)}$ and local vectors $F^{(e)}$ into their global counterparts was addressed in Section 4.7 in step (vi) of the general computational procedure. The assembly of B and G from $B^{(e)}$ and $G^{(e)}$ can be performed in the same way. In practice, it is worthwhile to assemble the global stiffness matrix

$$A = C + K + B$$

and the global load vector

$$Q = F - G$$

from the combined element contributions

$$A^{(e)} = C^{(e)} + K^{(e)} + B^{(e)},$$
$$Q^{(e)} = F^{(e)} - G^{(e)}.$$

The addition of matrices/vectors at the level of the local problem reduces the storage requirements and the amount of indirect addressing, i.e., the overhead cost of finding the entries of a sparse matrix stored in a compressed form. The savings may appear moderate in the context of 1D problems and tridiagonal matrices, but the efficient use of computer memory becomes an important issue in finite element codes for multidimensional problems.

Example 4.3. In Section 4.7, we discussed the process of global assembly for the 1D Dirichlet problem discretized using a uniform mesh of three linear finite elements. Let us now assemble the contribution of the Newton/Neumann boundary condition for the case of $N = 4$ mesh points. The local linear basis functions defined on the element E_3 are given by

$$\varphi_1^{(3)}(x) = \frac{-x}{h} + 3,$$
$$\varphi_2^{(3)}(x) = \frac{x}{h} - 2,$$

where $h = \frac{c}{3}$ is the constant mesh size. Inserting the nontrivial element contributions

$$B^{(3)} = h_c \begin{pmatrix} 0 & 0 \\ 0 & 1 \end{pmatrix}, \qquad G^{(3)} = g \begin{pmatrix} 0 \\ 1 \end{pmatrix}$$

into the 4×4 zero matrix B and the 4×1 zero vector G, we obtain

$$B = h_c \begin{pmatrix} 0 & 0 & 0 & 0 \\ 0 & 0 & 0 & 0 \\ 0 & 0 & 0 & 0 \\ 0 & 0 & 0 & 1 \end{pmatrix}, \qquad G = g \begin{pmatrix} 0 \\ 0 \\ 0 \\ 1 \end{pmatrix}.$$

Matrices C, K and vector F are the same as those presented in Section 4.7. ∎

(vii) Dirichlet Boundary Conditions

In light of the above, the coefficients of the global algebraic system

$$\begin{pmatrix} a_{11} & a_{12} & a_{13} & a_{14} \\ a_{21} & a_{22} & a_{23} & a_{24} \\ a_{31} & a_{32} & a_{33} & a_{34} \\ a_{41} & a_{42} & a_{43} & a_{44} \end{pmatrix} \begin{pmatrix} T_1 \\ T_2 \\ T_3 \\ T_4 \end{pmatrix} = \begin{pmatrix} Q_1 \\ Q_2 \\ Q_3 \\ Q_4 \end{pmatrix} \tag{4.64}$$

are given by

$$a_{ij} = c_{ij} + k_{ij} + b_{ij}, \qquad i,j = 1,\dots,4,$$

$$Q_i = F_i - G_i, \qquad i = 1,\dots,4.$$

Overwriting the redundant first equation by the Dirichlet boundary condition

$$T_1 = T_0,$$

we obtain the modified linear system

$$\begin{pmatrix} 1 & 0 & 0 & 0 \\ a_{21} & a_{22} & a_{23} & a_{24} \\ a_{31} & a_{32} & a_{33} & a_{34} \\ a_{41} & a_{42} & a_{43} & a_{44} \end{pmatrix} \begin{pmatrix} T_1 \\ T_2 \\ T_3 \\ T_4 \end{pmatrix} = \begin{pmatrix} T_0 \\ Q_2 \\ Q_3 \\ Q_4 \end{pmatrix}. \tag{4.65}$$

It consists of one algebraic constraint and three Galerkin equations. The degrees of freedom T_2, T_3, T_4 are associated with the test functions $\varphi_2, \varphi_3, \varphi_4$, which vanish at $x = 0$.

(viii) Solving the Algebraic System

The right-hand side of the linear system to be solved depends on the definition of the source term f in the governing equation (4.51). In our example, we set f equal to zero in the whole domain $(0,2)$. In this case, we have $Q = -G$. Using the parameter settings

$$c = 2, \qquad T_0 = 2, \qquad h_c = 1, \qquad g = -10$$

to define the Dirichlet–Newton boundary conditions, we end up with a linear system of the form (4.65). The unique solution to this linear system is given by

$$T_1 = 2.000, \quad T_2 = 2.533, \quad T_3 = 3.599, \quad T_4 = 5.733.$$

(x) Postprocessing

Fig. 4.16 displays the approximate solution T_h and the exact solution

$$T(x) = 2 + \frac{8}{2e^2 - 1}(e^x - 1).$$

Note that the Dirichlet boundary condition prescribed at the point $x = 0$ is satisfied exactly (in a *strong sense*), while the Newton boundary condition prescribed at the point $x = 2$ is satisfied approximately (in a *weak sense*). Since the slopes of the solutions T_h and T do not match exactly at $x = 2$, the corresponding heat fluxes are also slightly different.

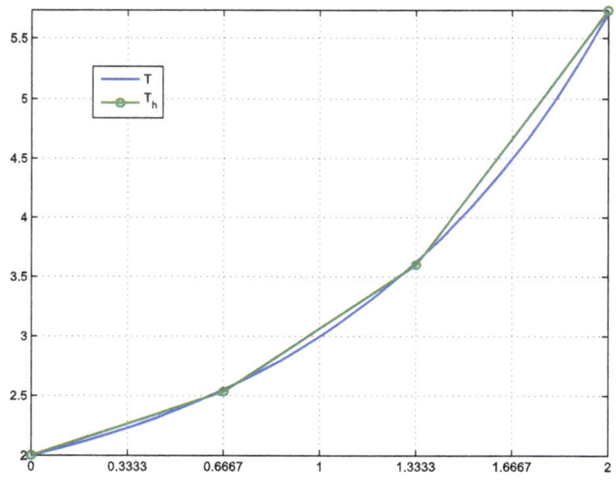

Figure 4.16. *Dirichlet–Newton problem:* T_h *vs.* T.

On the Implementation of the Radiative Boundary Condition

As a further example, let us replace the linear flux boundary condition (4.47) by a nonlinear radiative boundary condition of the form (4.50). Using the factorization

$$-k\frac{\partial T}{\partial x}\bigg|_{x=2} = \sigma\varepsilon(T^3 + T^2T_{\text{ext}} + TT_{\text{ext}}^2 + T_{\text{ext}}^3)(T - T_{\text{ext}})\bigg|_{x=2},$$

we define the temperature-dependent effective heat transfer coefficient

$$h_r(T) := \sigma\varepsilon(T^3 + T^2T_{\text{ext}} + TT_{\text{ext}}^2 + T_{\text{ext}}^3)\bigg|_{x=2}.$$

This leads us to the following representation of the radiative boundary condition:

$$-k\frac{\partial T}{\partial x}\bigg|_{x=2} = h_r(T(2))(T(2) - T_{\text{ext}}). \tag{4.66}$$

Since the boundary value $T(2)$ is unknown, so is the value of the heat flux across the boundary $x = 2$. Nonlinear boundary conditions of this kind can be implemented using lagged values of $T_4 \approx T(2)$ to evaluate h_r. Below we present an iterative algorithm which represents an extension of the finite element solution method we developed for the linear Dirichlet–Newton problem to nonlinear flux boundary conditions of the form (4.66).

Let $m = 0, 1, 2, \ldots$ denote the iteration index such that $T_i^{(m)}$ is the approximate nodal value of T_h at the grid point x_i at the end of the mth step. The iteration process begins with an initial guess $T^{(0)}$ which must satisfy the Dirichlet boundary condition $T_1^{(0)} = T_0$. In the absence of a better estimate, the remaining nodal values of T_h may be initialized by zeros. At the $(m+1)$st iteration step, the approximate solution is updated as follows.

Algorithm 4.2

1. Given the tentative nodal value $T_N^{(m)}$, calculate the parameters

$$h_r^{(m)}\bigg|_{x=2} := \sigma\varepsilon((T_N^{(m)})^3 + (T_N^{(m)})^2T_{\text{ext}} + (T_N^{(m)})T_{\text{ext}}^2 + T_{\text{ext}}^3)$$

and

$$g^{(m)} = -h_r^{(m)} T_{\text{ext}}.$$

2. Update the solution-dependent coefficients of the global algebraic system

$$(C + K + B^{(m)})T^{(m+1)} = F - G^{(m)}$$

using the linearized radiative boundary condition

$$-k \frac{\partial T}{\partial x}\bigg|_{x=2} = h_r^{(m)} T(2) + g^{(m)})$$

to calculate

$$B^{(m)} = h_r^{(m)} \begin{pmatrix} 0 & 0 & \dots & 0 \\ 0 & 0 & \dots & 0 \\ \vdots & \vdots & \ddots & \vdots \\ 0 & 0 & \dots & 1 \end{pmatrix}, \qquad G^{(m)} = g^{(m)} \begin{pmatrix} 0 \\ 0 \\ \vdots \\ 1 \end{pmatrix}.$$

3. Overwrite the first equation by the Dirichlet boundary condition $T_1 = T_0$ and solve the resulting linear system for the vector of nodal values:

$$T^{(m+1)} = (T_1^{(m+1)}, \dots, T_N^{(m+1)})^T.$$

4. If $||T^{(m+1)} - T^{(m)}|| < TOL$, where $||\cdot||$ is a suitably chosen vector norm and TOL is a prescribed tolerance, then set $T := T^{(m+1)}$ and stop. Otherwise go to step 1.

In practice, the components of the original solution vector $T^{(m)} = (T_1^{(m)}, \dots, T_N^{(m)})^T$ are overwritten after each iteration to avoid a waste of computer memory. The stopping criterion in step 4 terminates the iteration as soon as the difference between two successive iterates becomes as small as desired. It is also possible to monitor the relative changes

$$\frac{||T^{(m+1)} - T^{(m)}||}{||T^{(m)}||}$$

and/or the residuals of the nonlinear algebraic system evaluated using $T^{(m+1)}$.

4.9 ▪ Unsteady One-Dimensional Heat Equation

So far we have introduced the FEM in the context of the steady 1D heat equation. Let us now consider the time-dependent heat transport equation

$$\rho c_p \frac{\partial T(x,t)}{\partial t} + \rho c_p u \frac{\partial T(x,t)}{\partial x} - k \frac{\partial^2 T(x,t)}{\partial x^2} = f(x,t). \tag{4.67}$$

We will use this parabolic PDE to simulate the evolution of the temperature field $T(x,t)$ for $x \in (0,c)$ and $t \in (0,\tau)$, where $c > 0$ and $\tau > 0$ are positive constants. As before, the material parameters ρ, c_p, k and the velocity u are assumed to be constant. However, the heat source intensity $f(x,t)$ may depend on the space coordinate x and time t.

As in the case of the steady 1D heat equation, a (possibly time-dependent) boundary condition must be prescribed at each endpoint of the spatial domain $\Omega = (0,c)$. For example, we can use the general Dirichlet boundary condition

$$T(0,t) = T_0, \qquad t \in (0,\tau), \tag{4.68}$$

and the homogeneous Neumann boundary condition

$$-k\frac{\partial T}{\partial x}\bigg|_{x=c} = 0, \qquad t \in (0,\tau). \tag{4.69}$$

For simplicity, we assume that T_0 is constant, but the use of a time-dependent Dirichlet boundary condition $T_0 = T_0(t)$ may be appropriate in some applications.

In addition to the above boundary conditions, the initial temperature must be provided for the problem to be well-posed. We will use initial conditions of the form

$$T(x,0) = T^0(x), \qquad x \in (0,c), \tag{4.70}$$

where T^0 is a given function. The initial and boundary conditions are compatible if

$$T^0(0) = T_0,$$
$$-k\frac{\partial T^0}{\partial x}\bigg|_{x=c} = 0.$$

The solution $T(x,t)$ to the initial-boundary value problem is a function of two variables, which can be visualized using a 3D plot, as shown in Fig. 4.17. However, it is more common to plot the values of $T(x,t^*)$ versus the values of x for a fixed time instant t^*.

We will use the FEM to solve equation (4.67) subject to the boundary conditions (4.68), (4.69) and the initial condition (4.70). The discretization procedure to be presented below is based on a fairly straightforward generalization of the procedure introduced in Section 4.6. The main difference lies in the fact that the local time derivative also needs to be discretized using the nodal values of the approximate solution T_h at two consecutive time levels $t^n, t^{n+1} \in (0,\tau)$. In this section, we will focus on the aspects that have not been handled previously in the context of the steady 1D heat transfer model.

(ii) Variational Formulation

The construction of finite element approximations begins with the derivation of a variational formulation. Since we have prescribed a Dirichlet boundary condition at $x = 0$ and a Neumann boundary condition at $x = c$, the space of test functions is defined by

$$V = \{v \in H^1((0,c)) | v(0) = 0\}.$$

Using the method of weighted residuals and integration by parts over $(0,c)$, we find that

$$\rho c_p \int_0^c w\frac{\partial T}{\partial t}\mathrm{d}x + \rho c_p u \int_0^c w\frac{\partial T}{\partial x}\mathrm{d}x + k\int_0^c \frac{\partial w}{\partial x}\frac{\partial T}{\partial x}\,\mathrm{d}x - kw\frac{\partial T}{\partial x}\bigg|_0^c = \int_0^c wf\,\mathrm{d}x$$

for any $w \in V$ and any fixed time instant $t \in (0,\tau)$. The boundary term drops out because the test function w vanishes at $x = 0$, while the derivative $\frac{\partial T}{\partial x}$ vanishes at $x = c$ due to the homogeneous Neumann condition. Hence, the continuous variational problem reads

$$\rho c_p \int_0^c w\frac{\partial T}{\partial t}\mathrm{d}x + \rho c_p u \int_0^c w\frac{\partial T}{\partial x}\mathrm{d}x + k\int_0^c \frac{\partial w}{\partial x}\frac{\partial T}{\partial x}\,\mathrm{d}x = \int_0^c wf\,\mathrm{d}x \qquad \forall w \in V. \tag{4.71}$$

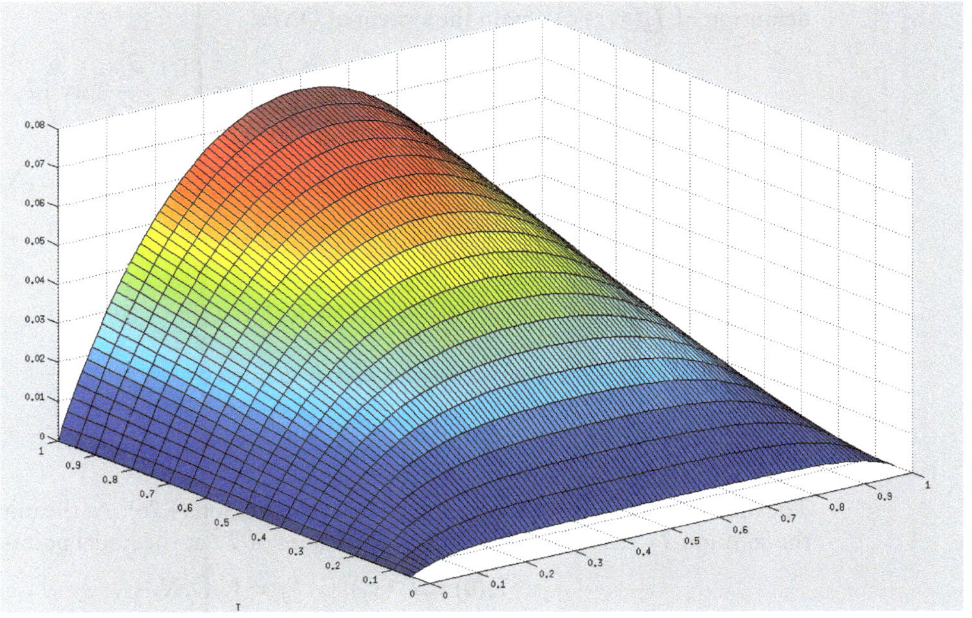

Figure 4.17. *Solution $T(x,t)$ to the initial-boundary value problem for the heat equation for $T(x,0) = T(0,t) = T(1,t) = 0$ and $f = 1$.*

The initial condition (4.70) can also be imposed weakly using the L^2 projection:

$$\int_0^c wT(x,0)\mathrm{d}x = \int_0^c wT^0(x)\mathrm{d}x \qquad \forall w \in V.$$

(iii) Discretization

Following the approach that we used to solve the steady 1D heat equation, we define the approximate solution $T_h(\cdot,t) \in V_h$ as a linear combination of global basis functions,

$$T_h(x,t) = \sum_{j=1}^{N} T_j(t)\varphi_j(x), \tag{4.72}$$

where the nodal values T_j depend on $t \in (0,\tau)$. The discrete variational problem becomes

$$\begin{aligned} \rho c_p \int_0^c w_h \frac{\partial T_h}{\partial t}\,\mathrm{d}x + \rho c_p u \int_0^c w_h \frac{\partial T_h}{\partial x}\,\mathrm{d}x + k \int_0^c \frac{\partial w_h}{\partial x}\frac{\partial T_h}{\partial x}\,\mathrm{d}x \\ = \int_0^c w_h f\,\mathrm{d}x, \qquad w_h \in V_h, \end{aligned} \tag{4.73}$$

where V_h denotes the finite-dimensional space of test functions. Neglecting the Dirichlet boundary condition for the time being, we substitute $w_h \in \{\varphi_1, \ldots, \varphi_N\}$ and the above definition of $T_h(x,t)$ to obtain the system of DAEs:

$$\sum_{j=1}^{N}\left(\rho c_p \int_0^c \varphi_i\varphi_j \,\mathrm{d}x\right)\frac{\mathrm{d}T_j}{\mathrm{d}t}+\sum_{j=1}^{N}\left(\rho c_p u \int_0^c \varphi_i\frac{\partial \varphi_j}{\partial x}\,\mathrm{d}x\right)T_j$$
$$+\sum_{j=1}^{N}\left(k\int_0^c \frac{\partial \varphi_i}{\partial x}\frac{\partial \varphi_j}{\partial x}\,\mathrm{d}x\right)T_j=\int_0^c \varphi_i f\,\mathrm{d}x, \qquad i=1,\dots,N. \tag{4.74}$$

The matrix form of this system is given by

$$M\frac{\mathrm{d}T}{\mathrm{d}t}+(C+K)T=F, \tag{4.75}$$

where $M=\{m_{ij}\}$ is the consistent *mass matrix* defined by

$$m_{ij}=\rho c_p\int_0^c \varphi_i\varphi_j\,\mathrm{d}x, \qquad i,j=1,\dots,N. \tag{4.76}$$

The matrices C,K and vector F are defined as in equation (4.26). At the initial time $t=0$, the solution T_h can be initialized using the values of T^0 at the nodal points,

$$T_j(0)=T^0(x_j), \qquad j=1,\dots,N.$$

Note that the function T^0 must be continuous for this definition to be mathematically correct.

Alternatively, the nodal values of T_h may be initialized using the discrete L^2 projection:

$$\int_0^c w_h T_h(x,0)\mathrm{d}x=\int_0^c w_h T^0(x)\mathrm{d}x \qquad \forall w_h\in\{\varphi_1,\dots,\varphi_N\}.$$

In this version, the initial nodal values $T_j(0)$ are obtained by solving the linear system

$$\sum_{j=1}^{N}\left(\int_0^c \varphi_i\varphi_j\mathrm{d}x\right)T_j(0)=\int_0^c \varphi_i T^0(x)\,\mathrm{d}x, \qquad i=1,\dots,N.$$

This initialization procedure is well-defined even for a discontinuous initial condition.

After the discretization in space and calculation of $T_j(0)$, the approximate solution can be advanced in time using one of the time-stepping schemes introduced in Section 2.3.3. Let T_j^n denote the approximate value of $T_j(t)$ at the time instant $t=t^n$. For a constant time step $\Delta t=\tau/M$, the time levels are defined by $t^n=n\Delta t$ for $n=0,1,2,\dots,M$.

In a two-level scheme, the time derivative is approximated by the finite difference

$$\frac{\mathrm{d}T}{\mathrm{d}t}=\frac{T^{n+1}-T^n}{\Delta t},$$

and the stiffness matrix $A=C+K$ is applied to a weighted average of T^n and T^{n+1}. The right-hand side F is assembled using the values of $f(x,t)$ at $t=t^n$ and/or $t=t^{n+1}$.

Forward Euler Method

In the forward Euler method, the term $(C+K)T$ is calculated using the known solution values from the previous time step. The components of the right-hand side F are also evaluated at the time level t^n. This approach leads to the linear system

$$M\frac{T^{n+1}-T^n}{\Delta t}+(C+K)T^n=F^n.$$

Moving all known terms to the right-hand side and multiplying by Δt, we obtain

$$MT^{n+1} = (M - \Delta t(C+K))T^n + \Delta t F^n.$$

Backward Euler Method

In the backward Euler method, the term $(C+K)T$ and the right-hand side F are defined at the new time level t^{n+1}. The fully discrete problem becomes

$$M\frac{T^{n+1}-T^n}{\Delta t} + (C+K)T^{n+1} = F^{n+1}$$

and can be written in the equivalent form

$$(M + \Delta t(C+K))T^{n+1} = MT^n + \Delta t F^{n+1}.$$

Crank–Nicolson Method

The Crank–Nicolson method corresponds to the arithmetic average of the forward and backward Euler approximations. The resulting linear system reads

$$M\frac{T^{n+1}-T^n}{\Delta t} + (C+K)\frac{T^{n+1}+T^n}{2} = \frac{F^{n+1}+F^n}{2}.$$

An equivalent form of the above discrete problem is given by

$$\left(M + \frac{\Delta t}{2}(C+K)\right)T^{n+1} = \left(M - \frac{\Delta t}{2}(C+K)\right)T^n + \Delta t\left(\frac{F^{n+1}+F^n}{2}\right).$$

In contrast to the first-order accurate forward and backward Euler methods, the Crank–Nicolson method is second-order accurate in time (see Section 2.3.3 for a more detailed presentation of the three time-stepping methods and further information).

(iv) Localization

The local matrices $C^{(e)}, K^{(e)}$ and the local vector $F^{(e)}$ are the same as in the case of a steady 1D heat equation discretized using linear finite elements. It remains to define the 2×2 element mass matrix $M^{(e)}$ for a generic element E_e, where $e = 1, \dots, N-1$.

Substituting the local basis functions $\varphi_1^{(e)}(x)$ and $\varphi_2^{(e)}(x)$ into the integrals

$$m_{ij}^{(e)} = \rho c_p \int_{E_e} \varphi_i^{(e)} \varphi_j^{(e)}\, \mathrm{d}x, \qquad i,j = 1,2,$$

one obtains the four entries of $M^{(e)}$. On a uniform mesh with $h = c/N$, this gives

$$M^{(e)} = \frac{\rho c_p h}{6}\begin{pmatrix} 2 & 1 \\ 1 & 2 \end{pmatrix}.$$

(vi) Assembly of the Global System

All of our two-level time discretizations lead to algebraic systems of the form

$$AT^{n+1} = BT^n + S,$$

where

$$
\begin{aligned}
A &= M + \theta \Delta t (C + K), \\
B &= M - (1-\theta) \Delta t (C + K), \\
S &= \Delta t \left(\theta F^{n+1} + (1-\theta) F^n \right).
\end{aligned}
$$

The parameter θ is used to select the time-stepping method (forward Euler: $\theta = 0$; backward Euler: $\theta = 1$; Crank–Nicolson: $\theta = \frac{1}{2}$). The global matrices A and B can be assembled element by element using the following modification of Algorithm 4.1.

Algorithm 4.3

For $e = 1, \ldots, N-1$, update the coefficients of the global matrices thus:

1. Calculate the entries of the 2×2 element matrices

$$
\begin{aligned}
A^{(e)} &= M^{(e)} + \theta \Delta t (C^{(e)} + K^{(e)}), \\
5pt]B^{(e)} &= M^{(e)} - (1-\theta) \Delta t (C^{(e)} + K^{(e)}).
\end{aligned}
$$

2. Determine the correspondence between the global and local node numbers:

$$
i = \text{node}(e, I), \qquad j = \text{node}(e, J), \qquad I, J = 1, 2.
$$

3. Update the appropriate entries of the global matrices

$$
\begin{aligned}
A_{ij} &= A_{ij} + A_{IJ}^{(e)}, \qquad I, J = 1, 2, \\
B_{ij} &= B_{ij} + B_{IJ}^{(e)}, \qquad I, J = 1, 2.
\end{aligned}
$$

Since the coefficients of the governing equation are independent of T, the matrices A and B need to be assembled just once, but a time-dependent source term vector S needs to be updated at each time step using the following assembly procedure for vectors.

Algorithm 4.4

For $e = 1, \ldots, N-1$, update the coefficients of the global vector thus:

1. Calculate the entries of the 2×1 element vector

$$
S^{(e)} = \Delta t \left(\theta F^{(e),n+1} + (1-\theta) F^{(e),n} \right).
$$

2. Determine the correspondence between the global and local node numbers:

$$
i = \text{node}(e, I), \qquad I = 1, 2.
$$

3. Update the appropriate components of the global vector

$$
S_i = S_i + S_I^{(e)}, \qquad I = 1, 2.
$$

If the source term $f = f(x)$ is independent of time, the assembly of global matrices and vectors can be performed in the same loop over mesh elements (cf. Algorithm 4.1.)

Many variations of the above assembly algorithms may be considered. For example, it is possible to assemble the global matrices M, C, and K element by element and use the entries of these matrices to determine the coefficients of A and B. This approach requires more computer memory but provides access to the discrete operators associated with different physical processes (M: accumulation; C: convection; K: heat conduction). On the other hand, the global right-hand side $Q^n = BT^n + S$ can be assembled directly from element contributions $Q^{(e)} = B^{(e)}T^{(e)} + S^{(e)}$. The best implementation strategy depends on the nature of the problem and on the amount of available computer memory.

(vii) Dirichlet Boundary Conditions

The end product of global assembly is a sparse linear system of the form

$$AT^{n+1} = Q^n.$$

If we discretize the interval $(0, c)$ using three linear elements, we end up with

$$\begin{pmatrix} a_{11} & a_{12} & a_{13} & a_{14} \\ a_{21} & a_{22} & a_{23} & a_{24} \\ a_{31} & a_{32} & a_{33} & a_{34} \\ a_{41} & a_{42} & a_{43} & a_{44} \end{pmatrix} \begin{pmatrix} T_1^{n+1} \\ T_2^{n+1} \\ T_3^{n+1} \\ T_4^{n+1} \end{pmatrix} = \begin{pmatrix} Q_1^n \\ Q_2^n \\ Q_3^n \\ Q_4^n \end{pmatrix},$$

where

$$\begin{pmatrix} Q_1^n \\ Q_2^n \\ Q_3^n \\ Q_4^n \end{pmatrix} = \begin{pmatrix} b_{11} & b_{12} & b_{13} & b_{14} \\ b_{21} & b_{22} & b_{23} & b_{24} \\ b_{31} & b_{32} & b_{33} & b_{34} \\ b_{41} & b_{42} & b_{43} & b_{44} \end{pmatrix} \begin{pmatrix} T_1^n \\ T_2^n \\ T_3^n \\ T_4^n \end{pmatrix} + \begin{pmatrix} S_1 \\ S_2 \\ S_3 \\ S_4 \end{pmatrix}.$$

Since the first component of T^{n+1} is known from the Dirichlet boundary condition

$$T_1^{n+1} = T_0,$$

the above system is overdetermined and must be modified using one of the three techniques (elimination of rows and columns, replacement of equations by constraints, penalty method) that we used to enforce Dirichlet boundary conditions for the steady model problem. For example, the replacement of the first equation by $T_1^{n+1} = T_0$ leads to the modified system

$$\begin{pmatrix} 1 & 0 & 0 & 0 \\ a_{21} & a_{22} & a_{23} & a_{24} \\ a_{31} & a_{32} & a_{33} & a_{34} \\ a_{41} & a_{42} & a_{43} & a_{44} \end{pmatrix} \begin{pmatrix} T_1^{n+1} \\ T_2^{n+1} \\ T_3^{n+1} \\ T_4^{n+1} \end{pmatrix} = \begin{pmatrix} T_0 \\ Q_2^n \\ Q_3^n \\ Q_4^n \end{pmatrix}.$$

The left-hand matrix needs to be modified just once, while the first component of the right-hand-side vector must be overwritten by T_0 at each time step before solving the linear system.

(viii) Solving the Algebraic System

On a uniform mesh of $N-1$ elements, the $N \times N$ tridiagonal global matrices are given by

$$M = \frac{\rho c_p h}{6} \begin{pmatrix} 2 & 1 & 0 & \dots & 0 \\ 1 & 4 & 1 & \dots & 0 \\ 0 & \ddots & \ddots & \ddots & 0 \\ \vdots & \vdots & 1 & 4 & 1 \\ 0 & 0 & \dots & 1 & 2 \end{pmatrix},$$

$$C = \rho c_p u \begin{pmatrix} -\frac{1}{2} & \frac{1}{2} & 0 & \dots & 0 \\ -\frac{1}{2} & 0 & \frac{1}{2} & \dots & 0 \\ 0 & \ddots & \ddots & \ddots & 0 \\ \vdots & \vdots & -\frac{1}{2} & 0 & \frac{1}{2} \\ 0 & 0 & \dots & -\frac{1}{2} & \frac{1}{2} \end{pmatrix},$$

$$K = \frac{k}{h} \begin{pmatrix} 1 & -1 & 0 & \dots & 0 \\ -1 & 2 & -1 & \dots & 0 \\ 0 & \ddots & \ddots & \ddots & 0 \\ \vdots & \vdots & -1 & 2 & -1 \\ 0 & 0 & \dots & -1 & 1 \end{pmatrix}.$$

When it comes to the implementation of the Dirichlet boundary condition, the first row of

$$A = M + \theta \Delta t (C + K)$$

is overwritten by the first row of the $N \times N$ identity matrix. That is, we set

$$a_{11} := 1, \qquad a_{1j} := 0, \qquad j = 2, \dots, N.$$

The right-hand side Q of each global linear system is an $N \times N$ vector of the form

$$Q = \begin{pmatrix} T_0 \\ Q_2^{(1)} + Q_1^{(2)} \\ \vdots \\ Q_2^{(N-2)} + Q_1^{(N-1)} \\ Q_2^{(N-1)} \end{pmatrix},$$

where we have replaced the first component Q_1 by the given Dirichlet boundary value T_0.

At each time step, the modified linear system can be solved efficiently, e.g., using the Thomas algorithm [69], a direct solver which exploits the structure of the tridiagonal $N \times N$ system matrix A and requires just $\mathcal{O}(N)$ arithmetic operations. It is also known as the *tridiagonal matrix algorithm* (TDMA) and can be implemented as follows.

Algorithm 4.5 (TDMA)

A linear system for N unknowns with a regular tridiagonal matrix can be written as

$$\begin{pmatrix} b_1 & c_1 & & & \\ a_2 & b_2 & c_2 & & \\ & a_3 & b_3 & c_3 & \\ & & & \dots & \\ & & & a_N & b_N \end{pmatrix} \begin{pmatrix} x_1 \\ x_2 \\ x_3 \\ \cdot \\ x_N \end{pmatrix} = \begin{pmatrix} d_1 \\ d_2 \\ d_3 \\ \cdot \\ d_N \end{pmatrix}.$$

The pseudocode of the TDMA solver for systems of this form is as follows:

1. Forward elimination

   for $k = 2, \ldots, N$ **do**

   $b_k = b_k - a_k \frac{c_{k-1}}{b_{k-1}}$

   $d_k = d_k - a_k \frac{d_{k-1}}{b_{k-1}}$

   end k-loop

2. Backward substitution

   $x_N = \frac{d_N}{b_N}$

   for $k = N-1, \ldots, 1$ **do**

   $x_k = \frac{d_k - c_k x_{k+1}}{b_k}$

   end k-loop

We encourage the reader to compare the CPU times for this fast direct solver and the standard Gaussian elimination method for general linear systems with $N \times N$ matrices.

4.10 ▪ Summary

In this chapter, we illustrated the implementation of the Galerkin FEM using simple model problems based on the 1D heat equation. As we will see in the next chapters, the design of multidimensional finite element approximations follows the same design principles, but the use of unstructured meshes leads to global matrices with irregular sparsity patterns and requires a careful choice of data structures. The design of efficient solvers for algebraic systems also becomes more involved. Moreover, many problems of practical interest (including the incompressible Navier–Stokes equations that we will use as a further model problem) are inherently nonlinear and require an iterative solution strategy like the one that we used to implement the radiative boundary condition in this chapter.

Chapter 5

The Galerkin Finite Element Method for Two-Dimensional Problems

In this chapter, we apply the Galerkin FEM to 2D models of heat transfer and fluid flow in general domains. The governing equations are transformed into algebraic systems following the generic algorithm introduced in Chapter 4. In particular, we address the derivation of the variational formulation for 2D model problems, construction of bilinear basis functions on quadrilateral meshes, the use of numerical integration on a reference element, and element-by-element assembly of the global algebraic system.

5.1 ▪ Two-Dimensional Heat Equation

Let $\Omega \subset \mathbb{R}^2$ be a bounded 2D domain, as shown in Fig. 5.1. The transport and production of heat in Ω are governed by the time-dependent general heat equation

$$\rho c_p \frac{\partial T}{\partial t} + \rho c_p \mathbf{u} \cdot \nabla T - \nabla \cdot (\bar{\bar{\kappa}} \nabla T) = f \qquad \text{in } \Omega, \tag{5.1}$$

where $T(\mathbf{x}, t)$ denotes the temperature at point $\mathbf{x} \in \Omega$ and time instant $t \in (0, \tau)$. The density ρ and specific heat capacity c_p are assumed to be constant. The rate of volumetric heating is given by $f = \rho h$, where h is the rate of heating per unit mass. The velocity

$$\mathbf{u} = (u_1, u_2)$$

and the source term f are given functions of $\mathbf{x} = (x_1, x_2)$ and t. In general, the thermal conductivity $\bar{\bar{\kappa}}$ is a 2×2 tensor in the 2D case and a 3×3 tensor in the 3D case. The material is called *isotropic* if there exists $k \in \mathbb{R}$ such that

$$\bar{\bar{\kappa}} = k \begin{pmatrix} 1 & 0 \\ 0 & 1 \end{pmatrix}.$$

Otherwise the material is called *anisotropic*. An isotropic material is called *homogeneous* if k is independent of the position $\mathbf{x}$. Otherwise the material is *nonhomogeneous*.

In our model, we assume the material is isotropic and homogeneous so that the instantaneous heat flux is proportional to the temperature gradient. This yields

$$\rho c_p \frac{\partial T}{\partial t} + \rho c_p \mathbf{u} \cdot \nabla T - \nabla \cdot (k \nabla T) = f \qquad \text{in } \Omega. \tag{5.2}$$

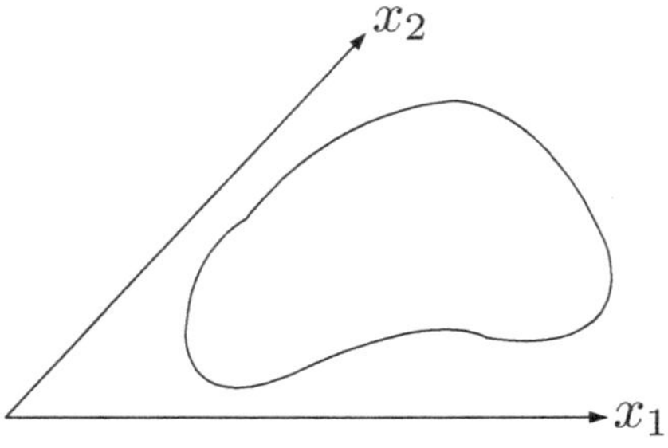

Figure 5.1. *A 2D domain* Ω.

Recalling the definition of the gradient and divergence operators given in Chapter 1, the simplified 2D heat equation can be written as follows:

$$\rho c_p \frac{\partial T}{\partial t} + \rho c_p \sum_{i=1}^{2} u_i \frac{\partial T}{\partial x_i} - \sum_{i=1}^{2}\sum_{j=1}^{2} \frac{\partial}{\partial x_i}\left(k \frac{\partial T}{\partial x_j}\right) = f \qquad \text{in } \Omega. \tag{5.3}$$

The solution to this equation is still defined up to an arbitrary constant $C \in \mathbb{R}$. If T is a solution to the temperature transport equation, then so will $T + C$. In order to formulate a well-posed problem, we must impose appropriate initial and boundary conditions.

5.1.1 ▪ Initial and Boundary Conditions

The general form of the initial condition for the parabolic 2D heat equation is given by

$$T(\mathbf{x}, 0) = T^0(\mathbf{x}), \qquad \mathbf{x} \in \Omega. \tag{5.4}$$

When it comes to the imposition of boundary conditions, we decompose the boundary $\Gamma = \partial\Omega$ of the domain Ω into two disjoint components Γ_D and Γ_N such that

$$\Gamma = \bar{\Gamma}_D \cup \bar{\Gamma}_N.$$

On the boundary component Γ_D, we prescribe the Dirichlet boundary condition

$$T = T_0 \quad \text{on } \Gamma_D. \tag{5.5}$$

On the boundary part Γ_N, we impose the generic flux boundary condition

$$-k\nabla T \cdot \mathbf{n} = h_c T + g \qquad \text{on } \Gamma_N, \tag{5.6}$$

where $\mathbf{n} = (n_1, n_2)^T$ denotes the unit outward normal to Γ (see Fig. 5.2). The parameters h_c and g have constant values in our model. The case $h_c = 0$ corresponds to a Neumann boundary condition, while $h_c \neq 0$ corresponds to a Newton boundary condition. Using the componentwise form of the vectors $\mathbf{n}$ and ∇T, we can write (5.6) in the form

$$-k \sum_{i=1}^{2} \frac{\partial T}{\partial x_i} n_i = h_c T + g \qquad \text{on } \Gamma_N.$$

The remainder of this section presents an extension of the finite element solution algorithm from Sections 4.7–4.9 to the 2D initial-boundary value problem. We intentionally consider the time-dependent heat transport model. The corresponding stationary boundary value problem can be obtained by omitting the time derivative and the initial condition. This model describes the equilibrium state which can be attained by marching the solution of the time-dependent problem to a steady state (see Chapter 2).

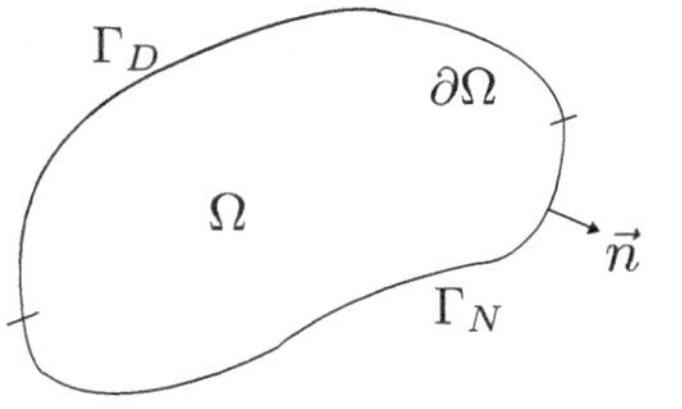

Figure 5.2. *Boundary components Γ_D and Γ_N.*

5.1.2 ▪ Variational Formulation

The space of test functions for the variational form of our model problem is given by

$$V = \{v \in H^1(\Omega) \,|\, v = 0 \text{ on } \Gamma_D\}.$$

Multiplying equation (5.2) by an arbitrary test function $w \in V$ and integrating the product over the domain Ω, one obtains the weighted residual formulation

$$\int_\Omega \rho c_p w \frac{\partial T}{\partial t}\,\mathrm{d}\mathbf{x} + \int_\Omega \rho c_p w \mathbf{u}\cdot\nabla T\,\mathrm{d}\mathbf{x} - \int_\Omega w\nabla\cdot(k\nabla T)\,\mathrm{d}\mathbf{x} = \int_\Omega wf\,\mathrm{d}\mathbf{x}. \tag{5.7}$$

The application of Green's theorem to the integral containing the diffusive term gives

$$\begin{aligned} -\int_\Omega w\nabla\cdot(k\nabla T)\,\mathrm{d}\mathbf{x} &= \int_\Omega k\nabla w\cdot\nabla T\,\mathrm{d}\mathbf{x} - \int_\Gamma kw\nabla T\cdot\mathbf{n}\,\mathrm{d}s \\ &= \int_\Omega k\nabla w\cdot\nabla T\,\mathrm{d}\mathbf{x} - \int_{\Gamma_D} kw\nabla T\cdot\mathbf{n}\,\mathrm{d}s - \int_{\Gamma_N} kw\nabla T\cdot\mathbf{n}\,\mathrm{d}s \\ &= \int_\Omega k\nabla w\cdot\nabla T\,\mathrm{d}\mathbf{x} + \int_{\Gamma_N} w(h_c T + g)\,\mathrm{d}s \qquad \forall w \in V. \end{aligned}$$

The integral over the Dirichlet boundary Γ_D vanishes because the test function $w \in V$ was chosen to have zero trace on this boundary. The integral over the Newton/Neumann boundary Γ_N is evaluated using the flux boundary condition (5.6). Substituting the result into (5.7), we obtain the following representation of the continuous variational problem:

$$\begin{aligned} &\int_\Omega \rho c_p w \frac{\partial T}{\partial t}\,\mathrm{d}\mathbf{x} + \int_\Omega \rho c_p w\mathbf{u}\cdot\nabla T\,\mathrm{d}\mathbf{x} + \int_\Omega k\nabla w\cdot\nabla T\,\mathrm{d}\mathbf{x} + \int_{\Gamma_N} h_c wT\,\mathrm{d}s \\ &\qquad = \int_\Omega wf\,\mathrm{d}\mathbf{x} - \int_{\Gamma_N} wg\,\mathrm{d}s \qquad \forall w \in V. \end{aligned} \tag{5.8}$$

Note that the integral over Γ_N has been split into two parts. The part that depends on the unknown temperature T belongs to the left-hand side of the integral equation, whereas the contribution of the given heat flux g belongs to the right-hand side.

5.1.3 ▪ Finite Element Discretization

Let V_h be a finite-dimensional space spanned by the basis functions $\varphi_1, \ldots, \varphi_N$. We define the approximate solution T_h as a linear combination of these basis functions:

$$T_h(\mathbf{x}, t) = \sum_{j=1}^{N} T_j(t)\varphi_j(\mathbf{x}). \tag{5.9}$$

Following the usual procedure, we consider the discretized variational problem

$$\int_\Omega \rho c_p w_h \frac{\partial T_h}{\partial t}\,\mathrm{d}\mathbf{x} + \int_\Omega \rho c_p w_h \mathbf{u}\cdot\nabla T_h\,\mathrm{d}\mathbf{x} + \int_\Omega k\nabla w_h\cdot\nabla T_h\,\mathrm{d}\mathbf{x} + \int_{\Gamma_N} h_c w_h T_h\,\mathrm{d}s$$
$$= \int_\Omega w_h f\,\mathrm{d}\mathbf{x} - \int_{\Gamma_N} w_h g\,\mathrm{d}s \qquad \forall w_h \in V_h. \tag{5.10}$$

Using $w_h \in \{\varphi_1,\dots,\varphi_N\}$ as test functions and substituting (5.9) into the integrals on the left-hand side, we obtain a system of N semidiscrete Galerkin equations

$$\sum_{j=1}^N \left(\int_\Omega \rho c_p \varphi_i\varphi_j\,\mathrm{d}\mathbf{x}\right)\frac{\mathrm{d}T_j}{\mathrm{d}t} + \sum_{j=1}^N \left(\int_\Omega \rho c_p \varphi_i \mathbf{u}\cdot\nabla\varphi_j\,\mathrm{d}\mathbf{x}\right)T_j$$
$$+\sum_{j=1}^N \left(\int_\Omega k\nabla\varphi_i\cdot\nabla\varphi_j\,\mathrm{d}\mathbf{x}\right)T_j + \sum_{j=1}^N \left(\int_{\Gamma_N} h_c\varphi_i\varphi_j\,\mathrm{d}s\right)T_j$$
$$= \int_\Omega \varphi_i f\,\mathrm{d}\mathbf{x} - \int_{\Gamma_N}\varphi_i g\,\mathrm{d}s, \qquad i=1,\dots,N. \tag{5.11}$$

Adopting the notation of Chapter 4, we write this system in the matrix form

$$M\frac{\mathrm{d}T}{\mathrm{d}t} + (C+K+B)T = F - G, \tag{5.12}$$

where T is the vector of time-dependent nodal values, M is the Galerkin mass matrix, C is the discrete convection operator, K is the discrete diffusion operator, and F is the source term. The contribution of the natural boundary conditions on Γ_N is represented by the matrix B and vector G. The coefficients of the semidiscrete problem are given by

$$m_{ij} = \int_\Omega \rho c_p\varphi_i\varphi_j\,\mathrm{d}\mathbf{x}, \qquad i,j=1,\dots,N,$$
$$c_{ij} = \int_\Omega \rho c_p\varphi_i\mathbf{u}\cdot\nabla\varphi_j\,\mathrm{d}\mathbf{x}, \qquad i,j=1,\dots,N,$$
$$k_{ij} = \int_\Omega k\nabla\varphi_i\cdot\nabla\varphi_j\,\mathrm{d}\mathbf{x}, \qquad i,j=1,\dots,N,$$
$$b_{ij} = \int_{\Gamma_N} h_c\varphi_i\varphi_j\,\mathrm{d}s, \qquad i,j=1,\dots,N,$$
$$F_i = \int_\Omega \varphi_i f\,\mathrm{d}\mathbf{x}, \qquad i=1,\dots,N,$$
$$G_i = \int_{\Gamma_N}\varphi_i g\,\mathrm{d}s, \qquad i=1,\dots,N.$$

The discretization in time can be performed, e.g., using the generic θ scheme

$$[M+\theta\Delta t(C+K+B)]T^{n+1} = [M-(1-\theta)\Delta t(C+K+B)]T^n$$
$$= \Delta t[\theta F^{n+1} + (1-\theta)F^n]$$
$$-\Delta t[\theta G^{n+1} + (1-\theta)G^n],$$

which combines the forward Euler ($\theta=0$), Crank–Nicolson ($\theta=\frac{1}{2}$), and backward Euler

($\theta = 1$) time-stepping methods. The nodal values may be initialized by setting

$$T_j^0 := T^0(\mathbf{x}_j), \qquad j = 1,\ldots,N,$$

or using the discrete L^2 projection

$$\sum_{j=1}^{N}\left(\int_\Omega \varphi_i\varphi_j\,\mathrm{d}\mathbf{x}\right)T_j^0 = \int_\Omega \varphi_i T^0\,\mathrm{d}\mathbf{x}, \qquad i = 1,\ldots,N,$$

which requires solving an additional linear system with a symmetric positive definite matrix.

5.1.4 ▪ Stabilization

The Galerkin discretization of the convective term $\rho c_p \mathbf{u}\cdot\nabla T$ is represented by the ill-behaved nonsymmetric matrix C, which may give rise to spurious oscillations and numerical instabilities at large values of the Péclet number $\mathrm{Pe} = \frac{\rho c_p}{k}$. We will explain the cause and cure of this phenomenon in detail in the next two chapters. The simplest way to stabilize the convective part is to add *anisotropic balancing dissipation* of the form

$$-\nabla\cdot(\tau\mathbf{u}\otimes\mathbf{u}\nabla T),$$

where τ is a suitably chosen stabilization parameter (see Section 7.3.3). The same stabilization effect can be achieved by using the modified test function

$$\tilde{\varphi}_i := \varphi_i + \tau\mathbf{u}\cdot\nabla\varphi_i$$

in the contribution of the convective term to the variational formulation. The discrete convection operator $\tilde{C}$ for the stabilized Galerkin scheme is defined by

$$\tilde{c}_{ij} = c_{ij} + \int_\Omega \rho c_p \tau(\mathbf{u}\cdot\nabla\varphi_i)(\mathbf{u}\cdot\nabla\varphi_j)\,\mathrm{d}\mathbf{x}, \qquad i,j = 1,\ldots,N.$$

Applying the modified test function to all components of the weighted residual leads to a strongly consistent streamline upwind Petrov–Galerkin (SUPG) method [23], one of the most prominent representatives of stabilized FEM for convection-dominated transport problems. The choice of stabilization parameters for SUPG and design of alternative stabilization techniques for convective terms will be addressed in Chapter 7. In the remainder of this section, we discuss the implementation of the standard Galerkin method ($\tau = 0$).

5.1.5 ▪ Localization

In Chapter 4, we constructed the approximate solution T_h by piecing together local interpolation polynomials $T_h^{(e)}$ expressed in terms of linear basis functions $\varphi_i^{(e)}$. In the 2D case, the computational mesh typically consists of triangular and/or quadrilateral elements. The construction of a linear shape function $T_h^{(e)}$ on a triangle E_e is straightforward. Each local basis function is a Lagrange basis polynomial of the form

$$\varphi_k^{(e)}(x_1,x_2) = a_i^{(e)}x_1 + b_i^{(e)}x_2 + c_i^{(e)}, \qquad i = 1,2,3,$$

where x_1 and x_2 denote the Cartesian coordinates of a point $\mathbf{x} = (x_1,x_2)^T$ in $\mathbb{R}^2$.

The interpolation conditions for a three-node linear triangle are given by

$$\varphi_i^{(e)}(\mathbf{x}_j) = \delta_{ij}, \qquad j = 1,2,3,$$

where $\mathbf{x}_j$ is a vertex of the triangular element and δ_{ij} is the Kronecker delta. This gives a system of three equations for the three unknown coefficients of the local basis function $\varphi_i^{(e)}$. For details regarding the construction of finite element approximations on triangular elements, we refer the reader to the FEM introductions by Cuvelier, Segal, and van Steenhoven [44] and Johnson [122].

In this section, we will use four-node quadrilateral elements. Since a linear polynomial is uniquely defined by its values at three points, we consider *bilinear* basis functions

$$\varphi_i^{(e)}(x_1,x_2) = a_i^{(e)} x_1 x_2 + b_i^{(e)} x_1 + c_i^{(e)} x_2 + d_i^{(e)}, \qquad i = 1,\dots,4.$$

Placing the nodal points at the corners of the quadrilateral element E_e, the values of the four unknown coefficients can be determined by imposing the interpolation conditions

$$\varphi_i^{(e)}(\mathbf{x}_j) = \delta_{ij}, \qquad j = 1,\dots,4.$$

Bilinear basis functions are particularly easy to construct on a rectangle. The generation of a rectangular mesh is possible for relatively simple domains like the backward facing step geometry in the example we considered in Section 4.2. Of course, the geometry is not always that simple. If the domain Ω is neither a rectangle nor a union of rectangular subdomains, it should be discretized using general quadrilaterals or triangles. Curved boundaries can be approximated using *isoparametric* finite elements to achieve optimal accuracy.

To get started, we will assume that the elements of the computational mesh are rectangular. When it comes to the design of bilinear finite element approximations for nonrectangular domains, we will extend the entire methodology to general quadrilaterals.

Bilinear Approximation on Rectangles

Let us consider a rectangular domain $\Omega = (0,c) \times (0,d)$. We discretize it using a uniform Cartesian mesh that consists of $N_{x_1} \times N_{x_2}$ grid points and $(N_{x_1} - 1) \times (N_{x_2} - 1)$ rectangular elements (see Fig. 5.3). The geometry of each element E_e is determined by the position of the four corners, which will serve as nodal points for constructing a local bilinear interpolation polynomial $T_h^{(e)}$ and the corresponding local basis functions $\varphi_i^{(e)}$, $i = 1,\dots,4$.

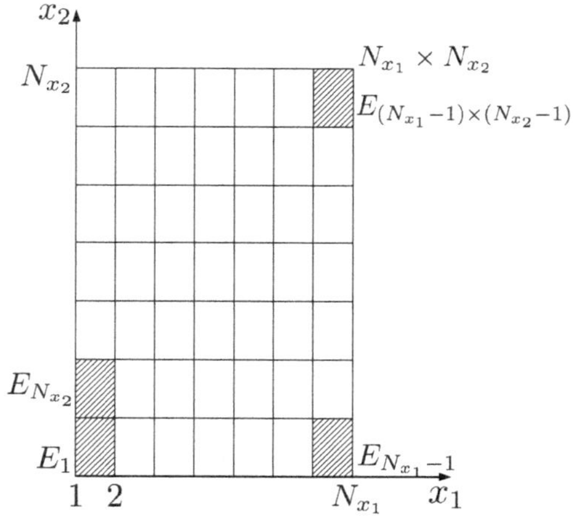

Figure 5.3. *Global numbering of the elements and nodal points.*

In our example, the vertices and elements of the mesh are numbered from left to right and from bottom to top, as illustrated in Fig. 5.3. We assume that all element edges parallel to the x_1 axis have the same length $h_{x_1} = \frac{c}{N_{x_1}-1}$ and all element edges parallel to the x_2 axis have the same length $h_{x_2} = \frac{d}{N_{x_2}-1}$. That is, each axis is discretized using a uniform 1D mesh like the one that we used in Chapter 4. The rectangular 2D mesh represents a Cartesian product of two 1D meshes. Each mesh vertex can be identified by two indices, $i = 1,\dots,N_{x_1}$ and $j = 1,\dots,N_{x_2}$, which correspond to the numbers of the vertical and horizontal grid lines intersecting at this vertex. The Cartesian coordinates of the nodal point with the global number $i+(j-1)N_{x_1}$ are $(i-1)h_{x_1}$ and $(j-1)h_{x_2}$.

Consider a single element E_e, where $e = 1,\dots,N_e := (N_{x_1}-1)(N_{x_2}-1)$. The four vertices of E_e are assigned local numbers in the range from 1 to 4. The enumeration is performed counterclockwise starting from the bottom left corner (see Fig. 5.4).

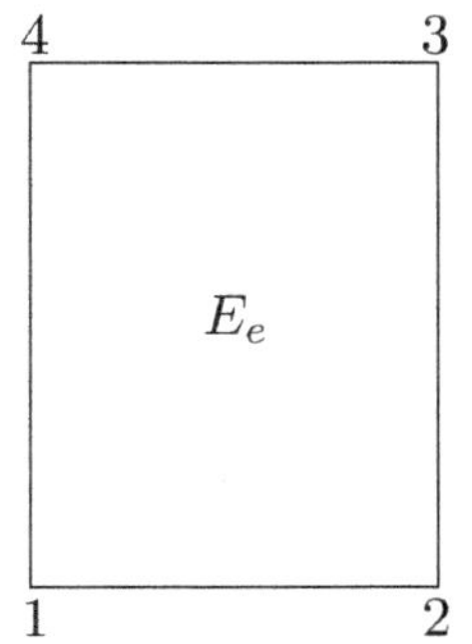

Figure 5.4. *Local node numbering for a rectangular element E_e.*

Let $j = 1,\dots,N_{x_2}-1$ denote the number of the horizontal grid line that contains the bottom edge of element E_e. Then the number of the vertical grid line containing the left edge of E_e is $i = e-(j-1)(N_{x_1}-1)$. The correspondence between the local and global numbers of the four vertices is summarized in the following table:

Local number	*Global number*	*Location*
1	$e+j-1$	bottom left corner
2	$e+j$	bottom right corner
3	$e+N_{x_1}+j$	top right corner
4	$e+N_{x_1}+j-1$	top left corner

The corresponding pairs of grid line numbers (i,j), $(i+1,j)$, $(i+1,j+1)$, $(i,j+1)$ determine the Cartesian coordinates of the four nodal points as shown in Fig. 5.5.

A 2D local basis function $\varphi_k^{(e)}$, $k = 1,\dots,4$, is a bilinear polynomial which equals one at the kth nodal point and zero at the remaining nodes of the element. The coefficients of this polynomial could be determined by solving a 2D interpolation problem. However, we will take advantage of the fact that our element E_e is rectangular and define $\varphi_k^{(e)}$ as a product of the 1D linear basis functions

$$\lambda_1^{(i)}(x,h) = -\frac{x}{h} + i,$$
$$\lambda_2^{(i)}(x,h) = \frac{x}{h} - i + 1.$$

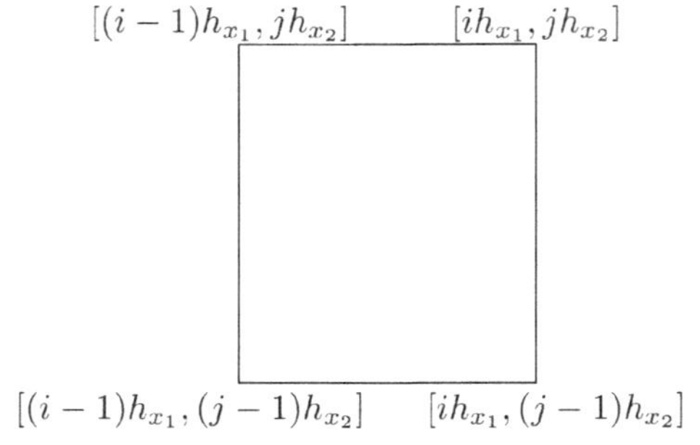

Figure 5.5. *Nodal point coordinates for a rectangular element E_e.*

Let the bottom left corner of E_e be located at the intersection of the ith vertical grid line and the jth horizontal grid line. It is easy to verify that the local basis functions

$$\begin{aligned}
\varphi_1^{(e)}(x_1,x_2) &= \lambda_1^{(i)}(x_1,h_{x_1})\lambda_1^{(j)}(x_2,h_{x_2}),\\
\varphi_2^{(e)}(x_1,x_2) &= \lambda_2^{(i)}(x_1,h_{x_1})\lambda_1^{(j)}(x_2,h_{x_2}),\\
\varphi_3^{(e)}(x_1,x_2) &= \lambda_2^{(i)}(x_1,h_{x_1})\lambda_2^{(j)}(x_2,h_{x_2}),\\
\varphi_4^{(e)}(x_1,x_2) &= \lambda_1^{(i)}(x_1,h_{x_1})\lambda_2^{(j)}(x_2,h_{x_2})
\end{aligned}$$

are bilinear polynomials such that $\varphi_k^{(e)}$ equals one at the kth nodal point and zero at the other three nodes. Substituting the above definition of $\lambda_1^{(i)}$ and $\lambda_2^{(i)}$, we obtain

$$\varphi_1^{(e)}(x_1,x_2) = \frac{1}{h_{x_1}h_{x_2}}x_1x_2 - \frac{j}{h_{x_1}}x_1 - \frac{i}{h_{x_2}}x_2 + ij, \tag{5.13}$$

$$\varphi_2^{(e)}(x_1,x_2) = -\frac{1}{h_{x_1}h_{x_2}}x_1x_2 + \frac{j}{h_{x_1}}x_1 + \frac{i-1}{h_{x_2}}x_2 + j(1-i), \tag{5.14}$$

$$\varphi_3^{(e)}(x_1,x_2) = \frac{1}{h_{x_1}h_{x_2}}x_1x_2 + \frac{1-j}{h_{x_1}}x_1 + \frac{1-i}{h_{x_2}}x_2 - i + j(i-1) + 1, \tag{5.15}$$

$$\varphi_4^{(e)}(x_1,x_2) = -\frac{1}{h_{x_1}h_{x_2}}x_1x_2 + \frac{j-1}{h_{x_1}}x_1 + \frac{i}{h_{x_2}}x_2 + i(1-j). \tag{5.16}$$

This representation enables us to write the basis functions in the canonical form

$$\varphi_k^{(e)}(x_1,x_2) = a_k^{(e)}x_1x_2 + b_k^{(e)}x_1 + c_k^{(e)}x_2 + d_k^{(e)}, \qquad k = 1,\ldots,4.$$

The coefficients of the four bilinear polynomials are listed in the following table:

k	$a_k^{(e)}$	$b_k^{(e)}$	$c_k^{(e)}$	$d_k^{(e)}$
1	$\frac{1}{h_{x_1}h_{x_2}}$	$-\frac{j}{h_{x_1}}$	$-\frac{i}{h_{x_2}}$	ij
2	$-\frac{1}{h_{x_1}h_{x_2}}$	$\frac{j}{h_{x_1}}$	$\frac{i-1}{h_{x_2}}$	$j(1-i)$
3	$\frac{1}{h_{x_1}h_{x_2}}$	$\frac{1-j}{h_{x_1}}$	$\frac{1-i}{h_{x_2}}$	$-i+j(i-1)+1$
4	$-\frac{1}{h_{x_1}h_{x_2}}$	$\frac{j-1}{h_{x_1}}$	$\frac{i}{h_{x_2}}$	$i(1-j)$

Fig. 5.6 shows the local basis function $\varphi_1^{(1)}$ defined on a unit square ($h_{x_1} = h_{x_2} = 1$).

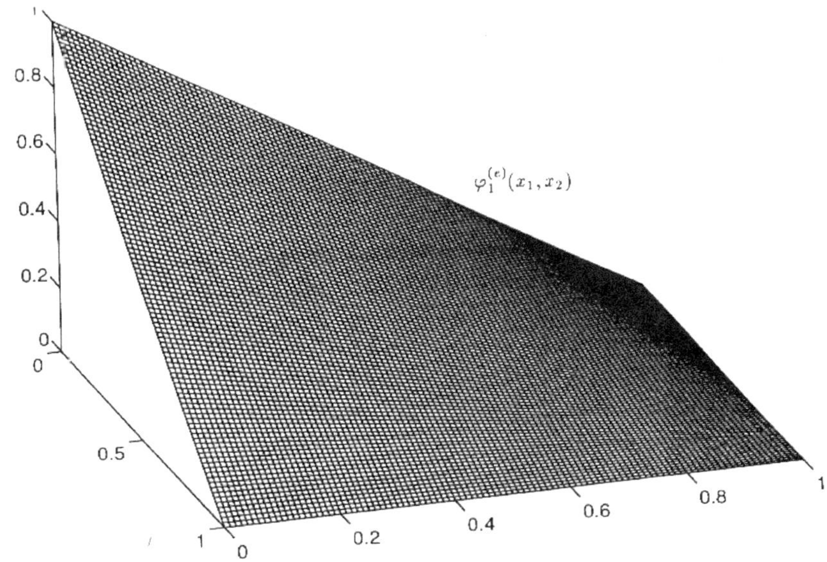

Figure 5.6. *Local basis function $\varphi_1^{(1)}(x_1, x_2)$.*

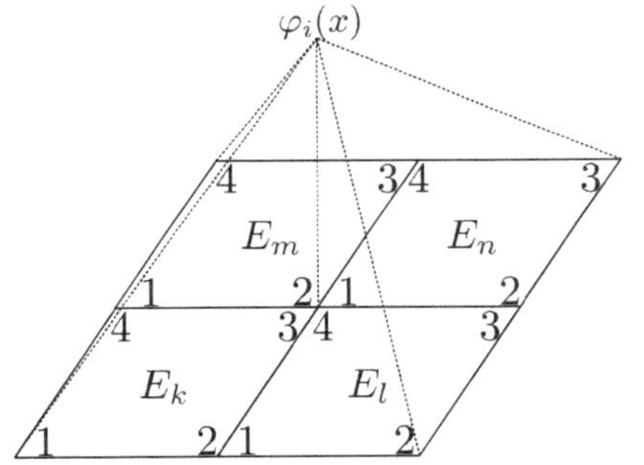

Figure 5.7. *Global basis function φ_i.*

As in the 1D case, the global basis function φ_i associated with a mesh point $\mathbf{x}_i$ can be constructed by piecing together local basis functions. Let $\mathbf{x}_i \in \Omega$ be an internal point that belongs to elements E_k, E_l, E_m, and E_n. The restriction of φ_i to the patch of these four elements is depicted in Fig. 5.7. The local basis functions associated with each element are referred to using the counterclockwise local numbering such that the point $\mathbf{x}_i$ is the first node of element E_n, the second node of element E_m, the third node of element E_k, and the fourth node of element E_l (see Fig. 5.7). The global basis function φ_i is defined in terms of the corresponding local bilinear basis functions as follows:

$$\varphi_i(x_1, x_2) = \begin{cases} \varphi_1^{(n)}(x_1, x_2), & (x_1, x_2) \in E_n, \\ \varphi_2^{(m)}(x_1, x_2), & (x_1, x_2) \in E_m, \\ \varphi_3^{(k)}(x_1, x_2), & (x_1, x_2) \in E_k, \\ \varphi_4^{(l)}(x_1, x_2), & (x_1, x_2) \in E_l, \\ 0 & \text{elsewhere.} \end{cases}$$

It follows that the restriction of the approximate solution T_h to each element is a bilinear polynomial that represents a linear combination of the four local basis functions.

The coefficients of the global system (5.12) are calculated using elementwise integration followed by summation in a loop over elements. Since there are four local degrees of freedom per element, the size of the element matrices $M^{(e)}, C^{(e)}, K^{(e)}, B^{(e)}$ is 4×4. The element vectors $F^{(e)}, G^{(e)}$ have four components. The contribution of E_e is defined by

$$m_{ij}^{(e)} = \int_{E_e} \rho c_p \varphi_i^{(e)} \varphi_j^{(e)} \,\mathrm{d}\mathbf{x}, \qquad i,j = 1,\ldots,4,$$

$$c_{ij}^{(e)} = \int_{E_e} \rho c_p \varphi_i^{(e)} \mathbf{u} \cdot \nabla \varphi_j^{(e)} \,\mathrm{d}\mathbf{x}, \qquad i,j = 1,\ldots,4,$$

$$k_{ij}^{(e)} = \int_{E_e} k \nabla \varphi_i^{(e)} \cdot \nabla \varphi_j^{(e)} \,\mathrm{d}\mathbf{x}, \qquad i,j = 1,\ldots,4,$$

$$b_{ij}^{(e)} = \int_{E_e \cap \Gamma_N} h_c \varphi_i^{(e)} \varphi_j^{(e)} \,\mathrm{d}s, \qquad i,j = 1,\ldots,4,$$

$$F_i^{(e)} = \int_{E_e} \varphi_i^{(e)} f \,\mathrm{d}\mathbf{x}, \qquad i = 1,\ldots,4,$$

$$G_i^{(e)} = \int_{E_e \cap \Gamma_N} \varphi_i^{(e)} g \,\mathrm{d}s, \qquad i = 1,\ldots,4,$$

where the local basis functions are defined by (5.13)–(5.16). If E_e has no edge on the boundary part Γ_N, we have $b_{ij}^{(e)} = 0$ for $i,j = 1,\ldots,4$ and $G_i = 0$ for $i = 1,\ldots,4$.

Example 5.1. By definition of the basis function $\varphi_1^{(e)}$, the first diagonal entry of the element mass matrix $M^{(e)}$ and the first component of the element vector $F^{(e)}$ are given by

$$\begin{aligned} m_{11}^{(e)} &= \int_{E_e} \rho c_p \varphi_1^{(e)} \varphi_1^{(e)} \,\mathrm{d}\mathbf{x} \\ &= \int_{(j-1)h_{x_2}}^{jh_{x_2}} \int_{(i-1)h_{x_1}}^{ih_{x_1}} \rho c_p \left(\frac{1}{h_{x_1} h_{x_2}} x_1 x_2 - \frac{j}{h_{x_1}} x_1 - \frac{i}{h_{x_2}} x_2 + ij \right)^2 \mathrm{d}x_1 \mathrm{d}x_2 \end{aligned}$$

and

$$\begin{aligned} F_1^{(e)} &= \int_{E_e} \varphi_1^{(e)} f \,\mathrm{d}\mathbf{x} \\ &= \int_{(j-1)h_{x_2}}^{jh_{x_2}} \int_{(i-1)h_{x_1}}^{ih_{x_1}} \left(\frac{1}{h_{x_1} h_{x_2}} x_1 x_2 - \frac{j}{h_{x_1}} x_1 - \frac{i}{h_{x_2}} x_2 + ij \right) f \,\mathrm{d}x_1 \mathrm{d}x_2. \end{aligned}$$

If exact integration is impossible or difficult, then numerical integration should be employed (see below). The remaining components of the element contribution are determined in a similar way. Note that the entries of the element matrices $C^{(e)}$ and $K^{(e)}$ depend on the (easily computable) partial derivatives of the local basis functions defined by (5.13)–(5.16). ∎

Example 5.2. Let us assume that the bottom edge of the element E_e lies on the boundary Γ_N, as shown in Fig. 5.8. The contribution of E_e to the boundary terms is given by

$$
\begin{aligned}
b_{ij}^{(e)} &= \int_{E_e \cap \Gamma_N} h_c \varphi_i^{(e)}(x_1,x_2)\varphi_j^{(e)}(x_1,x_2)\,\mathrm{d}s \\
&= \int_{(i-1)h_{x_1}}^{ih_{x_1}} h_c \varphi_i^{(e)}(x_1,0)\varphi_j^{(e)}(x_1,0)\,\mathrm{d}x_1, \qquad i,j=1,\dots,4, \\
G_i^{(e)} &= \int_{E_e \cap \Gamma_N} \varphi_i^{(e)}(x_1,x_2) g\,\mathrm{d}s \\
&= \int_{(i-1)h_{x_1}}^{ih_{x_1}} \varphi_i^{(e)}(x_1,0) g\,\mathrm{d}x_1, \qquad i=1,\dots,4,
\end{aligned}
$$

and should be evaluated using the local basis functions

$$
\begin{aligned}
\varphi_1^{(e)}(x_1,0) &= -\frac{x_1}{h_{x_1}} + e, \\
\varphi_2^{(e)}(x_1,0) &= \frac{x_1}{h_{x_1}} - e + 1, \\
\varphi_3^{(e)}(x_1,0) &= 0, \\
\varphi_4^{(e)}(x_1,0) &= 0.
\end{aligned}
$$

It follows that all integrals that depend on $\varphi_3^{(e)}$ or $\varphi_4^{(e)}$ will be equal to zero. ∎

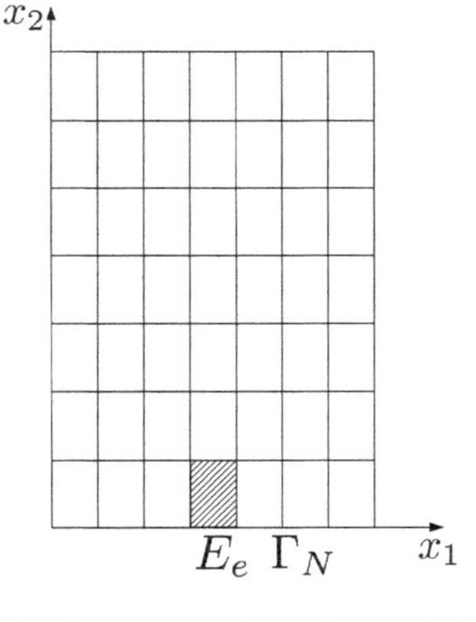

Figure 5.8. *An element E_e with an edge on the boundary Γ_N.*

Bilinear Approximation on Quadrilaterals

The generation of structured meshes for nonrectangular domains is typically performed using a global mapping to a rectangular reference domain [3] or local mappings to a square reference element [44]. The former approach is usually employed in the finite difference context, while the latter is ideally suited for finite element methods. Since a nonrectangular boundary-fitted mesh is topologically equivalent to a Cartesian grid, the elements

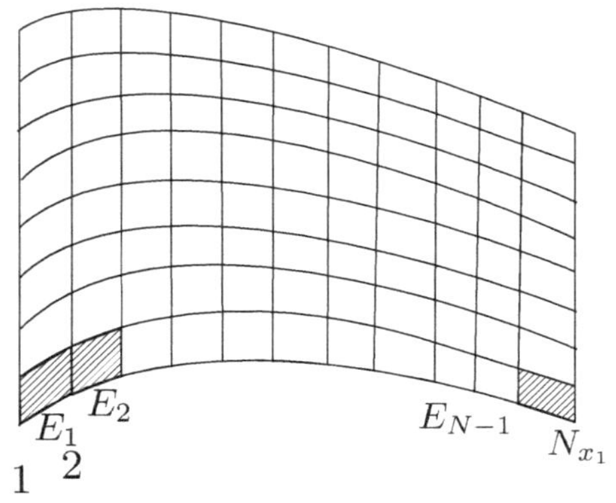

Figure 5.9. *Numbering of elements and nodes of a nonrectangular structured mesh.*

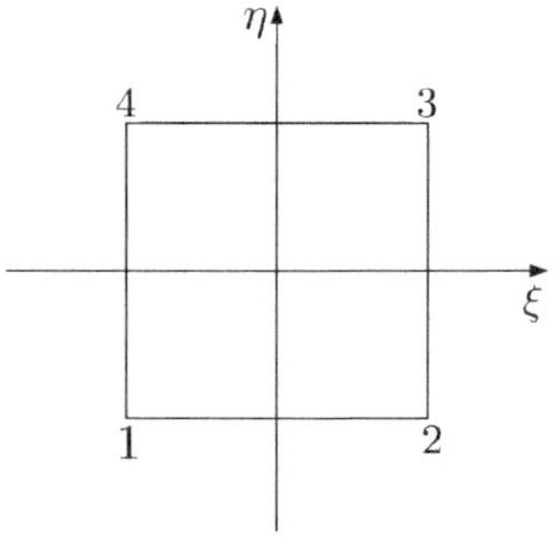

Figure 5.10. *The reference element and the local coordinate system.*

and vertices can be numbered in the same way (see Fig. 5.9), but the direct construction of local bilinear basis functions on a general quadrilateral is a tedious and laborious task. For this reason, the local basis functions are commonly defined on a reference element, and the corresponding transformation of variables is performed when it comes to integration/differentiation.

Each element of a general quadrilateral mesh represents an image of the square reference element $\hat{E} = [-1,1]\times[-1,1]$ centered at the origin of the Cartesian (ξ,η) coordinate system as shown in Fig. 5.10. In accordance with the numbering convention for rectangular elements, the four vertices of $\hat{E}$ are numbered counterclockwise starting from the bottom left corner. The bilinear basis functions associated with these nodal points are given by

$$\hat{\varphi}_1(\xi,\eta) = \frac{1}{4}(1-\xi)(1-\eta), \tag{5.17}$$

$$\hat{\varphi}_2(\xi,\eta) = \frac{1}{4}(1+\xi)(1-\eta), \tag{5.18}$$

$$\hat{\varphi}_3(\xi,\eta) = \frac{1}{4}(1+\xi)(1+\eta), \tag{5.19}$$

$$\hat{\varphi}_4(\xi,\eta) = \frac{1}{4}(1-\xi)(1+\eta) \tag{5.20}$$

and represent tensor products of 1D linear basis functions on the reference interval $[-1,1]$.

It is easy to verify that $\hat{\varphi}_1, \ldots, \hat{\varphi}_4$ are bilinear polynomials in ξ, η satisfying

$$\hat{\varphi}_i(\xi_j, \eta_j) = \delta_{ij}, \qquad i, j = 1, \ldots, 4,$$

where (ξ_j, r_{ij}) are the Cartesian coordinates of the jth nodal point. It follows that a linear combination of the four local basis functions will be a bilinear interpolation polynomial.

Let us now consider a quadrilateral element E_e and construct a one-to-one mapping

$$F_e : \hat{E} \mapsto E_e, \qquad F_e(\xi, \eta) = (x_1, x_2)$$

such that the vertices of the reference element $\hat{E}$ are mapped to the vertices of the physical element E_e. That is, the mapping F_e should satisfy the interpolation conditions

$$\begin{aligned} F_e(-1,-1) &= (x_{11}^{(e)}, x_{21}^{(e)}), \\ F_e(1,-1) &= (x_{12}^{(e)}, x_{22}^{(e)}), \\ F_e(1,1) &= (x_{13}^{(e)}, x_{23}^{(e)}), \\ F_e(-1,1) &= (x_{14}^{(e)}, x_{24}^{(e)}), \end{aligned}$$

where (x_{1i}, x_{2i}) are the Cartesian coordinates of the ith vertex of the physical element E_e.

In order to satisfy the above conditions, the mapping F_e should be a bilinear interpolation polynomial defined on the reference element in terms of the local basis functions:

$$F_e(\xi, \eta) = \sum_{j=1}^{4} \hat{\varphi}_j(\xi, \eta)(x_{1j}^{(e)}, x_{2j}^{(e)}) = (x_1, x_2),$$

where

$$\begin{aligned} x_1(\xi, \eta) &= \sum_{j=1}^{4} \hat{\varphi}_j(\xi, \eta) x_{1j}^{(e)}, \\ x_2(\xi, \eta) &= \sum_{j=1}^{4} \hat{\varphi}_j(\xi, \eta) x_{2j}^{(e)}. \end{aligned}$$

Fig. 5.11 illustrates the mapping of $\hat{E}$ onto the quadrilateral element $E_k = F_e(\hat{E})$.

If the same basis functions are used to construct the mapping F_e and the approximate solution of the governing equation, the element is called *isoparametric*. In this chapter,

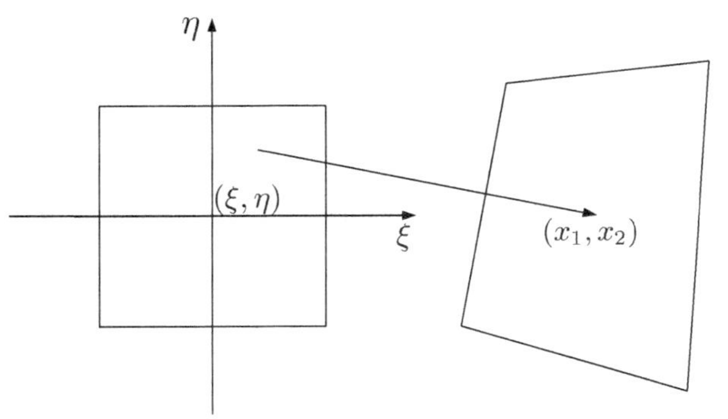

Figure 5.11. *Mapping of the reference element onto a general element.*

we will use isoparametric bilinear elements. The use of higher-order (e.g., biquadratic) isoparametric elements makes it possible to fit the shape of physical elements to curvilinear boundaries of the computational domain. For details, we refer the reader to Cuvelier, Segal, and van Steenhoven [44].

The bilinear mapping $(\xi,\eta) \mapsto (x_1,x_2)$ can be written in the matrix form

$$\begin{pmatrix} x_1(\xi,\eta) \\ x_2(\xi,\eta) \end{pmatrix} = X_e^T \hat{\Phi}(\xi,\eta), \tag{5.21}$$

where

$$X_e = \begin{pmatrix} x_{11}^{(e)} & x_{21}^{(e)} \\ x_{12}^{(e)} & x_{22}^{(e)} \\ x_{13}^{(e)} & x_{23}^{(e)} \\ x_{14}^{(e)} & x_{24}^{(e)} \end{pmatrix}, \qquad \hat{\Phi}(\xi,\eta) = \begin{pmatrix} \hat{\varphi}_1(\xi,\eta) \\ \hat{\varphi}_2(\xi,\eta) \\ \hat{\varphi}_3(\xi,\eta) \\ \hat{\varphi}_4(\xi,\eta) \end{pmatrix}.$$

Example 5.3. Consider a quadrilateral element, whose nodal point coordinates are given by

$$X_e = \begin{pmatrix} 0 & 0 \\ 2 & 1 \\ 2 & 3 \\ 1 & 4 \end{pmatrix}.$$

The center of the reference element is given by $(\xi,\eta) = (0,0)$ and maps to $(x_1,x_2) = (\frac{5}{4},2)$. Indeed, the evaluation of the local bilinear basis functions at the point $(0,0)$ yields

$$\hat{\Phi}(0,0) = \left(\frac{1}{4},\frac{1}{4},\frac{1}{4},\frac{1}{4}\right)^T,$$

whence

$$x_1(0,0) = \frac{1}{4}\cdot 0 + \frac{1}{4}\cdot 2 + \frac{1}{4}\cdot 2 + \frac{1}{4}\cdot 1 = \frac{5}{4},$$
$$x_2(0,0) = \frac{1}{4}\cdot 0 + \frac{1}{4}\cdot 1 + \frac{1}{4}\cdot 3 + \frac{1}{4}\cdot 4 = 2. \quad \blacksquare$$

When it comes to calculating the coefficients of our model problem, we will need not only the local basis functions $\hat{\varphi}_i$ but also their partial derivatives, which are given by

$$\begin{aligned}
\frac{\partial \hat{\varphi}_1(\xi,\eta)}{\partial \xi} &= -\frac{1}{4}(1-\eta), & \frac{\partial \hat{\varphi}_1(\xi,\eta)}{\partial \eta} &= -\frac{1}{4}(1-\xi), \\
\frac{\partial \hat{\varphi}_2(\xi,\eta)}{\partial \xi} &= \frac{1}{4}(1-\eta), & \frac{\partial \hat{\varphi}_2(\xi,\eta)}{\partial \eta} &= -\frac{1}{4}(1+\xi), \\
\frac{\partial \hat{\varphi}_3(\xi,\eta)}{\partial \xi} &= \frac{1}{4}(1+\eta), & \frac{\partial \hat{\varphi}_3(\xi,\eta)}{\partial \eta} &= \frac{1}{4}(1+\xi), \\
\frac{\partial \hat{\varphi}_4(\xi,\eta)}{\partial \xi} &= -\frac{1}{4}(1+\eta), & \frac{\partial \hat{\varphi}_4(\xi,\eta)}{\partial \eta} &= \frac{1}{4}(1-\xi).
\end{aligned} \tag{5.22}$$

We will store the values of these partial derivatives in the 2×4 matrix

$$\hat{\Phi}'(\xi,\eta) = \begin{pmatrix} \frac{\partial \hat{\varphi}_1(\xi,\eta)}{\partial \xi} & \frac{\partial \hat{\varphi}_2(\xi,\eta)}{\partial \xi} & \frac{\partial \hat{\varphi}_3(\xi,\eta)}{\partial \xi} & \frac{\partial \hat{\varphi}_4(\xi,\eta)}{\partial \xi} \\ \frac{\partial \hat{\varphi}_1(\xi,\eta)}{\partial \eta} & \frac{\partial \hat{\varphi}_2(\xi,\eta)}{\partial \eta} & \frac{\partial \hat{\varphi}_3(\xi,\eta)}{\partial \eta} & \frac{\partial \hat{\varphi}_4(\xi,\eta)}{\partial \eta} \end{pmatrix}. \tag{5.23}$$

The contribution of the element $E_e = F_e(\hat{E})$ to the global algebraic system is defined by the 4×4 matrices $M^{(e)}, C^{(e)}, K^{(e)}, B^{(e)}$ and 4×1 vectors $F^{(e)}, G^{(e)}$. Each entry of these element matrices and vectors is an integral over E_e which can be transformed into an integral over the reference element $\hat{E}$ using the corresponding substitution. We have

$$\int_{E_e} f(x_1, x_2)\, \mathrm{d}x_1 \mathrm{d}x_2 = \int_{\hat{E}} f(x_1(\xi,\eta), x_2(\xi,\eta)) |J^{(e)}(\xi,\eta)|\, \mathrm{d}\xi\, \mathrm{d}\eta, \tag{5.24}$$

where $f(x_1, x_2)$ is the function to be integrated. The factor $|J^{(e)}(\xi,\eta)|$ is the determinant of the element-dependent *Jacobian matrix*, which is defined by

$$J^{(e)}(\xi,\eta) = \begin{pmatrix} \frac{\partial x_1}{\partial \xi} & \frac{\partial x_2}{\partial \xi} \\ \frac{\partial x_1}{\partial \eta} & \frac{\partial x_2}{\partial \eta} \end{pmatrix}. \tag{5.25}$$

The entries of the Jacobian $J^{(e)}(\xi,\eta)$ can be readily obtained by differentiating the physical coordinates $x_1(\xi,\eta)$ and $x_2(\xi,\eta)$ w.r.t. the reference coordinates ξ and η. This gives

$$J_{11}^{(e)}(\xi,\eta) = \frac{\partial x_1}{\partial \xi} = \sum_{j=1}^{4} \frac{\partial \hat{\varphi}_j(\xi,\eta)}{\partial \xi} x_{1j}^{(e)},$$

$$J_{12}^{(e)}(\xi,\eta) = \frac{\partial x_2}{\partial \xi} = \sum_{j=1}^{4} \frac{\partial \hat{\varphi}_j(\xi,\eta)}{\partial \xi} x_{2j}^{(e)},$$

$$J_{21}^{(e)}(\xi,\eta) = \frac{\partial x_1}{\partial \eta} = \sum_{j=1}^{4} \frac{\partial \hat{\varphi}_j(\xi,\eta)}{\partial \eta} x_{1j}^{(e)},$$

$$J_{22}^{(e)}(\xi,\eta) = \frac{\partial x_2}{\partial \eta} = \sum_{j=1}^{4} \frac{\partial \hat{\varphi}_j(\xi,\eta)}{\partial \eta} x_{2j}^{(e)}.$$

It follows that

$$J^{(e)}(\xi,\eta) = \hat{\Phi}'(\xi,\eta) X_e,$$

where $\hat{\Phi}'(\xi,\eta)$ is the matrix containing the partial derivatives of the local basis functions at (ξ,η), and X_e is the matrix containing the physical coordinates of the nodal points.

In the 2D case, the determinant of the Jacobian matrix is given by

$$|J^{(e)}(\xi,\eta)| = J_{11}^{(e)}(\xi,\eta) J_{22}^{(e)}(\xi,\eta) - J_{21}^{(e)}(\xi,\eta) J_{12}^{(e)}(\xi,\eta). \tag{5.26}$$

The coordinate transformation is acceptable if $|J^{(e)}(\xi,\eta)| > 0$ for all points $(\xi,\eta) \in \hat{E}$.

Given the mapping F_e, we define the four local basis functions $\varphi_i^{(e)}$ as follows:

$$\varphi_i^{(e)}(x_1, x_2) := \hat{\varphi}_i(\xi,\eta), \qquad i = 1, \ldots, 4,$$
$$(\xi,\eta) = F_e^{-1}(x_1, x_2).$$

The entries of the element matrices $C^{(e)}$ and $K^{(e)}$ depend on the derivatives of $\varphi_i^{(e)}$, which are related to the derivatives of $\hat{\varphi}_i$ by the chain rule of calculus. We have

$$\begin{cases} \dfrac{\partial \hat{\varphi}_i}{\partial \xi} = \dfrac{\partial \varphi_i^{(e)}}{\partial x_1} \dfrac{\partial x_1}{\partial \xi} + \dfrac{\partial \varphi_i^{(e)}}{\partial x_2} \dfrac{\partial x_2}{\partial \xi}, \\[2ex] \dfrac{\partial \hat{\varphi}_i}{\partial \eta} = \dfrac{\partial \varphi_i^{(e)}}{\partial x_1} \dfrac{\partial x_1}{\partial \eta} + \dfrac{\partial \varphi_i^{(e)}}{\partial x_2} \dfrac{\partial x_2}{\partial \eta}. \end{cases}$$

Recalling that the partial derivatives of x_1 and x_2 w.r.t. ξ and η represent the entries of the Jacobian matrix $J^{(e)}(\xi,\eta)$, we write the above transformation rule in the matrix form

$$\begin{pmatrix} \frac{\partial \hat{\varphi}_i}{\partial \xi} \\ \frac{\partial \hat{\varphi}_i}{\partial \eta} \end{pmatrix} = \begin{pmatrix} \frac{\partial x_1}{\partial \xi} & \frac{\partial x_2}{\partial \xi} \\ \frac{\partial x_1}{\partial \eta} & \frac{\partial x_2}{\partial \eta} \end{pmatrix} \begin{pmatrix} \frac{\partial \varphi_i^{(e)}}{\partial x_1} \\ \frac{\partial \varphi_i^{(e)}}{\partial x_2} \end{pmatrix} = J^{(e)}(\xi,\eta) \begin{pmatrix} \frac{\partial \varphi_i^{(e)}}{\partial x_1} \\ \frac{\partial \varphi_i^{(e)}}{\partial x_2} \end{pmatrix}.$$

This yields

$$\begin{pmatrix} \frac{\partial \varphi_i^{(e)}}{\partial x_1} \\ \frac{\partial \varphi_i^{(e)}}{\partial x_2} \end{pmatrix} = (J^{(e)}(\xi,\eta))^{-1} \begin{pmatrix} \frac{\partial \hat{\varphi}_i}{\partial \xi} \\ \frac{\partial \hat{\varphi}_i}{\partial \eta} \end{pmatrix},$$

where

$$(J^{(e)}(\xi,\eta))^{-1} = \frac{1}{|J^{(e)}(\xi,\eta)|} \begin{pmatrix} \frac{\partial x_2}{\partial \eta} & -\frac{\partial x_2}{\partial \xi} \\ -\frac{\partial x_1}{\partial \eta} & \frac{\partial x_1}{\partial \xi} \end{pmatrix} \tag{5.27}$$

is the inverse of $J^{(e)}(\xi,\eta)$. The four entries of $(J^{(e)}(\xi,\eta))^{-1}$ are given by

$$\left(J^{(e)}(\xi,\eta)\right)^{-1}_{11} = \frac{1}{|J^{(e)}(\xi,\eta)|}\frac{\partial x_2}{\partial \eta} = \frac{1}{|J^{(e)}(\xi,\eta)|}\sum_{j=1}^{4}\frac{\partial \hat{\varphi}_j(\xi,\eta)}{\partial \eta}x_{2j}^{(e)},$$

$$\left(J^{(e)}(\xi,\eta)\right)^{-1}_{12} = -\frac{1}{|J^{(e)}(\xi,\eta)|}\frac{\partial x_2}{\partial \xi} = -\frac{1}{|J^{(e)}(\xi,\eta)|}\sum_{j=1}^{4}\frac{\partial \hat{\varphi}_j(\xi,\eta)}{\partial \xi}x_{2j}^{(e)},$$

$$\left(J^{(e)}(\xi,\eta)\right)^{-1}_{21} = -\frac{1}{|J^{(e)}(\xi,\eta)|}\frac{\partial x_1}{\partial \eta} = -\frac{1}{|J^{(e)}(\xi,\eta)|}\sum_{j=1}^{4}\frac{\partial \hat{\varphi}_j(\xi,\eta)}{\partial \eta}x_{1j}^{(e)},$$

$$\left(J^{(e)}(\xi,\eta)\right)^{-1}_{22} = \frac{1}{|J^{(e)}(\xi,\eta)|}\frac{\partial x_1}{\partial \xi} = \frac{1}{|J^{(e)}(\xi,\eta)|}\sum_{j=1}^{4}\frac{\partial \hat{\varphi}_j(\xi,\eta)}{\partial \xi}x_{1j}^{(e)}.$$

Putting the partial derivatives of the local basis functions $\varphi_i^{(e)}$ into the 2×4 matrix

$$\Phi'(x_1,x_2) = \begin{pmatrix} \frac{\partial \varphi_1^{(e)}}{\partial x_1} & \frac{\partial \varphi_2^{(e)}}{\partial x_1} & \frac{\partial \varphi_3^{(e)}}{\partial x_1} & \frac{\partial \varphi_4^{(e)}}{\partial x_1} \\ \frac{\partial \varphi_1^{(e)}}{\partial x_2} & \frac{\partial \varphi_2^{(e)}}{\partial x_2} & \frac{\partial \varphi_3^{(e)}}{\partial x_2} & \frac{\partial \varphi_4^{(e)}}{\partial x_2} \end{pmatrix}, \tag{5.28}$$

we obtain

$$\Phi'(x_1,x_2) = \left(J^{(e)}(\xi,\eta)\right)^{-1}\hat{\Phi}'(\xi,\eta).$$

This substitution should be performed whenever the partial derivatives of $\varphi_i^{(e)}$ are needed.

The change of variables in the derivatives and integrals makes it possible to perform all local computations on the reference element $\hat{E}$. The exact evaluation of 2D integrals over $\hat{E}$ is often difficult or even impossible. For this reason, it is worthwhile to use numerical quadrature rules. This approach corresponds to exact integration of an interpolation polynomial and replaces the integral by a weighted sum of function values at a certain number of quadrature points. The general form of an n-point quadrature rule for numerical integration on the reference element $\hat{E} = [-1,1] \times [-1,1]$ is as follows:

$$\int_{\hat{E}} f(\xi,\eta)\,\mathrm{d}\xi\,\mathrm{d}\eta \approx \sum_{k=1}^{n}\hat{\omega}_k f(\xi_k,\eta_k), \tag{5.29}$$

where $\hat{\omega}_k$ are the weights associated with the integration points (ξ_k,η_k), $k = 1,\dots,n$.

The weights of a numerical quadrature rule are chosen so that polynomials of certain degree are integrated exactly. The best accuracy is attained by *Gaussian quadrature* rules in which the integration points are defined as roots of Legendre polynomials. In the 1D case, an n-point Gaussian quadrature rule is exact for polynomials of degree $2n-1$ or less. The weights and nodes of a formula for numerical integration on squares and rectangles can be easily derived using 1D quadrature rules and the Fubini theorem.

Example 5.4. The nodes of the four-point Gaussian quadrature rule for numerical integration on the reference element $\hat{E} = [-1,1] \times [-1,1]$ are given by (see Fig. 5.12)

$$(\xi_1, \eta_1) = \left(-\frac{1}{\sqrt{3}}, -\frac{1}{\sqrt{3}}\right),$$
$$(\xi_2, \eta_2) = \left(\frac{1}{\sqrt{3}}, -\frac{1}{\sqrt{3}}\right),$$
$$(\xi_3, \eta_3) = \left(\frac{1}{\sqrt{3}}, \frac{1}{\sqrt{3}}\right),$$
$$(\xi_4, \eta_4) = \left(-\frac{1}{\sqrt{3}}, \frac{1}{\sqrt{3}}\right).$$

The corresponding weights are given by $\hat{\omega}_k = 1$, $k = 1, \ldots, 4$. The so-defined Gaussian formula is exact for products of up to third-degree polynomials in ξ and η. ∎

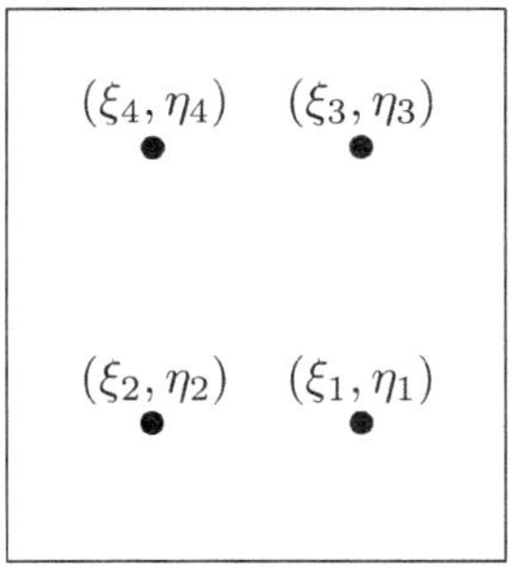

Figure 5.12. *Gaussian integration points on $\hat{E}$.*

Equipped with the definition of the local basis functions, transformation formulas for derivatives and integrals, and numerical quadrature rules, we will now generate the element matrices $M^{(e)}$, $C^{(e)}$, $K^{(e)}$ and the element vector $F^{(e)}$ using the following algorithm.

Algorithm 5.1 (Integration on four-node bilinear elements)

- Initialize all components of the element matrices and vectors by zeros. Save the vertex coordinates of the physical element E_e in the matrix X_e. Choose a numerical integration formula, for example, the four-point Gaussian rule. Save the coordinates of the integration points (ξ_k, η_k) and the weights $\hat{\omega}_k$ in auxiliary arrays.
- In a loop over the integration points (ξ_k, η_k), $k = 1, \ldots, n$, do the following:
  1. Evaluate the local basis functions $\hat{\varphi}_1, \ldots, \hat{\varphi}_4$ and their derivatives at the point (ξ_k, η_k). Store the results in the vector $\hat{\Phi}$ and matrix $\hat{\Phi}'$.

2. Calculate the physical coordinates (x_1, x_2) of the point (ξ_k, η_k)

$$\begin{pmatrix} x_1 \\ x_2 \end{pmatrix} = X_e^T \hat{\Phi}.$$

3. Calculate the four entries of the Jacobian matrix

$$J = \hat{\Phi}' X_e.$$

4. Calculate the determinant $|J|$ and the scaled weight

$$\omega_k := \hat{\omega}_k |J|.$$

5. Calculate the four entries of the inverse Jacobian

$$J^{-1} = \frac{1}{|J|} \begin{pmatrix} J_{22} & -J_{12} \\ -J_{21} & J_{11} \end{pmatrix}.$$

6. Calculate the matrix of transformed derivatives

$$\Phi' = J^{-1} \hat{\Phi}'.$$

7. Evaluate the source term and velocity components

$$f^* = \sum_{j=1}^{4} f(x_{1j}^{(e)}, x_{2j}^{(e)}) \hat{\Phi}_j,$$

$$u_1^* = \sum_{j=1}^{4} u_1(x_{1j}^{(e)}, x_{2j}^{(e)}) \hat{\Phi}_j,$$

$$u_2^* = \sum_{j=1}^{4} u_2(x_{1j}^{(e)}, x_{2j}^{(e)}) \hat{\Phi}_j.$$

8. Update the entries of local matrices and vectors:

for $i = 1, \ldots, 4$ **do**

  for $j = 1, \ldots, 4$ **do**

  $m_{ij}^{(e)} := m_{ij}^{(e)} + \rho c_p \hat{\Phi}_i \hat{\Phi}_j \omega_k$ *(mass matrix)*

  $c_{ij}^{(e)} := c_{ij}^{(e)} + \rho c_p \hat{\Phi}_i (u_1^* \Phi'_{1j} + u_2^* \Phi'_{2j}) \omega_k$ *(convective term)*

  $k_{ij}^{(e)} := k_{ij}^{(e)} + k(\Phi'_{1i} \Phi'_{1j} + \Phi'_{2i} \Phi'_{2j}) \omega_k$ *(diffusive term)*

  end j-loop

  $F_i^{(e)} := F_i^{(e)} + \hat{\Phi}_i f^* \omega_k$ *(source term)*

end i-loop

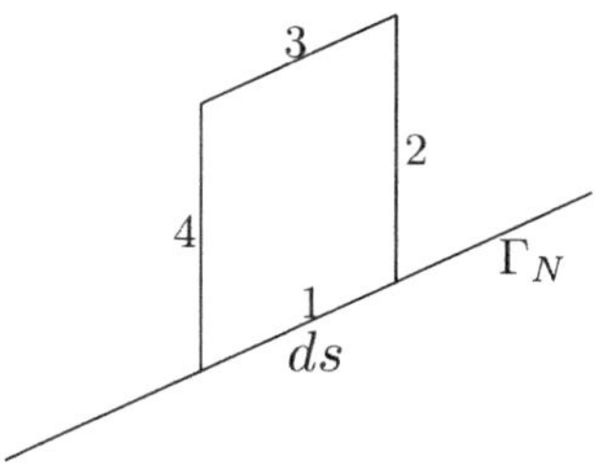

Figure 5.13. *Numbering of the edges of a boundary element.*

If the element E_e has an edge on the boundary part Γ_N, where the heat flux is given by a natural boundary condition of Newton type, we also need to calculate the surface integrals that constitute the element matrix $B^{(e)}$ and element vector $G^{(e)}$. The element depicted in Fig. 5.13 has one edge on the boundary Γ_N and three interior edges. The edges are numbered counterclockwise starting from the one that lies on the boundary Γ_N. The differential $\mathrm{d}s$ that appears in the boundary integrals over Γ_N is given by

$$\mathrm{d}s = \sqrt{(\mathrm{d}x_1)^2+(\mathrm{d}x_2)^2} = \sqrt{\left(\frac{\partial x_1}{\partial \xi}\mathrm{d}\xi + \frac{\partial x_1}{\partial \eta}\mathrm{d}\eta\right)^2 + \left(\frac{\partial x_2}{\partial \xi}\mathrm{d}\xi + \frac{\partial x_2}{\partial \eta}\mathrm{d}\eta\right)^2}$$
$$= \sqrt{\left(J_{11}^{(e)}(\xi,\eta)\mathrm{d}\xi + J_{21}^{(e)}(\xi,\eta)\mathrm{d}\eta\right)^2 + \left(J_{12}^{(e)}(\xi,\eta)\mathrm{d}\xi + J_{22}^{(e)}(\xi,\eta)\mathrm{d}\eta\right)^2}.$$

If we assume that the boundary edge of element E_e corresponds to the bottom edge of the reference element $\hat{E}$, then $\eta = -1$ and $\mathrm{d}\eta = 0$ along this edge. It follows that

$$\mathrm{d}s = J_{\Gamma_N}^{(e)}(\xi,-1)\mathrm{d}\xi,$$

where

$$J_{\Gamma_N}^{(e)}(\xi,-1) = \sqrt{\left(J_{11}^{(e)}(\xi,-1)\mathrm{d}\xi\right)^2 + \left(J_{12}^{(e)}(\xi,-1)\mathrm{d}\xi\right)^2}.$$

This yields

$$\begin{aligned} b_{ij}^{(e)} &= \int_{E_e\cap\Gamma_N} h_c \varphi_i^{(e)}(x_1,x_2)\varphi_j^{(e)}(x_1,x_2)\,\mathrm{d}s \\ &= \int_{-1}^{1} h_c \hat{\varphi}_i(\xi,-1)\hat{\varphi}_j(\xi,-1)J_{\Gamma_N}^{(e)}(\xi,-1)\mathrm{d}\xi, \qquad i,j = 1,\dots,4, \end{aligned}$$

and

$$G_i^{(e)} = \int_{E_e\cap\Gamma_N} \varphi_i^{(e)} g\,\mathrm{d}s = \int_{-1}^{1} \hat{\varphi}_i(\xi,-1) g J_{\Gamma_N}^{(e)}(\xi,-1)\mathrm{d}\xi, \qquad i = 1,\dots,4,$$

where

$$\begin{cases} \hat{\varphi}_1(\xi,-1) = \frac{1}{2}(1-\xi), \\ \hat{\varphi}_2(\xi,-1) = \frac{1}{2}(1+\xi), \\ \hat{\varphi}_3(\xi,-1) = 0, \\ \hat{\varphi}_4(\xi,-1) = 0. \end{cases}$$

The components of the Jacobian matrix $J^{(e)}_{\Gamma_N}$ are given by

$$J^{(e)}_{11}(\xi,-1)=\frac{\partial x_1}{\partial \xi}(\xi,-1)=\sum_{j=1}^{4}\frac{\partial \hat{\varphi}_j}{\partial \xi}(\xi,-1)x^{(e)}_{1j},$$

$$J^{(e)}_{12}(\xi,-1)=\frac{\partial x_2}{\partial \xi}(\xi,-1)=\sum_{j=1}^{4}\frac{\partial \hat{\varphi}_j}{\partial \xi}(\xi,-1)x^{(e)}_{2j}.$$

Applying a 1D two-point Gaussian integration formula (interval: $[-1,1]$; quadrature points: $\xi_1=-\frac{1}{\sqrt{3}}$ and $\xi_2=\frac{1}{\sqrt{3}}$; weights: $\hat{\omega}_1=\hat{\omega}_2=1$), we obtain

$$\begin{aligned} b^{(e)}_{ij} &= \int_{-1}^{1} h_c\hat{\varphi}_i(\xi,-1)\hat{\varphi}_j(\xi,-1)J^{(e)}_{\Gamma_N}(\xi,-1)\mathrm{d}\xi \\ &\approx \sum_{k=1}^{2}\omega_k h_c\hat{\varphi}_i(\xi_k,-1)\hat{\varphi}_j(\xi_k,-1)J^{(e)}_{\Gamma_N}(\xi_k,-1), \qquad i,j=1,\dots,4, \\ G^{(e)}_{i} &= \int_{-1}^{1} \hat{\varphi}_i(\xi,-1)gJ^{(e)}_{\Gamma_N}(\xi,-1)\mathrm{d}\xi \\ &\approx \sum_{k=1}^{2}\omega_k\hat{\varphi}_i(\xi_k,-1)gJ^{(e)}_{\Gamma_N}(\xi_k,-1), \qquad i=1,\dots,4. \end{aligned}$$

Thus, the evaluation of $b^{(e)}_{ij}$ and $G^{(e)}_{i}$ reduces to 1D integration. The Gaussian quadrature may be replaced by Simpson's rule or another numerical integration formula.

5.1.6 ▪ Assembly of the Global System

The element-by-element assembly of global matrices and vectors is performed as in the 1D case. The correspondence between the local and global numbers of nodal points must be known to insert the entries of local matrices and vectors in the appropriate rows/columns of their global counterparts. The following algorithm illustrates the process of global assembly for the bilinear finite element discretization of the 2D heat transport equation on a mesh consisting of N_e four-node quadrilateral elements.

Algorithm 5.2 (Global assembly for four-node bilinear elements)

for $e=1,\dots,N_e$ do

- Retrieve the global numbers and coordinates of the four nodes of element E_e.
- Save the node numbers in the vector IND_e and the coordinates in the matrix X_e.
- Calculate the element matrices $M^{(e)}$, $C^{(e)}$, $K^{(e)}$ and the element vector $F^{(e)}$.
- Update the global matrices M,C,K and the global load vector F as follows:

  for $I=1,\dots,4$ **do**

  $i=\mathrm{IND}_e(I)$

  for $J=1,\dots,4$ **do**

$j = \mathrm{IND}_e(J)$

$m_{ij} := m_{ij} + m_{IJ}^{(e)}$ *(mass matrix)*

$c_{ij} := c_{ij} + c_{IJ}^{(e)}$ *(convective term)*

$k_{ij} := k_{ij} + k_{IJ}^{(e)}$ *(diffusive term)*

end J-loop

$F_i := F_i + F_I^{(e)}$ *(source term)*

end I-loop

end e-loop

The contribution of natural boundary conditions is calculated and inserted in an additional loop over N_s element edges lying on Γ_N. The assembly procedure is as follows:

for $s = 1, \dots, N_s$ **do**

- Retrieve the number e of the element E_e containing the sth boundary edge.
- Retrieve the global numbers and coordinates of the four nodes of element E_e.
- Save the node numbers in the vector IND_e and the coordinates in the matrix X_e.
- Calculate the element matrix $B^{(e)}$ and the element vector $G^{(e)}$.
- Update the global matrix B and the global load vector G:

for $I = 1, \dots, 4$ **do**

$i = \mathrm{IND}_e(I)$

for $J = 1, \dots, 4$ **do**

$j = \mathrm{IND}_e(J)$

$b_{ij} := b_{ij} + b_{IJ}^{(e)}$ *(Newton flux)*

end J-loop

$G_i := G_i + G_I^{(e)}$ *(Neumann flux)*

end I-loop

end s-loop

Remark 5.1. The assembly of B and G may be separated in a computer implementation.

5.1.7 ▪ Dirichlet Boundary Conditions

The Dirichlet boundary conditions can be incorporated into a global system of the form

$$AT^{n+1} = Q$$

using one of the three implementation techniques (row-column elimination, overwriting Galerkin equations by Dirichlet constraints, penalty method) introduced in Chapter 4.

Let $\mathbf{x}_i$ be a nodal point on the boundary part Γ_D. The overwriting method replaces the Galerkin equation associated with the basis function φ_i by the constraint

$$T_i^{n+1} = T_0(\mathbf{x}_i).$$

Thus, the matrix A and right-hand side Q must be modified as follows:

$$a_{ij} := \delta_{ij}, \quad j = 1, \dots, N,$$

$$Q_i := T_0(\mathbf{x}_i).$$

The penalty method adds a large number α to the diagonal element a_{ii} and the product $\alpha T_0(\mathbf{x}_i)$ to the right-hand side of the ith algebraic equation:

$$a_{ii} := a_{ij} + \alpha, \qquad Q_i := Q_i + \alpha T_0(\mathbf{x}_i), \qquad \alpha \gg 1.$$

The above modifications must be performed for each node on the Dirichlet boundary, so it is essential to use data structures that provide fast access to diagonal matrix entries.

5.1.8 ▪ Solving the Algebraic System

The global matrix A of a linear system produced by the Galerkin discretization on a structured mesh of bilinear finite elements exhibits the following block-tridiagonal structure:

$$A = \begin{pmatrix} A_{11} & A_{12} & & & & \\ A_{21} & A_{22} & A_{23} & & & \\ & & \dots & & \dots & \\ & & & A_{N_{x_2}-1,N_{x_2}-2} & A_{N_{x_2}-1,N_{x_2}-1} & A_{N_{x_2}-1,N_{x_2}} \\ & & & & A_{N_{x_2},N_{x_2}-1} & A_{N_{x_2}N_{x_2}} \end{pmatrix},$$

where each diagonal block is an $N_{x_1} \times N_{x_1}$ tridiagonal matrix associated with the nodal points lying on the same horizontal grid line. The off-diagonal matrix blocks arise due to the coupling with the nodes lying on the neighboring horizontal grid lines.

The TDMA algorithm that we used to solve tridiagonal systems in one dimension can be extended to block-tridiagonal matrices, but the cost of the resulting method is $\mathcal{O}(N^2)$. As an alternative, one may consider using an iterative *alternating direction implicit* (ADI) method. This approach is based on solving a sequence of tridiagonal systems associated with the horizontal and vertical grid lines [3]. The use of unstructured quadrilateral meshes leads to matrices with irregular sparsity patterns, and their properties depend on the shape of mesh elements. Therefore, more work needs to be invested in the design of efficient solvers for linear systems. General-purpose CFD codes typically provide a large selection of direct and iterative solution techniques which can be configured using problem-specific fine-tuning. For problems of moderate size, we recommend UMFPACK [235], a GNU GPL software tool for solving sparse linear systems using the unsymmetric multifrontal method.

The software package *Elmer* introduced in Section 2.6 also features a collection of direct and iterative solvers (including multigrid methods) for linear algebraic systems. The default direct solution method utilizes the well-known LAPACK collection of subroutines for band matrices. This method is recommended for small linear systems only. Another direct solver employs the UMFPACK routines to solve sparse linear systems. In practice, it may be the most efficient method for solving 2D problems as long as there is enough memory available. *ElmerSolver* may also be compiled with Mumps, SuperLU, and Pardiso packages, but due to licensing restrictions they are not included in precompiled binaries.

In addition to direct solvers, *ElmerSolver* provides a set of Krylov subspace methods for iterative solution of linear systems. In particular, the conjugate gradient (CG), conjugate gradient squared (CGS), biconjugate gradient stabilized (BiCGStab), BiCGStab(l), transpose-free quasi-minimal residual (TFQMR), generalized minimal residual (GMRES), and generalized conjugate residual (GCR) methods are available.

A class of iterative methods based on a multilevel solution strategy provides an efficient way of solving large linear systems. For certain problems, they perform nearly optimally, that is, nearly $\mathcal{O}(N)$. Two different kinds of multilevel method are available in *ElmerSolver*: geometric multigrid (GMG) and algebraic multigrid (AMG) methods.

For a more detailed documentation of the solution methods used in *ElmerSolver*, we refer the reader to the *Elmer* web pages [62] (see "ElmerSolver Manual" there).

5.1.9 ▪ Elmer Tutorials for the Heat Equation

In this section, we present two tutorials which illustrate the numerical solution of the heat equation using the *Elmer* package. In the first tutorial, we consider a 2D steady-state problem with a heat source. In the second one, we simulate 3D heat transfer in a solid structure of a complex geometry. These "learning by doing" tutorials can be found on the *Elmer* web site [62] along with some other tutorials and recommended materials describing, e.g., the PDE models included in *Elmer*, as well as the use of the graphical user interface (GUI), generation of finite element meshes, and solver properties.

We recommend that the reader download the *Elmer* software from the web page [63] and go through the tutorials presented in this chapter step by step. All the needed input files are available in the `ElmerGUI/samples` directory that should come with the installation. In what follows, the instructions written in `verbatim` refer to operations with the GUI, and indentation means a step in the menu hierarchy. Any feedback to the team developing *Elmer* is greatly appreciated (`elmeradm@csc.fi`).

Two-Dimensional Steady-State Heat Source Problem

We begin with a 2D heat transfer tutorial, in which an L-shaped structure Ω (see Fig. 5.14) is heated by an internal heat source at the rate h =1 W/m^3. The density of the solid material is ρ =1 kg/m^3 and the heat conductivity is $\varkappa$ =1 W/mK. All the boundaries Γ are kept at the constant temperature of 0 K. The objective is to calculate the temperature distribution in the structure, that is, to solve the mathematical problem

$$\begin{cases} -\varkappa\Delta T = \rho h & \text{in } \Omega, \\ T = 0 & \text{on } \Gamma. \end{cases} \tag{5.30}$$

To set up the finite element simulation, start *ElmerGUI* from the command line or by clicking the icon on your desktop. Here we explain the setup of the simulation run

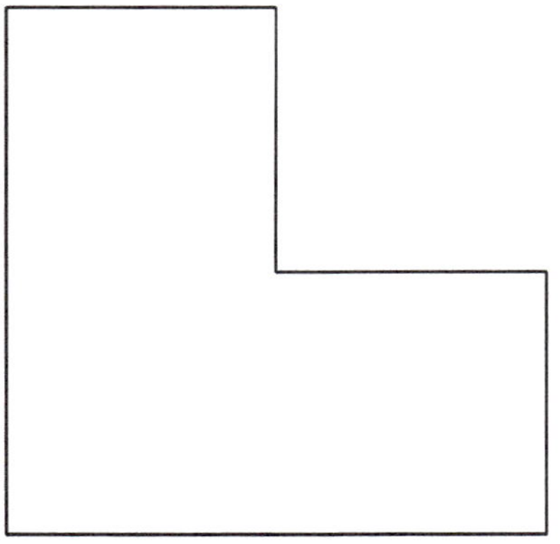

Figure 5.14. *The L-shaped domain.*

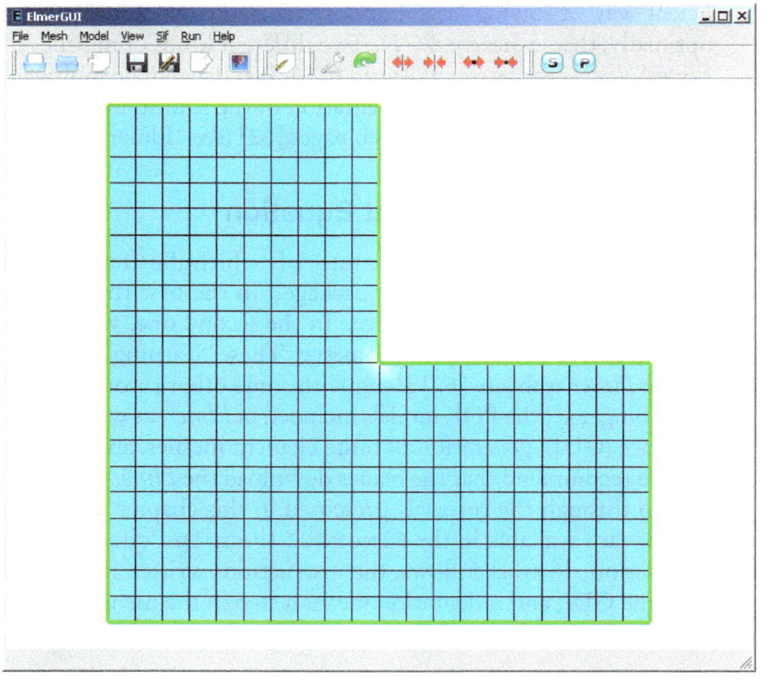

Figure 5.15. *The finite element mesh in ElmerGUI.*

in *ElmerGUI* by describing the clicking procedure in words. Tabulation generally means that selections are made within the window chosen at the higher level.

The mesh is stored in the *ElmerGrid* format in the file `angle.grd` in the samples directory of *ElmerGUI*. Load this file:

```
File
  Open -> angle.grd
```

You should obtain the mesh and may check in the `Model summary` window that it consists of 341 nodes and 300 bilinear elements. If the mesh was successfully imported, your window should look like the one in Fig. 5.15.

After loading the mesh, you can go through the `Model` menu from the top to bottom. Under `Setup`, you can choose parameters related to the simulation such as file names, time stepping, constants, etc. The steady-state simulation is carried out in the 2D Cartesian coordinates. Since the problem is linear, just one steady-state iteration is needed.

```
Model
  Setup
    Simulation Type = Steady state
    Steady state max. iter = 1
```

Choose `Accept` to close the window.

In the equation section, you can choose the relevant equations and parameters related to their solution. In this tutorial case, we have only one equation: the heat equation.

When it comes to the definition of equations and materials, it is possible to assign them to bodies (domains) right away, or use mouse selection to assign them later. So far we have just one domain and one boundary. Therefore, it is easier to select `Equation` and `Material` directly.

For the linear system solver, we recommend the defaults. You may, however, try out different preconditioners (ILU1,...) or the direct UMFPACK solver, for example.

```
Model
  Equation
    Add
      Name = Heat Equation
      Apply to bodies = 1
      Heat Equation
        Active = on
  Apply
  OK
```

The `Material` section is used to specify all material parameters. They include generic parameters which specify direct properties of the material, such as density, without making any assumptions on the physical or mathematical model. Other properties, such as the heat conductivity in Fourier's law, specify the parameters of a given model.

```
Model
  Material
    Add
      Name = Ideal
      Apply to bodies = 1
      General
        Density = 1.0
      Heat Equation
        Heat Conductivity = 1.0
    Apply
    OK
```

The `Body Force` represents the right-hand side of the equation to be solved. In this test case, it models the heat source.

```
Model
  Body Force
    Add
      Name = Heating
      Heat Source = 1.0
      Apply to bodies = 1
    Apply
    OK
```

Since the problem at hand is stationary, no initial conditions need to be imposed. In this test case, we have only one boundary, and the boundary value is set to zero.

```
Model
  BoundaryCondition
    Add
      Heat Equation
        Temperature = 0.0
      Name = Zero
      Apply to boundaries = 1
    Apply
    OK
```

For the actual simulation, *ElmerSolver* needs the mesh files and the command file. In the previous steps, you have defined all the information that *ElmerGUI* needs to create the latter. After writing it to a hard drive, you may also visually inspect the command file.

```
Sif
  Generate
  Edit -> look how your command file came out
```

Before running the solver, you should back up all the files needed to restart the simulation. The command `Save Project` saves these files to the destination directory.

```
File
  Save Project
```

After you have successfully saved the files, you may start the solver.

```
Run
  Start solver
```

A convergence monitoring window automatically pops up showing the relative changes of the solution at each iteration. Since the governing equation is linear in this test case, only one iteration is required to obtain the solution; the second one is just needed to check the convergence. The resulting output log is shown in Fig. 5.16.

Note that if you face problems at the solution phase, you may need to edit the settings and restart the code. Always remember to save the project before execution.

To view the results, you may start the internal VTK widget, or use *ElmerPost* for postprocessing, as is done here:

```
Run
  Start postprocessor
```

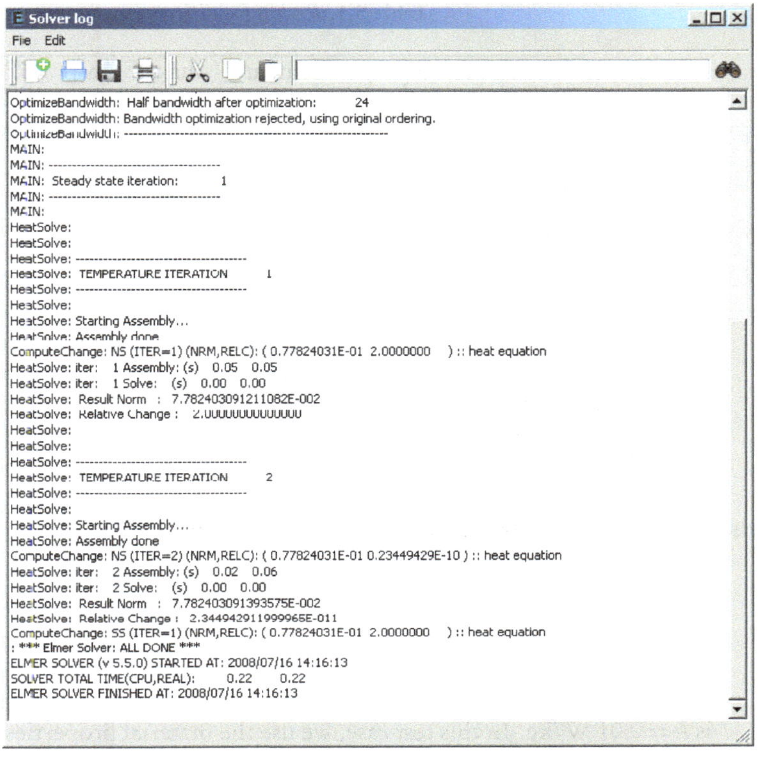

Figure 5.16. *The output log of ElmerSolver when used under ElmerGUI.*

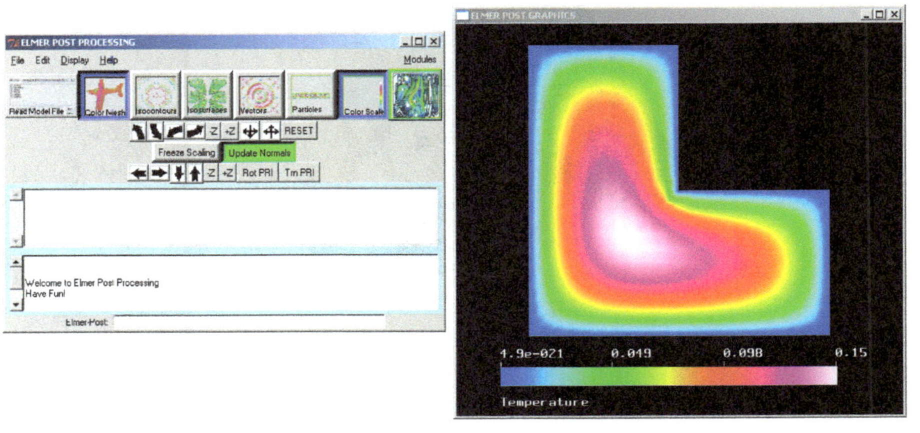

Figure 5.17. *The temperature distribution visualized in ElmerPost.*

The default configuration shows the surface mesh colored with temperature as in Fig. 5.17. You may test different visualization options. For example, by choosing in the `Color Mesh Edit` window the `Mesh Style` to be `Line` and `Edge Style` to be `All` and pressing `Apply`, the wireframe of the computational mesh is displayed.

As a reference quantity, the maximum temperature in the structure is calculated. For a comparison, the same problem was solved six times with different mesh sizes using the

Table 5.1. *Results for different mesh sizes.*

Model mesh h [m]	Elements	max(T) [K]	CPU time [s]
0.2	75	0.1452	0.14
0.1	300	0.1477	0.19
0.05	1200	0.1489	0.41
0.025	5120	0.14925	1.22
0.01	30800	0.14937	7.77

`-relh` flag of `ElmerGrid` format located in the `Mesh` menu under `Configure`. After choosing `Remesh` and saving the mesh, simulation may be repeated on the modified mesh.

The maximum temperature obtained by using different meshes is listed in Table 5.1. From these results, one can see that the result converges. With a denser mesh, the result is more accurate, but solving the problem takes more computing time. For comparison, the CPU time used in each case (on a Lenovo 60 laptop) is also shown in the table.

Heat Transfer in a Complex Three-Dimensional Geometry

The second tutorial deals with a 3D problem. We solve the same heat equation as in the previous 2D heat source tutorial but the domain Ω has a complex 3D geometry (see the finite element mesh in Fig. 5.18). Here, a generic solid object is heated internally by a heat source and the temperature is fixed on selected parts of the boundary Γ. We assume that the boundary value of the temperature is $T_0 = 293$ K and the internal heat generation rate is $h = 0.01$ W/kg. In this test case, we use the material properties of aluminium.

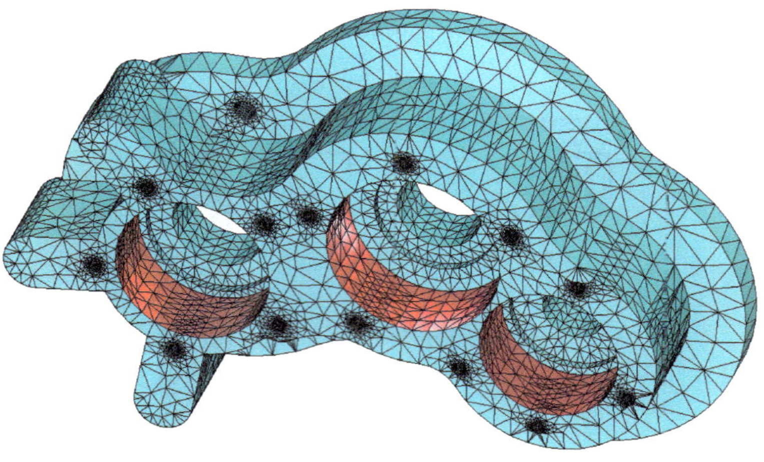

Figure 5.18. *The computational mesh showing the three joined boundaries.*

The geometry is given in the step format in the file `pump_carter_sup.stp` in the `samples/step` directory of *ElmerGUI*. This file is kindly provided at the AIM@SHAPE Shape Repository by INRIA. If you lack OpenCascade, you might try to solve a similar problem with the `grd` files `angle3d.grd`, `angles3d.grd`, `bench.grd`, or `cooler.grd`, for example.

The CAD geometry defined by the step file is transformed on-the-fly by the OpenCascade library into a `stl` file for which nglib creates a tetrahedral computational mesh. You may also use the tetlib library (tetgen) if you have installed it as a plug-in.

Load the input file:

```
File
  Open -> pump_carter_sup.stp
```

The meshing will take a minute or two. You should obtain your mesh and may check the number of elements in the `Model summary`. With netgen the default setting generates 8,371 nodes and 36,820 tetrahedral elements. Visual inspection reveals that the mesh is not quite satisfactory in terms of geometric accuracy. We choose to modify the mesh by altering the settings in the following way:

```
View -> Cad model...
  Model -> Preferences...
    Restrict mesh size on surfaces by STL density = on
    Apply
Mesh -> Remesh
```

The modified mesh should include 16,159 nodes and 65,689 tetrahedral elements and is more appealing to the eye. In order to control the mesh density, please study the command-line options of the netgen manual. Here we continue with the default mesh.

We want to set the temperature boundary condition at the inside of the holes. For this purpose, you may join the three boundaries (see Fig. 5.18). Choose the six pieces that constitute the boundaries as shown in the picture by pressing the `Ctrl` key.

```
Mesh
  Unify Surface
```

After loading the mesh, you can go through the `Model` menu from top to bottom. Similarly to the previous tutorial problem, in the `Setup` you can choose things related to the simulation, such as file names, time stepping, constants, and others. Since the problem is linear, only one steady-state iteration is needed.

```
Model
  Setup
    Simulation Type = Steady state
    Steady state max. iter = 1
```

Choose `Apply` to close the window.

In the equation section, you need to choose the relevant equations and parameters related to their solution. Again, we have only one equation, that is, the heat equation. Similarly to the 2D tutorial, you can now assign Equations and Materials to bodies (domains), and you might test different solver parameters.

```
Model
  Equation
    Add
      Name = Heat Equation
      Apply to bodies = Body 1
      Heat Equation
        Active = on
      Add
      OK
```

The `Material` section is used to specify all material parameters. They include generic parameters which specify direct properties of the material, such as density, without making any assumptions on the physical or mathematical model. Other properties, such as the heat conductivity in Fourier's law, specify the parameters of a given model.

We choose `Aluminium` from the `Material library`, which automatically sets the needed material properties for our simulation.

```
Model
  Material
    Add
      Material library
        Aluminium
      Apply to bodies = Body 1
      Add
      OK
```

Again, the `Body Force` represents the heat source in the right-hand side of our equation.

```
Model
  Body Force
    Add
      Name = Heating
      Heat Source = 0.01
      Apply to bodies = Body 1
      Add
      OK
```

In this test case, we have only one boundary component, and you may prescribe the room temperature there. After selecting the corresponding boundary condition

```
Model
  BoundaryCondition
    Add
      Heat Equation
        Temperature = 293.0
      Name = RoomTemp
      Add
      OK
```

you may set the boundary properties:

```
Model
  Set boundary properties
```

Choose the previously defined union of three boundaries by clicking with the mouse and apply the room temperature condition to this boundary.

```
Boundary condition
  RoomTemp
```

You have now defined all the information for *ElmerGUI* to write the command file for *ElmerSolve*. After saving it, you may also visually inspect the command file.

```
Sif
  Generate
  Edit -> look how your command file came out
```

Before running the solver, you should back up the project files. The command `Save Project` saves these files to the destination directory.

```
File
  Save Project
```

After you have successfully saved the files, you may start the solver:

```
Run
  Start solver
```

To view the results, you may use the *ElmerPost* postprocessor or start the internal VTK widget as is done here:

```
Run
  Postprocessor (VTK)
```

The default configuration shows just the object. To color the surface with the temperature, choose

```
Surfaces
  Surface: Temperature
  Apply
```

The maximum temperature should be about 586.5 K.You may turn on opacity in order to see through the object (10–20% is a good value). This will enable you to see the isosurfaces that you might want to define. Some examples of visualizations are displayed in Fig. 5.19.

We will continue with more *Elmer* tutorials after introducing the Galerkin finite element discretization of the incompressible Navier–Stokes equations in two dimensions.

5.2 ▪ Navier–Stokes Equations

The previous section dealt with a scalar linear transport model, the heat equation. In more realistic models, the velocity field is unknown a priori and must be determined by solving the incompressible Navier–Stokes equations derived in Chapter 3. These equations represent a nonlinear saddle-point problem for the vector-valued velocity $\mathbf{u}$ and scalar-valued pressure p. The number of velocity components equals the number of space dimensions. In the 2D case, the Navier–Stokes system consists of two momentum equations and one continuity equation. In Chapter 8, we will discuss the properties of this model in detail.

The incompressible Navier–Stokes equations belong to the most widely used flow models in CFD. During the past three decades, considerable research effort has been expended on the development of unstructured grid finite element methods for incompressible flow problems. In this book, we give just a short introduction to the field. For a more focused and detailed presentation of different aspects of finite element methods for the incompressible Navier–Stokes equations, we refer the reader to the books by Cuvelier, Segal, and van Steenhoven [44], Donea and Huerta [50], Elman, Silvester, and Wathen [61], Girault and Raviart [79], Glowinski [80], Gresho and Sani [87], Layton [155], Lewis, Nithiarasu, and Seetharamu [160], Pironneau [188], Reddy and Gartling [191], Quartapelle [192], Quarteroni and Valli [195], and Turek [230].

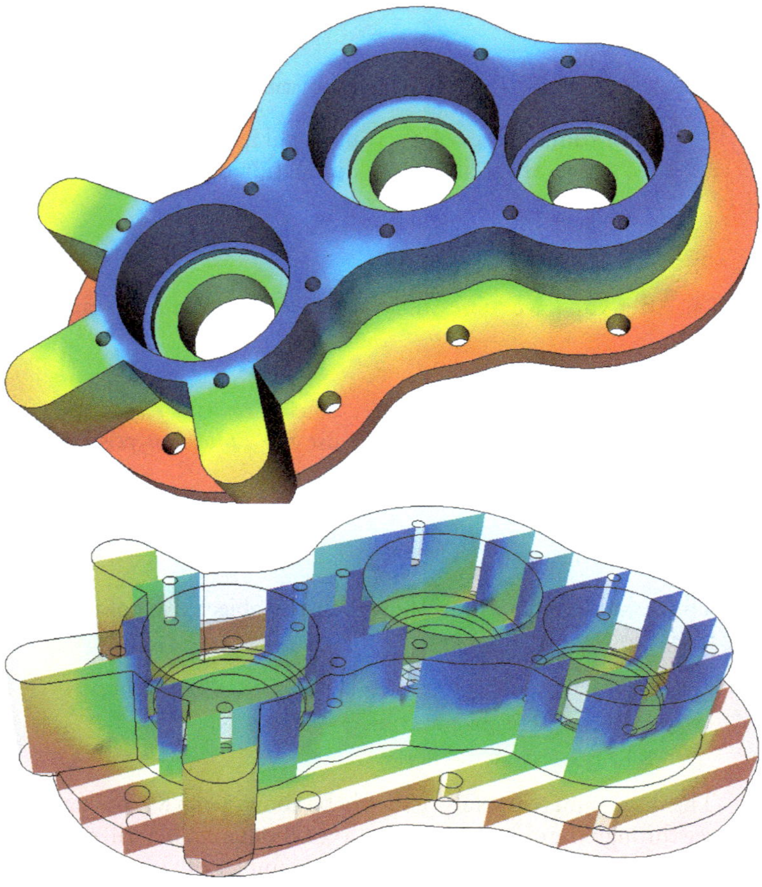

Figure 5.19. *The temperature distribution of the solid object visualized by using the VTK-based postprocessor.*

In this section, we discretize the steady Navier–Stokes equations using the standard Galerkin method. In particular, we address the linearization of the convective term, the derivation of the variational formulation, the finite element discretization, generation of local matrices/vectors, and assembly of the global algebraic system. Stabilization techniques and efficient solvers for the discrete saddle-point problem are discussed in Chapters 6–8.

5.2.1 ▪ Linearization of the Convective Term

At low Reynolds numbers, the flow is laminar and can be modeled by the steady incompressible Navier–Stokes equations, which will be our model problem in this section. The numerical treatment of the time-dependent Navier–Stokes equations and the implementation of the k–ε turbulence model for high Reynolds number flows are addressed in Chapter 8.

The dimensionless form of the steady incompressible Navier–Stokes equations reads

$$\begin{cases} \mathbf{u}\cdot\nabla\mathbf{u} - \dfrac{1}{\mathrm{Re}}\Delta\mathbf{u} + \nabla p = \mathbf{g}, \\ \qquad\qquad\qquad\quad \nabla\cdot\mathbf{u} = 0. \end{cases}$$

The momentum equation is nonlinear because the velocity $\mathbf{u}$ is convected by itself. A finite element discretization of a nonlinear governing equation inevitably leads to a nonlinear algebraic system such that the coefficients of the global matrix depend on the unknown solution and must be updated in an iterative way until convergence is achieved.

To keep things simple, we will linearize the convection term $\mathbf{u}\cdot\nabla\mathbf{u}$ *before* discretizing the Navier–Stokes equations. Consider a sequence of successive approximations

$$\mathbf{u}^0, \mathbf{u}^1, \ldots, \mathbf{u}^{n-1}$$

to the unknown velocity field $\mathbf{u}$. At the nth iteration step, we obtain

$$\mathbf{u}^n = \mathbf{u}^{n-1} + \delta\mathbf{u}^{n-1}, \tag{5.31}$$

where $\delta\mathbf{u}^{n-1}$ is the correction to be applied to the old iterate $\mathbf{u}^{n-1}$. We have

$$\begin{aligned}\mathbf{u}^n\cdot\nabla\mathbf{u}^n &= (\mathbf{u}^{n-1}+\delta\mathbf{u}^{n-1})\cdot\nabla(\mathbf{u}^{n-1}+\delta\mathbf{u}^{n-1})\\ &= \mathbf{u}^{n-1}\cdot\nabla\mathbf{u}^{n-1}\\ &+ \delta\mathbf{u}^{n-1}\cdot\nabla\mathbf{u}^{n-1} + \mathbf{u}^{n-1}\cdot\nabla(\delta\mathbf{u}^{n-1}) + \delta\mathbf{u}^{n-1}\cdot\nabla(\delta\mathbf{u}^{n-1}).\end{aligned}$$

If all correction terms are neglected, we obtain the linearized Navier–Stokes system

$$\begin{cases} -\dfrac{1}{\mathrm{Re}}\Delta\mathbf{u}^n + \nabla p = \mathbf{g} - \mathbf{u}^{n-1}\cdot\nabla\mathbf{u}^{n-1}, \\ \nabla\cdot\mathbf{u}^n = 0\,. \end{cases} \tag{5.32}$$

Thus, each update boils down to solving a linear Stokes problem of the form

$$\begin{cases} -\dfrac{1}{\mathrm{Re}}\Delta\mathbf{u}^n + \nabla p = \mathbf{g}^*(\mathbf{u}^{n-1}), \\ \nabla\cdot\mathbf{u}^n = 0, \end{cases}$$

where the right-hand side depends on the velocity from the previous iteration,

$$\mathbf{g}^*(\mathbf{u}^{n-1}) := \mathbf{g} - \mathbf{u}^{n-1}\cdot\nabla\mathbf{u}^{n-1}.$$

In practice, this method converges only for very low Reynolds number flows. Therefore, a more accurate approximation to the convective term $\mathbf{u}^n\cdot\nabla\mathbf{u}^n$ is to be employed.

Clearly, a better convergence behavior can be expected if we use the approximation

$$\mathbf{u}^n\cdot\nabla\mathbf{u}^n \approx \mathbf{u}^{n-1}\cdot\nabla\mathbf{u}^n.$$

This approach corresponds to a lagged evaluation of the convective velocity in the momentum transport equations. The linearized Navier–Stokes system becomes

$$\begin{cases} \mathbf{u}^{n-1}\cdot\nabla\mathbf{u}^n - \dfrac{1}{\mathrm{Re}}\Delta\mathbf{u}^n + \nabla p = \mathbf{g}, \\ \nabla\cdot\mathbf{u}^n = 0\,. \end{cases} \tag{5.33}$$

This first-order linearization is more robust than (5.32). It is known as the *fixed-point iteration* or the *Picard method.* The convergence rates are linear at best, but the method is easy to implement and its performance is satisfactory for practical purposes.

Second-order convergence can be achieved using the *Newton linearization*

$$\mathbf{u}^n \cdot \nabla \mathbf{u}^n \approx \mathbf{u}^{n-1} \cdot \nabla \mathbf{u}^n + \mathbf{u}^n \cdot \nabla \mathbf{u}^{n-1} - \mathbf{u}^{n-1} \cdot \nabla \mathbf{u}^{n-1},$$

which leads to

$$\begin{cases} \mathbf{u}^{n-1} \cdot \nabla \mathbf{u}^n + \mathbf{u}^n \cdot \nabla \mathbf{u}^{n-1} - \dfrac{1}{\mathrm{Re}} \Delta \mathbf{u}^n + \nabla p = \mathbf{g} + \mathbf{u}^{n-1} \cdot \nabla \mathbf{u}^{n-1}, \\ \nabla \cdot \mathbf{u}^n = 0. \end{cases} \tag{5.34}$$

We remark that the presence of the "source" term $\mathbf{u}^n \cdot \nabla \mathbf{u}^{n-1}$ on the left-hand side of the linearized system may cause convergence problems at the discrete level [233].

The convergence behavior of all iterative methods introduced in this section depends on the Reynolds number and, of course, on the nature of the flow phenomena under investigation. As the Reynolds number increases, convergence tends to slow down and may even stall, making further iterations useless. The choice of the initial guess $\mathbf{u}^0$ is also an important issue. In particular, the Newton method requires a good initial guess. For this reason, it is often recommended to start with a first-order iteration method. When the solution begins to converge, the solver can switch to a second-order method to speed up convergence.

If convergence problems are encountered, one may limit the changes of the solution by using *underrelaxation*, i.e., a weighted average of the old and new iterates:

$$\mathbf{u}^n := \omega \mathbf{u}^n + (1-\omega)\mathbf{u}^{n-1},$$

where $\omega \in (0,1]$ is the relaxation parameter that determines the shares of the two iterates and may be chosen adaptively. In fact, an optimal linear combination of more than two iterates may be used to accelerate fixed-point iterations by taking the convergence history into account. This approach is known as *Anderson acceleration* [239].

Note that nested iterations may occur at different levels of the solution process. In addition to the iterative procedure that we used to handle the nonlinear convective term, iterative methods may be used, e.g., for coupling the Navier–Stokes equations with the heat equation or with the equations of the k–ε turbulence model (see Chapter 8), for the numerical implementation of boundary conditions, and for solving the algebraic systems.

5.2.2 ▪ Variational Formulation

The derivation of the variational formulation for the Navier–Stokes equations relies on certain assumptions regarding the domain Ω and functional spaces. For further reading on those theoretical aspects, we refer the reader to Ladyzhenskaya [147], Temam [220], and Layton [155].

We begin with the linearized Navier–Stokes equations written in the form

$$\begin{cases} \rho \mathbf{v} \cdot \nabla \mathbf{u} - \nabla \cdot (2\mu \bar{\bar{\varepsilon}}) + \nabla p = \rho \mathbf{g}, \\ \nabla \cdot \mathbf{u} = 0, \end{cases}$$

where $\mathbf{v}$ is the velocity from the previous iteration and $\bar{\bar{\varepsilon}} = \frac{1}{2}(\nabla \mathbf{u} + \nabla \mathbf{u}^T)$ is the strain rate tensor. The dynamic viscosity μ is constant so that $\nabla \cdot (2\mu \bar{\bar{\varepsilon}}) = \mu \Delta \mathbf{u}$. We intentionally use the general form of the viscous term to facilitate further extensions.

In the 2D case, the componentwise form of this system reads

$$\begin{cases} \rho\mathbf{v}\cdot\nabla u_1 - \nabla\cdot(2\mu\boldsymbol{\varepsilon}_1) + \dfrac{\partial p}{\partial x_1} = \rho g_1, \\ \rho\mathbf{v}\cdot\nabla u_2 - \nabla\cdot(2\mu\boldsymbol{\varepsilon}_2) + \dfrac{\partial p}{\partial x_2} = \rho g_2, \\ \nabla\cdot\mathbf{u} = 0. \end{cases}$$

On the boundary $\Gamma = \Gamma_0 \cup \Gamma_F$ of the domain $\Omega \subset \mathbb{R}^2$, we prescribe the boundary conditions

$$\begin{cases} \mathbf{u} = \mathbf{u}_0 \quad \text{on } \Gamma_0 & \text{(Dirichlet)}, \\ 2\mu\boldsymbol{\varepsilon}_k\cdot\mathbf{n} - pn_k + \alpha_k u_k = s_k \quad \text{on } \Gamma_F & \text{(Newton)}, \end{cases} \tag{5.35}$$

where $\mathbf{n}$ is the unit outward normal vector and $\mathbf{u}_0, \alpha_k, s_k$ are given functions of $\mathbf{x}$.

Let $V = H_0^1(\Omega)$ be the space of scalar-valued test functions which have zero trace on the Dirichlet boundary component Γ_0. Multiplying the kth momentum equation ($k = 1,2$) by a test function $w_k \in V$ and integrating over the domain Ω, we obtain

$$\rho\int_\Omega w_k(\mathbf{v}\cdot\nabla u_k)\,d\mathbf{x} - \int_\Omega w_k\nabla\cdot(2\mu\boldsymbol{\varepsilon}_k)\,d\mathbf{x} + \int_\Omega w_k\frac{\partial p}{\partial x_k}\,d\mathbf{x} = \rho\int_\Omega w_k g_k\,d\mathbf{x}.$$

Invoking Green's formula, we perform integration by parts in the viscous term,

$$\begin{aligned} -\int_\Omega w_k\nabla\cdot(2\mu\boldsymbol{\varepsilon}_k)\,d\mathbf{x} &= -\int_\Omega w_k\left(\frac{\partial(2\mu\varepsilon_{k1})}{\partial x_1} + \frac{\partial(2\mu\varepsilon_{k2})}{\partial x_2}\right)d\mathbf{x} \\ &= \int_\Omega 2\mu\left(\frac{\partial w_k}{\partial x_1}\varepsilon_{k1} + \frac{\partial w_k}{\partial x_2}\varepsilon_{k2}\right)d\mathbf{x} - \int_\Gamma 2\mu w_k(\varepsilon_{k1}n_1 + \varepsilon_{k2}n_2)\,ds \\ &= \int_\Omega 2\mu(\nabla w_k\cdot\boldsymbol{\varepsilon}_k)\,d\mathbf{x} - \int_\Gamma 2\mu w_k(\boldsymbol{\varepsilon}_k\cdot\mathbf{n})\,ds, \end{aligned}$$

and in the pressure term,

$$\int_\Omega w_k\frac{\partial p}{\partial x_k}\,d\mathbf{x} = -\int_\Omega\frac{\partial w_k}{\partial x_k}p\,d\mathbf{x} + \int_\Gamma w_k p n_k\,ds.$$

This leads us to the following variational formulation of the kth momentum equation:

$$\begin{aligned} &\rho\int_\Omega w_k(\mathbf{v}\cdot\nabla u_k)\,d\mathbf{x} + \int_\Omega 2\mu(\nabla w_k\cdot\boldsymbol{\varepsilon}_k)\,d\mathbf{x} - \int_\Omega\frac{\partial w_k}{\partial x_k}p\,d\mathbf{x} \\ &-\int_\Gamma w_k(2\mu\boldsymbol{\varepsilon}_k\cdot\mathbf{n} - pn_k)\,ds = \rho\int_\Omega w_k g_k\,d\mathbf{x} \qquad \forall w_k \in \Omega. \end{aligned} \tag{5.36}$$

The integral over the boundary component $\Gamma_0 \subseteq \Gamma$ vanishes since $w_k = 0$ on Γ_0. Substituting the Newton boundary condition into the integral over Γ_F, we obtain

$$\begin{aligned} -\int_\Gamma w_k(2\mu\boldsymbol{\varepsilon}_k\cdot\mathbf{n} - pn_k)\,ds &= -\int_{\Gamma_F} w_k(s_k - \alpha_k u_k)\,ds \\ &= \int_{\Gamma_F}\alpha_k w_k u_k\,ds - \int_{\Gamma_F} w_k s_k\,ds. \end{aligned}$$

Hence, the variational formulation of the kth momentum equation becomes

$$\rho\int_\Omega w_k(\mathbf{v}\cdot\nabla u_k)\,d\mathbf{x}+\int_\Omega 2\mu(\nabla w_k\cdot\boldsymbol{\varepsilon}_k)\,d\mathbf{x}-\int_\Omega \frac{\partial w_k}{\partial x_k}p\,d\mathbf{x}+\int_{\Gamma_F}\alpha_k w_k u_k\,ds$$
$$=\rho\int_\Omega w_k g_k\,d\mathbf{x}+\int_{\Gamma_F} w_k s_k\,ds \qquad \forall w_k\in V, \qquad k=1,2.$$

By definition of the strain rate tensor, the final form of the two systems is given by

$$\rho\int_\Omega w_1(\mathbf{v}\cdot\nabla u_1)\,d\mathbf{x}+\int_\Omega 2\mu\frac{\partial w_1}{\partial x_1}\frac{\partial u_1}{\partial x_1}\,d\mathbf{x}+\int_\Omega \mu\frac{\partial w_1}{\partial x_2}\left(\frac{\partial u_1}{\partial x_2}+\frac{\partial u_2}{\partial x_1}\right)d\mathbf{x}$$
$$+\int_\Omega \frac{\partial w_1}{\partial x_1}p\,d\mathbf{x}+\int_{\Gamma_F}\alpha_1 w_1 u_1\,ds=\rho\int_\Omega w_1 g_1\,d\mathbf{x}+\int_{\Gamma_F} w_1 s_1\,ds \tag{5.37}$$

and

$$\rho\int_\Omega w_2(\mathbf{v}\cdot\nabla u_2)\,d\mathbf{x}+\int_\Omega \mu\frac{\partial w_2}{\partial x_1}\left(\frac{\partial u_1}{\partial x_2}+\frac{\partial u_2}{\partial x_1}\right)d\mathbf{x}+\int_\Omega 2\mu\frac{\partial w_2}{\partial x_2}\frac{\partial u_2}{\partial x_2}\,d\mathbf{x}$$
$$+\int_\Omega \frac{\partial w_2}{\partial x_2}p\,d\mathbf{x}+\int_{\Gamma_F}\alpha_2 w_2 u_2\,ds=\rho\int_\Omega w_2 g_2\,d\mathbf{x}+\int_{\Gamma_F} w_2 s_2\,ds. \tag{5.38}$$

This formulation corresponds to the first-order linearization. The Newton linearization

$$\mathbf{u}\cdot\nabla\mathbf{u}\approx\underbrace{\mathbf{v}\cdot\nabla\mathbf{u}}_{\text{a}}+\underbrace{\mathbf{u}\cdot\nabla\mathbf{v}}_{\text{b}}-\underbrace{\mathbf{v}\cdot\nabla\mathbf{v}}_{\text{c}}$$

produces two additional terms which must be included in the variational formulation.

Adding the Newton terms to the kth momentum transport equation, we obtain

$$\rho\int_\Omega w_k\underbrace{(\mathbf{v}\cdot\nabla u_k)}_{\text{a}}\,d\mathbf{x}+\rho\int_\Omega w_k\underbrace{(\mathbf{u}\cdot\nabla\mathbf{v}_k)}_{\text{b}}\,d\mathbf{x}+\int_\Omega 2\mu(\nabla w_k\cdot\boldsymbol{\varepsilon}_k)\,d\mathbf{x}$$
$$+\int_\Omega \frac{\partial w_k}{\partial x_k}p\,d\mathbf{x}+\int_{\Gamma_F}\alpha_k w_k u_k\,ds=\rho\int_\Omega w_k g_k\,d\mathbf{x}+\int_{\Gamma_F} w_k s_k\,ds$$
$$+\rho\int_\Omega w_k\underbrace{(\mathbf{v}\cdot\nabla\mathbf{v}_k)}_{\text{c}}\,d\mathbf{x} \qquad \forall w_k\in V, \quad k=1,2. \tag{5.39}$$

The continuity equation must also be written in a variational form using test functions from a space $W\subseteq L^2(\Omega)$, which is generally different from the space V. The reason for this lies in the fact that the velocity components must belong to $H^1(\Omega)$, while the pressure is a (possibly discontinuous) function from $L^2(\Omega)$. The difference between the functional spaces for different variables is reflected in the choice of the test functions.

Multiplying the incompressibility constraint $\nabla\cdot\mathbf{u}=0$ by a test function $w_p\in W$ and integrating over the domain Ω, we obtain the system of integral relations

$$\int_\Omega w_p\nabla\cdot\mathbf{u}\,d\mathbf{x}=0 \qquad \forall w_p\in W. \tag{5.40}$$

By definition of the divergence, the componentwise form of this system is given by

$$\int_\Omega w_p\frac{\partial u_1}{\partial x_1}\,d\mathbf{x}+\int_\Omega w_p\frac{\partial u_2}{\partial x_2}\,d\mathbf{x}=0 \qquad \forall w_p\in W. \tag{5.41}$$

5.2.3 ▪ Finite Element Discretization

We will begin with the kth momentum equation and transform its variational form into an algebraic system using the Galerkin method to find the best approximation to u_k in a finite-dimensional space $V_h \subset H^1(\Omega)$ spanned by the basis functions $\varphi_1,\ldots,\varphi_N$.

The finite element approximation to the velocity component u_k is defined by

$$u_{kh}(\mathbf{x}) = \sum_{j=1}^{N} u_{kj}\varphi_j(\mathbf{x}), \qquad k=1,2, \tag{5.42}$$

where u_{kj} denotes the nodal value of the approximate solution u_{kh} at the grid point $\mathbf{x}_j$.

The test function space $W_h \subset W$ for the discrete continuity equations is defined as the span of basis functions $\psi_1,\ldots,\psi_M$ leading to pressure approximations of the form

$$p_h(\mathbf{x}) = \sum_{j=1}^{M} p_j\psi_j(\mathbf{x}). \tag{5.43}$$

For the standard Galerkin method to produce a well-posed discrete problem, we must have $M < N$. Moreover, an *inf-sup stability condition* restricts the choice of the spaces V_h and W_h for the finite element approximation of the velocity and pressure unknowns. An inf-sup stable pair of approximation spaces can be constructed, e.g., by using lower-degree basis functions or a coarser mesh for the pressure. The use of equal-order interpolations for the velocity and pressure may be admissible, provided that stabilization terms are added to the discrete continuity equation. This issue is discussed in more detail in Chapter 8.

Replacing u_1, u_2, p by the finite element approximations u_{1h}, u_{2h}, p_h and using test functions from the finite-dimensional spaces, we obtain the desired system of algebraic equations. The discrete counterparts of equations (5.37) and (5.38) are given by

$$\begin{aligned}
&\sum_{j=1}^{N} u_{1j}\left(\rho\int_\Omega \varphi_i(\mathbf{v}\cdot\nabla\varphi_j)\,\mathrm{d}\mathbf{x} + 2\int_\Omega \mu\frac{\partial\varphi_i}{\partial x_1}\frac{\partial\varphi_j}{\partial x_1}\,\mathrm{d}\mathbf{x} + \int_\Omega \mu\frac{\partial\varphi_i}{\partial x_2}\frac{\partial\varphi_j}{\partial x_2}\,\mathrm{d}\mathbf{x}\right.\\
&\left.+\int_{\Gamma_F}\alpha_1\varphi_i\varphi_j\,\mathrm{d}s\right) + \sum_{j=1}^{N} u_{2j}\int_\Omega \mu\frac{\partial\varphi_i}{\partial x_2}\frac{\partial\varphi_j}{\partial x_1}\,\mathrm{d}\mathbf{x} + \sum_{j=1}^{M} p_j\int_\Omega \frac{\partial\varphi_i}{\partial x_1}\psi_j\,\mathrm{d}\mathbf{x}\\
&= \rho\int_\Omega \varphi_i g_1\,\mathrm{d}\mathbf{x} + \int_{\Gamma_F}\varphi_i s_1\,\mathrm{d}s, \qquad i=1,\ldots,N,
\end{aligned}$$

and

$$\begin{aligned}
&\sum_{j=1}^{N} u_{2j}\left(\rho\int_\Omega \varphi_i(\mathbf{v}\cdot\nabla\varphi_j)\,\mathrm{d}\mathbf{x} + \int_\Omega \mu\frac{\partial\varphi_i}{\partial x_1}\frac{\partial\varphi_j}{\partial x_1}\,\mathrm{d}\mathbf{x} + 2\int_\Omega \mu\frac{\partial\varphi_i}{\partial x_2}\frac{\partial\varphi_j}{\partial x_2}\,\mathrm{d}\mathbf{x}\right.\\
&\left.+\int_{\Gamma_F}\alpha_2\varphi_i\varphi_j\,\mathrm{d}s\right) + \sum_{j=1}^{N} u_{1j}\int_\Omega \mu\frac{\partial\varphi_i}{\partial x_1}\frac{\partial\varphi_j}{\partial x_2}\,\mathrm{d}\mathbf{x} + \sum_{j=1}^{M} p_j\int_\Omega \frac{\partial\varphi_i}{\partial x_2}\psi_j\,\mathrm{d}\mathbf{x}\\
&= \rho\int_\Omega \varphi_i g_2\,\mathrm{d}\mathbf{x} + \int_{\Gamma_F}\varphi_i s_2\,\mathrm{d}s, \qquad i=1,\ldots,N.
\end{aligned}$$

This is a system of $2N$ linear algebraic equations. Using $\psi_1,\ldots,\psi_M$ as test functions in the weak form (5.41) of the continuity equation, we obtain M additional equations,

$$\sum_{j=1}^{N} u_{1j}\int_\Omega \psi_i\frac{\partial\varphi_j}{\partial x_1}\,\mathrm{d}\mathbf{x} + \sum_{j=1}^{N} u_{2j}\int_\Omega \psi_i\frac{\partial\varphi_j}{\partial x_2}\,\mathrm{d}\mathbf{x} = 0, \qquad i=1,\ldots,M, \tag{5.44}$$

which gives us a total of $2N+M$ Galerkin equations for $2N+M$ degrees of freedom.

The above system of algebraic equations can be written in the block matrix form

$$\underbrace{\begin{pmatrix} A_{11} & A_{12} & -B_1^T \\ A_{21} & A_{22} & -B_2^T \\ B_1 & B_2 & 0 \end{pmatrix}}_{A} \underbrace{\begin{pmatrix} u_1 \\ u_2 \\ p \end{pmatrix}}_{U} = \underbrace{\begin{pmatrix} F_1 \\ F_2 \\ 0 \end{pmatrix}}_{F}, \tag{5.45}$$

where the vectors $u_1 = (u_{11}, \ldots, u_{1N})^T$ and $u_2 = (u_{21}, \ldots, u_{2N})^T$ contain the nodal values of the two velocity components. The pressure degrees of freedom are stored in the vector $p = (p_1, \ldots, p_M)^T$. The blocks of the global system $AU = F$ are defined by

$$\begin{aligned} A_{11} &= C + 2K_{11} + K_{22} + R_1, \\ A_{12} &= K_{12}, \\ A_{21} &= K_{21}, \\ A_{22} &= C + K_{11} + 2K_{22} + R_2, \\ F_1 &= G_1 + S_1, \\ F_2 &= G_2 + S_2, \end{aligned}$$

where C is the discrete convection operator, $K_{11}, K_{12}, K_{21}, K_{22}$ are the blocks of the viscous dissipation term, and F_1, F_2 is the contribution of the body force. The natural boundary conditions are taken into account via the matrices R_1, R_2 and vectors G_1, G_2.

The entries of the above $N \times N$ matrices and $N \times 1$ vectors are given by

$$\begin{aligned} C_{ij} &= \rho \int_\Omega \varphi_i (\mathbf{v} \cdot \nabla \varphi_j)\, \mathrm{d}\mathbf{x}, \\ (K_{kl})_{ij} &= \int_\Omega \mu \frac{\partial \varphi_i}{\partial x_l} \frac{\partial \varphi_j}{\partial x_k}\, \mathrm{d}\mathbf{x}, \quad k, l = 1, 2, \\ (F_k)_i &= \rho \int_\Omega \varphi_i g_k\, \mathrm{d}\mathbf{x}, \quad k = 1, 2, \\ (R_k)_{ij} &= \int_{\Gamma_F} \alpha_k \varphi_i \varphi_j\, \mathrm{d}s, \quad k = 1, 2, \\ (S_k)_i &= \int_{\Gamma_F} \varphi_i s_k\, \mathrm{d}s, \quad k = 1, 2, \end{aligned} \tag{5.46}$$

for $i, j = 1, \ldots, N$. The entries of the $M \times N$ blocks B_1 and B_2 are given by

$$(B_k)_{ij} = \int_\Omega \psi_i \frac{\partial \varphi_j}{\partial x_k}\, \mathrm{d}\mathbf{x}, \quad k = 1, 2, \tag{5.47}$$

for $i = 1, \ldots, M$ and $j = 1, \ldots, N$. It is easy to verify that all matrix dimensions agree.

If Newton linearization is employed, the blocks of the algebraic system become

$$\begin{aligned} A_{11} &= C + C_{11} + 2K_{11} + K_{22} + R_1, \\ A_{12} &= C_{12} + K_{12}, \\ A_{21} &= C_{21} + K_{21}, \\ A_{22} &= C + C_{22} + K_{11} + 2K_{22} + R_2, \\ F_1 &= C_1 + G_1 + S_1, \\ F_2 &= C_2 + G_2 + S_2. \end{aligned}$$

The definition of the additional $N \times N$ matrices and $N \times 1$ vectors is as follows:

$$(C_{kl})_{ij} = \rho \int_\Omega \frac{\partial \mathbf{v}_k}{\partial x_l} \varphi_i \varphi_j \, d\mathbf{x}, \qquad k,l = 1,2,$$

$$(C_k)_i = \rho \int_\Omega \varphi_i (\mathbf{v} \cdot \nabla \mathbf{v}_k) \, d\mathbf{x}, \qquad k = 1,2,$$

for $i,j = 1,\dots,N$. Note that the blocks C_{kl} have the structure of a mass matrix.

5.2.4 ▪ Localization

The blocks of the global algebraic system introduced in the previous section are generated using the usual elementwise integration procedure. The element matrix

$$A^{(e)} = \begin{pmatrix} A_{11}^{(e)} & A_{12}^{(e)} & -(B_1^{(e)})^T \\ A_{21}^{(e)} & A_{22}^{(e)} & -(B_2^{(e)})^T \\ B_1^{(e)} & B_2^{(e)} & 0 \end{pmatrix}$$

and the element load vector

$$F^{(e)} = \begin{pmatrix} F_1^{(e)} \\ F_2^{(e)} \\ 0 \end{pmatrix}$$

exhibit the same structure as their global counterparts. The size of $A^{(e)}$ and $F^{(e)}$ depends on the number of local degrees of freedom for the velocity and pressure.

Let $N^{(e)}$ denote the number of nodal points for the velocity approximation on element E_e. The number of nodal points for the pressure will be denoted by $M^{(e)}$. Then the size of the element matrix $A^{(e)}$ is given by $(2N^{(e)} + M^{(e)}) \times (2N^{(e)} + M^{(e)})$, and the element vector $F^{(e)}$ will have $2N^{(e)} + M^{(e)}$ elements. The size of the auxiliary element matrices/vectors can also be determined by replacing N by $N^{(e)}$ and M by $M^{(e)}$. We have

$$\begin{aligned} A_{11}^{(e)} &= C^{(e)} + 2K_{11}^{(e)} + K_{22}^{(e)} + R_1^{(e)}, \\ A_{12}^{(e)} &= K_{12}^{(e)}, \\ A_{21}^{(e)} &= K_{21}^{(e)}, \\ A_{22}^{(e)} &= C^{(e)} + K_{11}^{(e)} + 2K_{22}^{(e)} + R_2^{(e)}, \\ F_1^{(e)} &= G_1^{(e)} + S_1^{(e)}, \\ F_2^{(e)} &= G_2^{(e)} + S_2^{(e)}. \end{aligned}$$

The element matrix $A^{(e)}$ and element vector $F^{(e)}$ are generated using (numerical) integration on element $E^{(e)}$ to calculate the entries of the auxiliary matrices and vectors

$$C_{ij}^{(e)} = \rho \int_{E^{(e)}} \varphi_i^{(e)} (\mathbf{v} \cdot \nabla \varphi_j^{(e)})\, \mathrm{d}\mathbf{x},$$

$$(K_{kl}^{(e)})_{ij} = \int_{E^{(e)}} \mu \frac{\partial \varphi_i^{(e)}}{\partial x_l} \frac{\partial \varphi_j^{(e)}}{\partial x_k}\, \mathrm{d}\mathbf{x}, \qquad k,l = 1,2,$$

$$(B_k^{(e)})_{ij} = \int_{E^{(e)}} \psi_i^{(e)} \frac{\partial \varphi_j^{(e)}}{\partial x_k}\, \mathrm{d}\mathbf{x}, \qquad k = 1,2,$$

$$(G_k^{(e)})_i = \int_{E^{(e)}} \varphi_i^{(e)} g_k\, \mathrm{d}\mathbf{x}, \qquad k = 1,2,$$

$$(R_k^{(e)})_{ij} = \int_{\partial E^{(e)} \cap \Gamma_F} \alpha_k \varphi_i^{(e)} \varphi_j^{(e)}\, \mathrm{d}s, \qquad k = 1,2,$$

$$(S_k^{(e)})_i = \int_{\partial E^{(e)} \cap \Gamma_F} \varphi_i^{(e)} s_k\, \mathrm{d}s, \qquad k = 1,2.$$

The boundary integrals are calculated only if element E_e has an edge on the boundary part Γ_F, that is, if $\partial E^{(e)} \cap \Gamma_F \neq \emptyset$. Otherwise, we set $R_k^{(e)} = 0$ and $S_k^{(e)} = 0$ for $k = 1,2$.

On a mesh of rectangular elements, the local basis functions $\varphi_i^{(e)}$ and $\psi_i^{(e)}$ are defined as tensor products of 1D Lagrange basis polynomials. Note that the use of a piecewise-constant pressure approximation is admissible in finite element methods for the Navier–Stokes equations. The above-mentioned inf-sup stability condition must be taken into account if the discretization is performed using the standard Galerkin method (see Chapter 8). For a review of inf-sup stable finite element pairs and available stabilization techniques, we refer the reader to the books by Cuvelier, Segal, and van Steenhoven [44] and Quarteroni and Valli [195].

5.2.5 ▪ Assembly of the Global System

For simplicity, we assume that the velocity and pressure are approximated using the same basis functions ($M = N$). In this case, three degrees of freedom (two velocity components and pressure) are associated with each nodal point. Hence, the element-by-element assembly procedure differs from Algorithm 5.2 designed for the scalar 2D heat equation.

The bandwidth of the global matrix A depends on the numbering of the global degrees of freedom. In our partition of the global system $AU = F$, we assumed that

$$U = (\underbrace{u_{11}, u_{12}, \ldots, u_{1N}}_{u_1}, \underbrace{u_{21}, u_{22}, \ldots, u_{2N}}_{u_2}, \underbrace{p_1, \ldots, p_N}_{p})^T.$$

Clearly, the bandwidth is smaller if the following node-by-node numbering is adopted:

$$U = (\underbrace{u_{11}, u_{21}, p_1}_{U_1}, \underbrace{u_{12}, u_{22}, p_2}_{U_2} \ldots, \underbrace{u_{1N}, u_{2N}, p_N}_{U_N})^T.$$

In the case of the four-node quadrilateral element depicted in Fig. 5.20, the correspondence between the local and global numbers of nodal points is given by

$$\mathrm{IND}_e = (12, 27, 28, 13).$$

If $I \in \{1, \ldots, 4\}$ is the local number of a nodal point, its global number is $\mathrm{IND}_e(I)$.

The following pseudocode can be used to assemble the global algebraic system from the $N^{(e)} \times N^{(e)}$ element matrices $A_{kl}^{(e)}, B_k^{(e)}$ and $N^{(e)} \times 1$ element vectors $F_k^{(e)}$.

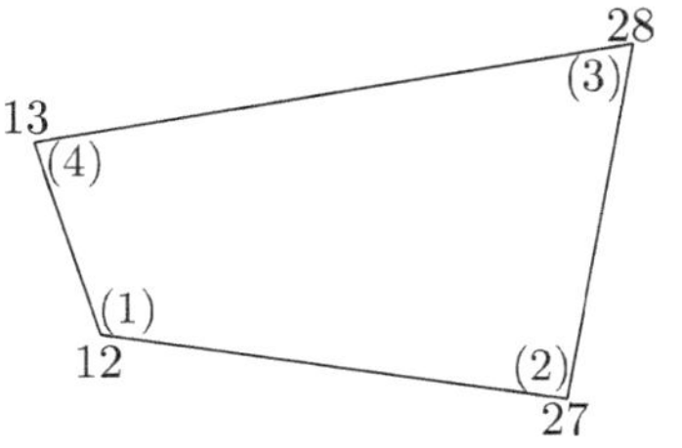

Figure 5.20. *Global vs. local node numbering for a quadrilateral element.*

Algorithm 5.3 (Global assembly for a generic quadrilateral element)

for $e = 1,\ldots,N_e$ **do**

Generate the element matrices/vectors for element E_e.

for $I = 1,\ldots,N^{(e)}$ **do**

$i = \mathrm{IND}_e(I)$

for $J = 1,\ldots,N^{(e)}$ **do**

$j = \mathrm{IND}_e(J)$

$$\begin{aligned}
A(3i-2,3j-2) &= A(3i-2,3j-2) + A_{11}^{(e)}(I,J)\\
A(3i-2,3j-1) &= A(3i-2,3j-1) + A_{12}^{(e)}(I,J)\\
A(3i-1,3j-2) &= A(3i-1,3j-2) + A_{21}^{(e)}(I,J)\\
A(3i-1,3j-1) &= A(3i-1,3j-1) + A_{22}^{(e)}(I,J)\\
A(3i\quad,3j-2) &= A(3i\quad,3j-2) + B_1^{(e)}(I,J)\\
A(3i\quad,3j-1) &= A(3i\quad,3j-1) + B_2^{(e)}(I,J)\\
A(3i-2,3j\quad) &= A(3i-2,3j\quad) + B_1^{(e)}(J,I)\\
A(3i-1,3j\quad) &= A(3i-1,3j\quad) + B_2^{(e)}(J,I)\\
A(3i\quad,3j\quad) &= 0
\end{aligned}$$

end J-loop

$$\begin{aligned}
F(3i-2) &= F(3i-2) + F_1^{(e)}(I)\\
F(3i-1) &= F(3i-1) + F_2^{(e)}(I)\\
F(3i-0) &= 0
\end{aligned}$$

end I-loop

end e-loop

Note that the blocks $B_1^{(e)}$ and $B_2^{(e)}$ are transposed by swapping the indices I and J.

After visiting all elements, one obtains the matrix A and the right-hand side F of the global system. Due to the presence of the zero block in the bottom right corner of

$$A = \begin{pmatrix} A_{11} & A_{12} & -B_1^T \\ A_{21} & A_{22} & -B_2^T \\ B_1 & B_2 & 0 \end{pmatrix},$$

the numerical solution of $AU = F$ is a challenging task. The design of advanced solution techniques for the discretized Navier–Stokes equations is discussed in Chapter 8.

5.2.6 ▪ Stabilized FEM

The standard Galerkin discretization of the steady incompressible Navier–Stokes equations may produce spurious oscillations in the velocity and/or pressure fields. In addition to convective instabilities that arise at high Reynolds numbers, an inappropriate combination of the finite element spaces for the velocity and pressure approximations may result in nonphysical pressure modes. We will explain the reasons for this in Chapter 8.

The basic stabilization technique to be presented in this section follows the idea of the SUPG method [23] and is utilized in the *Elmer* software. Let $\mathbf{w} = (w_1, w_2)$ and $q = w_p$ be test functions for the 2D momentum and continuity equations, respectively. The variational form of the problem at hand reads

$$(\mathbf{w}, \mathbf{v}\cdot\nabla\mathbf{u}) + (2\nu\bar{\bar{\varepsilon}}(\mathbf{w}), \bar{\bar{\varepsilon}}(\mathbf{u})) - (\nabla\cdot\mathbf{w}, \tilde{p}) - (q, \nabla\cdot\mathbf{u}) = (\mathbf{w}, \mathbf{f}),$$

where $(\cdot,\cdot)$ denotes the L^2 scalar product, $\nu = \frac{\mu}{\rho}$ is the kinematic viscosity, and $\tilde{p} = \frac{p}{\rho}$.

In order to stabilize the velocity and pressure, we replace $\mathbf{w}$ and q by

$$\tilde{\mathbf{w}} = \mathbf{w} + \tau(\mathbf{v}\cdot\nabla\mathbf{w} + \nabla q),$$
$$\tilde{q} = q + \delta\nabla\cdot\mathbf{w},$$

where τ and δ are the stabilization parameters. This gives

$$\begin{aligned}&(\mathbf{w}, \mathbf{v}\cdot\nabla\mathbf{u}) + (2\nu\bar{\bar{\varepsilon}}(\mathbf{w}), \bar{\bar{\varepsilon}}(\mathbf{u})) - (\nabla\cdot\mathbf{w}, \tilde{p}) - (q, \nabla\cdot\mathbf{u}) + (\delta\nabla\cdot\mathbf{w}, \nabla\cdot\mathbf{u})\\ &+ (\tau(\mathbf{v}\cdot\nabla\mathbf{w} + \nabla q), \mathbf{v}\cdot\nabla\mathbf{u} + \nabla\tilde{p}) = (\mathbf{w} + \tau(\mathbf{v}\cdot\nabla\mathbf{w} + \nabla q), \mathbf{f}).\end{aligned} \tag{5.48}$$

The stabilization parameters δ and τ are defined as follows [74, 95]:

$$\delta = \lambda\|\mathbf{v}\|h_K\xi(Re_h),$$
$$\tau = \frac{h_K}{2\|\mathbf{v}\|}\xi(Re_h),$$

where the mesh Reynolds number Re_h is given by

$$Re_h = \frac{m_k\|\mathbf{v}\|h_K}{2(2\nu)} = \frac{\rho\, m_k\|\mathbf{v}\|h_K}{4\mu}$$

and

$$\xi(Re_h) = \begin{cases} Re_h & \text{if } 0 \le Re_h < 1,\\ 1 & \text{if } Re_h \ge 1.\end{cases}$$

The parameters λ and m_k are positive constants. Typical values are $\lambda = 1$ and $m_k = \frac{1}{3}$. The parameter h_K denotes the local mesh size.

Multiplying (5.48) by the density ρ and using the test functions $\mathbf{w} = (w_1, 0)$ and $q = 0$, we obtain the variational formulation of the first momentum equation,

$$\begin{aligned}&(w_1 + \tau\mathbf{v}\cdot\nabla w_1, \rho\mathbf{v}\cdot\nabla u_1) + (\nabla w_1, 2\mu\bar{\bar{\varepsilon}}(u_1)) - \left(\frac{\partial w_1}{\partial x_1}, p\right)\\ &+ \left(\delta\frac{\partial w_1}{\partial x_1}, \rho\nabla\cdot\mathbf{u}\right) + \left(\tau\mathbf{v}\cdot\nabla w_1, \frac{\partial p}{\partial x_1}\right) = (w_1 + \tau\mathbf{v}\cdot\nabla w_1, \rho f_1).\end{aligned}$$

Using $\mathbf{w} = (0, w_2)$ and $q = 0$, we obtain the second variational momentum equation,

$$(w_2 + \tau \mathbf{v}\cdot\nabla w_2, \rho\mathbf{v}\cdot\nabla u_2) + (\nabla w_2, 2\mu\bar{\bar{\varepsilon}}(u_2)) - \left(\frac{\partial w_2}{\partial x_2}, p\right)$$
$$+\left(\delta\frac{\partial w_2}{\partial x_2}, \rho\nabla\cdot\mathbf{u}\right) + \left(\tau\mathbf{v}\cdot\nabla w_2, \frac{\partial p}{\partial x_2}\right) = (w_2 + \tau\mathbf{v}\cdot\nabla w_2, \rho f_2).$$

The choice $\mathbf{w} = (0, 0)$ yields the stabilized variational form of the continuity equation

$$(q, \rho\nabla\cdot\mathbf{u}) + (\tau\nabla q, \rho\mathbf{v}\cdot\nabla\mathbf{u}) + (\tau\nabla q, \nabla p) = (\tau\nabla q, \rho\mathbf{f}).$$

The finite element discretization of the stabilized variational formulation can now be performed similarly to the standard Galerkin method. We define

$$u_{1h} = \sum_{j=1}^{N} u_{1j}\varphi_j, \qquad u_{2h} = \sum_{j=1}^{N} u_{2j}\varphi_j, \qquad p_h = \sum_{j=1}^{N} p_j\varphi_j,$$

where the same basis functions are used to interpolate the two velocity components and the pressure. The resultant global algebraic system has the following structure:

$$\left(\begin{pmatrix} A_{11} & A_{12} & A_{13} \\ A_{21} & A_{22} & A_{23} \\ A_{31} & A_{32} & 0 \end{pmatrix} + \begin{pmatrix} A_{11}^s & A_{12}^s & A_{13}^s \\ A_{21}^s & A_{22}^s & A_{23}^s \\ A_{31}^s & A_{32}^s & A_{33}^s \end{pmatrix}\right)\begin{pmatrix} u_1 \\ u_2 \\ p \end{pmatrix} = \begin{pmatrix} F_1 \\ F_2 \\ 0 \end{pmatrix} + \begin{pmatrix} F_1^s \\ F_2^s \\ F_3^s \end{pmatrix}.$$

The matrices $A_{kl} \in \mathbb{R}^{N\times N}$ and vectors $F_k \in \mathbb{R}^N$ represent the standard Galerkin discretization of the Navier–Stokes equations, whereas A_{kl}^s and F_k^s represent the contribution of the stabilization terms. The coefficients of the Galerkin part are given by

$$A_{11}(i,j) = \int_\Omega \rho\varphi_i(\mathbf{v}\cdot\nabla\varphi_j)\,\mathbf{dx} + \int_\Omega \mu\left(2\frac{\partial\varphi_i}{\partial x_1}\frac{\partial\varphi_j}{\partial x_1} + \frac{\partial\varphi_i}{\partial x_2}\frac{\partial\varphi_j}{\partial x_2}\right)\mathbf{dx},$$
$$A_{12}(i,j) = \int_\Omega \mu\frac{\partial\varphi_i}{\partial x_2}\frac{\partial\varphi_j}{\partial x_1}\,\mathbf{dx},$$
$$A_{13}(i,j) = -\int_\Omega \frac{\partial\varphi_i}{\partial x_1}\varphi_j\,\mathbf{dx},$$
$$A_{21}(i,j) = \int_\Omega \mu\frac{\partial\varphi_i}{\partial x_1}\frac{\partial\varphi_j}{\partial x_2}\,\mathbf{dx},$$
$$A_{22}(i,j) = \int_\Omega \rho\varphi_i(\mathbf{v}\cdot\nabla\varphi_j)\,\mathbf{dx} + \int_\Omega \mu\left(\frac{\partial\varphi_i}{\partial x_1}\frac{\partial\varphi_j}{\partial x_1} + 2\frac{\partial\varphi_i}{\partial x_2}\frac{\partial\varphi_j}{\partial x_2}\right)\mathbf{dx},$$
$$A_{23}(i,j) = -\int_\Omega \frac{\partial\varphi_i}{\partial x_2}\varphi_j\,\mathbf{dx},$$
$$A_{31}(i,j) = \int_\Omega \rho\varphi_i\frac{\partial\varphi_j}{\partial x_1}\,\mathbf{dx},$$
$$A_{32}(i,j) = \int_\Omega \rho\varphi_i\frac{\partial\varphi_j}{\partial x_2}\,\mathbf{dx},$$

and

$$F_1(i) = \int_\Omega \rho \varphi_i f_1 \, \mathbf{dx},$$

$$F_2(i) = \int_\Omega \rho \varphi_i f_2 \, \mathbf{dx},$$

where $i, j = 1, \ldots, N$. The stabilization terms are given by the matrices

$$A_{11}^s(i,j) = \int_\Omega \rho \tau (\mathbf{v} \cdot \nabla \varphi_i)(\mathbf{v} \cdot \nabla \varphi_j) \, \mathbf{dx} + \int_\Omega \rho \delta \frac{\partial \varphi_i}{\partial x_1} \frac{\partial \varphi_j}{\partial x_1} \, \mathbf{dx},$$

$$A_{12}^s(i,j) = \int_\Omega \rho \delta \frac{\partial \varphi_i}{\partial x_1} \frac{\partial \varphi_j}{\partial x_2} \, \mathbf{dx},$$

$$A_{13}^s(i,j) = \int_\Omega \tau (\mathbf{v} \cdot \nabla \varphi_i) \frac{\partial \varphi_j}{\partial x_1} \, \mathbf{dx},$$

$$A_{21}^s(i,j) = \int_\Omega \rho \delta \frac{\partial \varphi_i}{\partial x_2} \frac{\partial \varphi_j}{\partial x_1} \, \mathbf{dx},$$

$$A_{22}^s(i,j) = \int_\Omega \rho \tau (\mathbf{v} \cdot \nabla \varphi_i)(\mathbf{v} \cdot \nabla \varphi_j) \, \mathbf{dx} + \int_\Omega \rho \delta \frac{\partial \varphi_i}{\partial x_2} \frac{\partial \varphi_j}{\partial x_2} \, \mathbf{dx},$$

$$A_{23}^s(i,j) = \int_\Omega \tau (\mathbf{v} \cdot \nabla \varphi_i) \frac{\partial \varphi_j}{\partial x_2} \, \mathbf{dx},$$

$$A_{31}^s(i,j) = \int_\Omega \rho \tau \frac{\partial \varphi_i}{\partial x_1} (\mathbf{v} \cdot \nabla \varphi_j) \, \mathbf{dx},$$

$$A_{32}^s(i,j) = \int_\Omega \rho \tau \frac{\partial \varphi_i}{\partial x_2} (\mathbf{v} \cdot \nabla \varphi_j) \, \mathbf{dx},$$

$$A_{33}^s(i,j) = \int_\Omega \tau_{33} \left(\frac{\partial \varphi_i}{\partial x_1} \frac{\partial \varphi_j}{\partial x_1} + \frac{\partial \varphi_i}{\partial x_2} \frac{\partial \varphi_j}{\partial x_2} \right) \mathbf{dx}$$

and vectors

$$F_1^s(i) = \int_\Omega \rho \tau (\mathbf{v} \cdot \nabla \varphi_i) f_1 \, \mathbf{dx},$$

$$F_2^s(i) = \int_\Omega \rho \tau (\mathbf{v} \cdot \nabla \varphi_i) f_2 \, \mathbf{dx},$$

$$F_3^s(i) = \int_\Omega \rho \tau \left(\frac{\partial \varphi_i}{\partial x_1} f_1 + \frac{\partial \varphi_i}{\partial x_2} f_2 \right) \mathbf{dx}.$$

To illustrate the arising of spurious pressure oscillations, we consider a classical example in which the instability is triggered by a small disturbance of a single mesh node. The original and perturbed meshes, as well as the velocity and pressure fields calculated using the standard Galerkin method with 8×8 elements, are shown in Figs. 5.21 and 5.22, respectively. In this numerical example, the velocity is approximated using bilinear (Q_1) basis functions, whereas the pressure approximation is piecewise constant (Q_0). A usable approximation of the velocity field is obtained on both meshes but the instability of the Q_1Q_0 finite element pair gives rise to checkerboard pressure modes on the perturbed mesh. The inclusion of stabilization terms makes it possible to avoid spurious oscillations.

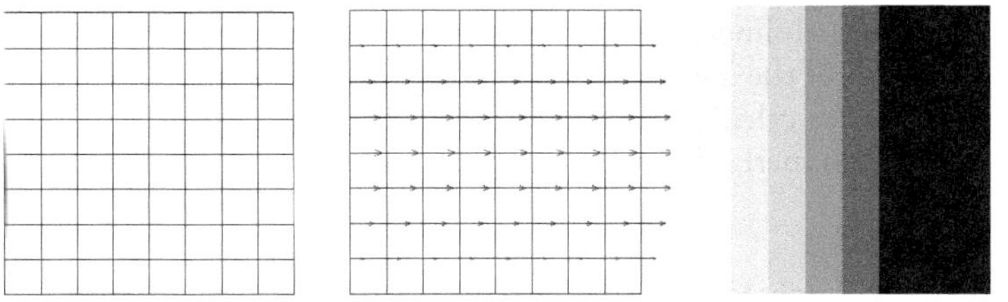

Figure 5.21. *Rectangular mesh, velocity, and pressure—standard FEM.*

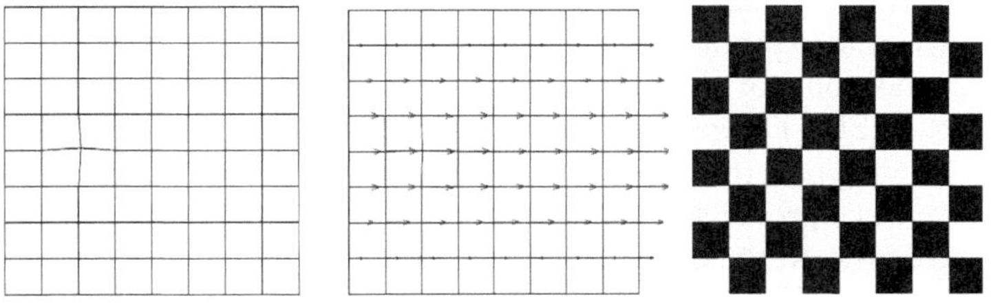

Figure 5.22. *Slightly disturbed mesh, velocity, and pressure—standard FEM.*

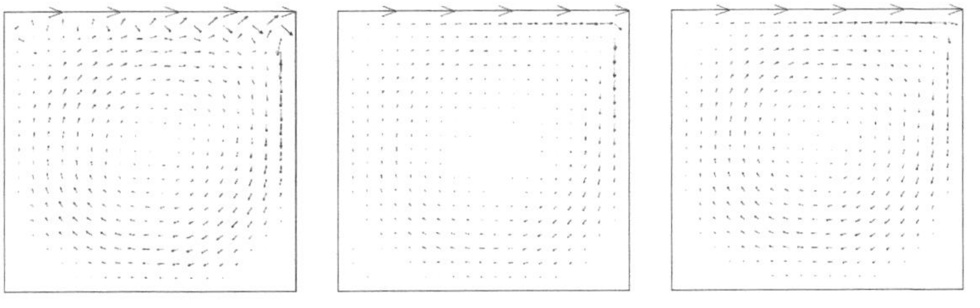

Figure 5.23. *Driven cavity: velocity fields calculated using the* Q_1Q_0 *pair without stabilization (left), SUPG (middle), and the* Q_1Q_1 *pair with full stabilization (right).*

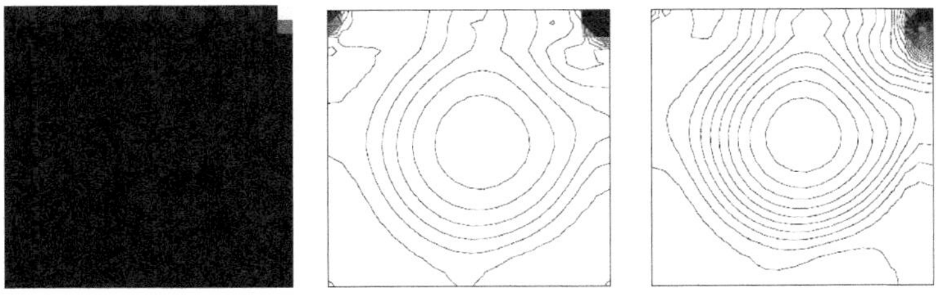

Figure 5.24. *Driven cavity: pressure fields calculated using the* Q_1Q_0 *pair without stabilization (left), SUPG (middle), and the* Q_1Q_1 *pair with full stabilization (right).*

Another classical example is the 2D incompressible flow in a lid-driven cavity. The computational domain is a unit square. The tangential velocity is prescribed on the upper wall and the no-slip condition on the other walls. The left diagrams in Figs. 5.23 and 5.24 show the oscillatory velocity and pressure fields calculated using the Q_1Q_0 finite element pair without any stabilization. The use of SUPG stabilization ($\tau \neq 0$) is sufficient to obtain an accurate velocity field (see Fig. 5.23, middle) but the pressure is still oscillating if the stabilization parameter δ is set to zero. Fig. 5.24 displays the isolines of the piecewise-bilinear function obtained by averaging the piecewise-constant pressure at the vertices of the mesh. The right diagrams in Figs. 5.23 and 5.24 show the velocity and pressure fields calculated using the Q_1Q_1 finite element pair and the stabilized version of the Galerkin method. It can be seen that there are no oscillations in the velocity or pressure fields.

5.2.7 ▪ Elmer Tutorials for the Navier–Stokes Equations

Laminar Flow over a Backward Facing Step

In this tutorial, the incompressible flow solver of *Elmer* is applied to the famous backward facing step problem which we used in Chapter 3 to illustrate the imposition of boundary conditions for the Navier–Stokes equations and in Chapter 4 to illustrate the usage of the FEM. The geometry of the computational domain is depicted in Fig. 5.25. The fluid flowing past the step has the density $\rho = 1$ kg/m^3 and viscosity $\mu = 0.01$ kg/ms. The velocity profile at the inlet is parabolic with the mean velocity $\bar{u}_1 = 1.0$ m/s and $\bar{u}_2 = 0.0$ m/s. At the outlet, the vertical velocity component u_2 and the outlet pressure are set to zero. At all the solid walls, the no-slip boundary condition, $\mathbf{u} = 0$, is applied. The height of the inlet section and that of the channel behind the step are equal to 1 m and 2 m, respectively. The Reynolds number of the flow is 100 based on the 2D cross-sections of the geometry and the corresponding average velocities.

Figure 5.25. *Backward facing step: Geometry of the 2D domain.*

The mesh is stored in the *ElmerGrid* format in the file `step.grd`. Load this file:

```
File
  Open -> step.grd
```

You should obtain your mesh and may check that it consists of 9,696 nodes and 9,442 bilinear elements.

```
Model
  Summary...
```

After loading the mesh, you can start going through the `Model` menu from the top to bottom. Under `Setup`, you can choose parameters related to the simulation. The steady-state simulation is carried out in the 2D Cartesian coordinates, which is the default.

```
Model
  Setup
    Simulation Type = Steady state
    Coordinate system = Cartesian
```

In the equation section, you can choose the relevant equations and parameters related to their solution. In this test case, the Navier–Stokes equations are needed.

When it comes to the definition of equations and materials, it is possible to assign them to bodies (domains) right away, or use mouse selection to assign them later. In this test case, we have just one domain and, therefore, it is easier to select `Equation` and `Material` directly. You could also edit the solver settings in order to try different strategies for solving the nonlinear or linear system. Initially, the Navier–Stokes solver uses the more robust Picard iteration, which is changed to Newton iteration after a few steps. The default values are fine for the given viscosity but may require tuning when going to higher Reynolds number flows.

```
Model
  Equation
    Name = Navier--Stokes
    Apply to Bodies = Body 1
    Navier--Stokes
      Active = on
      Edit Solver Setting
        Nonlinear System
          Max. iterations = 20
          Newton after iterations = 3
    Add
    OK
```

The `Material` section is used to specify all material properties. They include generic parameters which specify direct properties of the material, such as its density, without making any assumptions on the physical and mathematical model. Other properties may represent model parameters in a problem-specific constitutive law. For example, the effective viscosity of the fluid can be non-Newtonian or given by a turbulence model.

```
Model
  Material
    Name = Ideal
    General
      Density = 1.0
    Navier--Stokes
      Viscosity = 0.01
    Apply to Bodies = Body 1
    Add
    OK
```

In the current case study, there are no body forces and the flow is stationary. Convergence to a steady-state solution can be obtained using the default initial condition, which sets the velocity field to zero. Hence, there is no need to prescribe an initial condition.

The parabolic inlet velocity profile is generated using the MATC environment. To edit the content of the inlet profile, click `Enter` to open an edit box for `Velocity 1`. The given expression will be interpreted at run-time producing $u_1(x_1,x_2)=6(x_2-1)(2-x_2)$, $x_2\in[1,2]$, and thereby creating a parabolic velocity profile with a mean velocity of unity.

```
Model
  BoundaryCondition
    Name = Inlet
    Navier--Stokes
      Velocity 1 = Variable Coordinate 2; Real MATC
                                            "6*(tx-1)*(2-tx)"
```

```
    Velocity 2 = 0.0
  Add
  New

  Name = Outlet
  Navier--Stokes
    Velocity 2 = 0.0
  Add
  New

  Name = Walls
  Navier--Stokes
    Noslip wall BC = on
  Add
  OK
```

The boundary conditions may also be assigned to boundaries in the `Boundary condition` menu, or by clicking with the mouse. Here we use the latter approach, which spares us the need to know the indices of each boundary.

```
Model
  Set boundary properties
    Choose Inlet -> set boundary condition Inlet
    Choose Outlet -> set boundary condition Outlet
    Choose Walls -> set boundary condition Walls
```

After writing the files to a hard drive, the problem is ready for handling by *ElmerSolver*. As usual, you may visually inspect the command file.

```
Sif
  Generate
  Edit -> look how your command file came out
```

Before running the solver, you should save the project files needed to restart the simulation. Create a new directory for this test case if you have not done so yet.

```
File
  Save Project
```

After you have successfully saved the files, you may start the solver.

```
Run
  Start solver
```

A convergence monitoring window automatically pops up showing the relative changes of the solution at each iteration. The code should converge in about ten iterations to a tolerance threshold of 0.4347 visible in the output.

When there are some results to view, you may also start the postprocessor.

```
Run
  Start postprocessor
```

The results may be viewed using the postprocessor as shown in Figs. 5.26 and 5.27. The analyst may also be interested in the values of certain derived quantities. For example,

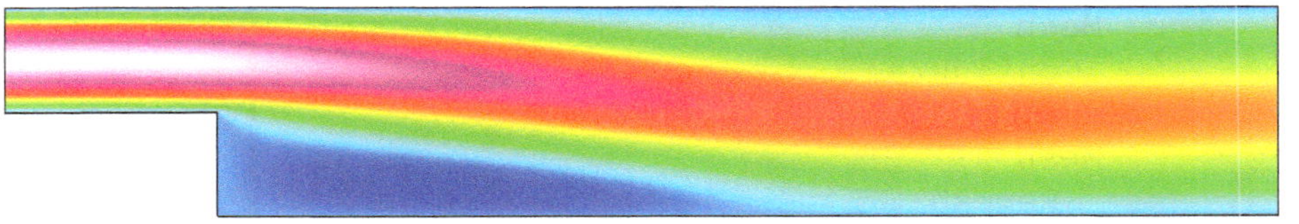

Figure 5.26. *Backward facing step: velocity magnitude.*

Figure 5.27. *Backward facing step: pressure distribution.*

the pressure difference is 0.388 Pa, and the minimum and maximum lateral velocities are −0.1666 m/s and 1.5 m/s, respectively. A further quantity of particular interest is the location of the so-called reattachment point, that is, the point on the x-axis at which the direction of the flow changes. In this case, its position is about 5.0 m behind the step.

The fluid dynamics of the problem at hand strongly depends on the Reynolds number, and it affects the numerics as well. For instance, you may check what happens if the viscosity is further decreased by a factor of 10, which increases the Reynolds number up to 1,000. In this case, convergence of the iterative solution method for the nonlinear Navier–Stokes equations may be difficult to achieve. Some tricks that may be tested include

- introducing an underrelaxation factor (typically in the range 0.5–0.7);
- increasing the maximum number of nonlinear iterations;
- favoring the Picard iteration over Newton's method;
- refining the mesh and increasing the length of the domain.

For a large Reynolds number, you should resolve an additional flow separation at the upper wall downstream of the step position. Note that at high Reynolds numbers the flow becomes turbulent and you cannot find a converged solution without using turbulence models.

Vortex Shedding: Von Karman Instability

In the second tutorial illustrating the application of *Elmer* to the incompressible Navier–Stokes equations, we simulate a transient laminar flow over an obstacle, a cylinder, in a 2D channel (see [208] for a detailed description of this benchmark). This tutorial illustrates the vortex shedding phenomenon which is known as the *von Karman instability*.

The mesh is stored in the 2d netgen format in the file `circle_in_channel.in2d`. Load this file:

```
File
  Open -> circle_in_channel.in2d
```

You should get a mesh consisting of 749 nodes and 1,328 triangles. This is a rather coarse mesh. You may want to have a denser grid, which can be achieved by refining the mesh size and remeshing the domain

```
Mesh
  Configure
    nglib / Max H: 0.02
Mesh
  Remesh
```

The new mesh includes 3,464 nodes and 6,506 triangles. This mesh is presented in Fig. 5.28.

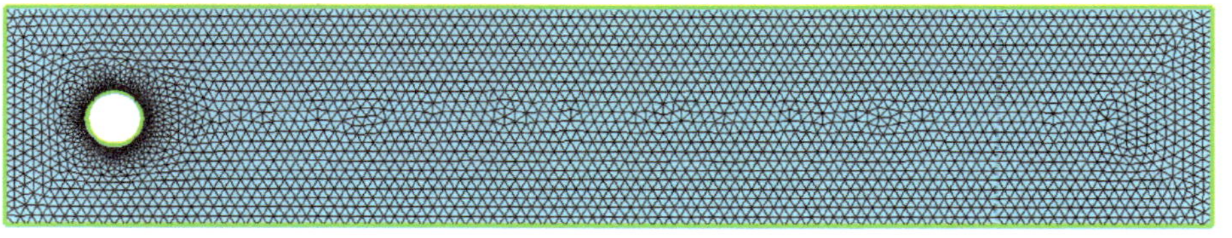

Figure 5.28. *Computational mesh for the flow around a cylinder problem.*

If you are satisfied with the mesh, you may go through the `Model` menu from top to bottom. Now that we have to deal with a transient flow problem, you need to select the time discretization method. For example, you may want to use the second-order backward difference formula (BDF) with 200 time steps. To determine the time step size, divide the total simulation time (e.g., 8 seconds) by the number of time steps.

```
Model
  Setup
    Simulation Type = Transient
    Steady state max. iter = 1
    Time Stepping Method = bdf
    BDF Order = 2
    Time Step Intervals = 200
    Time Step Sizes = $ 8/200
```

You may control the convergence tolerances in the solver-specific settings.

```
Model
  Equation
    Name = Navier--Stokes
    Apply to Bodies = 1
    Navier--Stokes
      Active = on
    Edit Solver Settings
      Nonlinear system
        Convergence tol. = 1.0e-4
      Linear System
        Convergence tol. = 1.0e-6
    Add
    OK
```

The `Material` section is used to specify the values of all material parameters. In this test case, you can choose the following parameter settings:

```
Model
  Material
    Name = Ideal
    General
      Density = 1
    Navier Stokes
      Viscosity = 0.001
    Apply to Bodies = 1
    Add
    OK
```

As in the previous example, we assume that there are no body forces and use the zero velocity field as an initial solution for the time-stepping scheme.

When it comes to the imposition of boundary conditions, we need to distinguish between three different kinds of boundaries: inlet, no-slip walls, and outlet. At the inlet, the horizontal inlet velocity has a parabolic fully developed laminar profile with a maximum speed of 1.5 m/s. The vertical velocity component of the inflow velocity is set to zero. The no-slip condition is imposed on the boundary of the cylinder as well as on the lower and upper walls of the channel. At the outlet, the vertical velocity component is set to zero. If the normal velocity component is not defined, the variational formulation of the problem imposes a zero pressure condition in a weak sense.

```
Model
  BoundaryCondition
    Name = Inlet
    Navier--Stokes
      Velocity 1 = Variable Coordinate 2;
                   Real MATC "4*1.5*tx*(0.41-tx)/0.41^2"
      Velocity 2 = 0.0
    Add
    New

    Name = Walls
    Navier--Stokes
      Velocity 1 = 0.0
      Velocity 2 = 0.0
    Add
    New

    Name = Outlet
    Navier--Stokes
      Velocity 2 = 0.0
    Add
    Ok
```

The boundary conditions may also be assigned to boundaries in the `Boundary condition` menu, or by clicking with the mouse. Here you may use the latter approach since it spares you the need to know the indices of each boundary.

```
Model
  Set boundary properties
    Choose inlet -> set boundary condition Inlet
```

```
Choose both horizontal walls and circle
                         -> set boundary condition Walls
Choose outlet -> set boundary condition Outlet
```

Everything is ready now for generating the mesh files and the command file for *ElmerSolver*. You may inspect the command file to double-check the settings.

```
Sif
  Generate
  Edit -> look how your command file came out
```

Before running the solver, save your project files in a new directory.

```
File
  Save Project
```

After you have successfully saved the files, you may start the solver.

```
Run
  Start solver
```

A convergence monitoring window automatically pops up showing the relative changes of the solution at each iteration. The norm should be around 0.695 after the first time step and around 0.749 after the last one.

When there are some results to view, you may start the postprocessor.

```
Run
  Start postprocessor
```

Due to a large number of time steps, the simulation will take a few minutes. You may inspect the results with *ElmerPost* as the time steps are completed, or wait until the entire simulation run is over. When opening the result file using *ElmerGUI*, *ElmerPost* displays the solution from the first time step only. Therefore, it is important to reopen the file and load all the time steps of your interest. Pressing the button `All` selects all the calculated time steps. A video animation of the results can be viewed by selecting the option `Timestep Control` and pressing the button `Loop` under the `Edit` menu.

In Fig. 5.29, snapshots of the velocity magnitude are presented for three different time steps. The maximum velocity should be about 2.1724 m/s in this simulation.

The Reynolds number is around 100 in this test case. The transition to an unsteady flow behavior occurs at Reynolds numbers around 90. Reducing the flow velocity so that the Reynolds number becomes as small as, say, 20 leads to a stationary solution. On the other hand, a further increase of the velocity would make the von Karman vortices even more pronounced until they break into a fully chaotic motion and the flow becomes turbulent. Remember that a turbulence model needs to be activated for higher Reynolds number flows. Moreover, you may need a refined mesh for the flow to be resolved accurately enough.

Coupled Flow and Heat Transfer: Rayleigh–Benard Instability

The last tutorial is an example of heat transfer in fluid flow. Therefore, the mathematical model of the problem at hand consists of the incompressible Navier–Stokes equations coupled with the heat equation. This test case illustrates the development of the Rayleigh–Benard instability in a rectangular domain (see Fig. 5.30). The height and length of the domain are given by 0.01 m and 0.06 m, respectively. The simulation is performed with

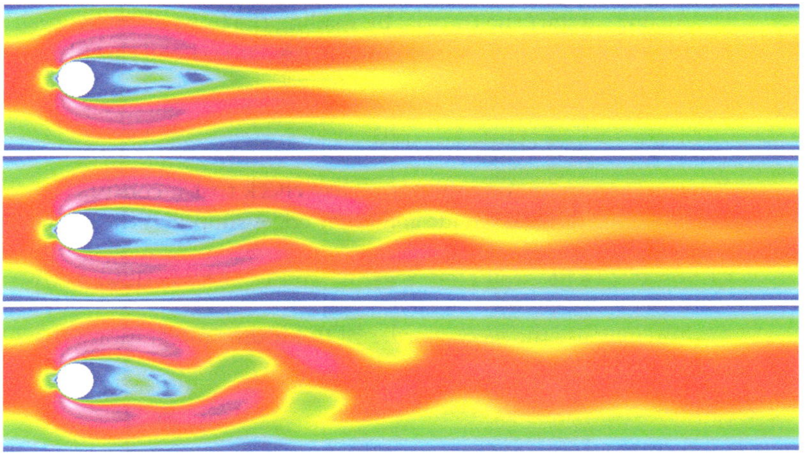

Figure 5.29. *Vortex shedding: velocity magnitude at steps 20, 100, and 200.*

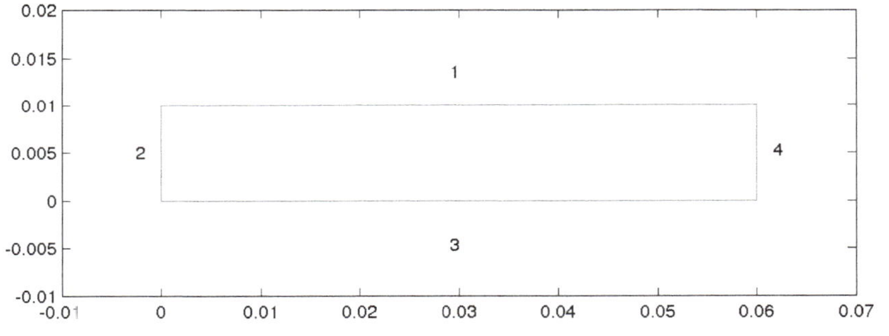

Figure 5.30. *Computational domain for the Rayleigh–Benard problem.*

Table 5.2. *Material parameters of water.*

Parameter	Value
density	998.3 kg/m^3
viscosity	1040e-6 Ns/m^2
heat capacity	4183 J/(kg·K)
heat conductivity	0.58 W/(m·K)
heat expansion coefficient	2.07e-4 K^{-1}
reference temperature	293 K

water. The material parameters are presented in Table 5.2. The temperature difference between the upper and lower boundary is set to 0.5 K. The lower wall has a temperature of 293.5 K and the upper wall has a temperature of 293 K.

The density of water is inversely proportional to its temperature. Due to gravity, warmer/colder water tends to flow upward/downward. In this test case, we assume that the Boussinesq approximation introduced in Chapter 3 is valid for the incompressible fluid flow with thermal effects. That is, the density in the body force term is given by

$$\rho = \rho_0(1 - \beta(T - T_0)),$$

where β is the heat expansion coefficient and the subscript 0 refers to a reference state.

Again, we start the tutorial by loading the finite element mesh which is stored in the *ElmerGrid* format in the file `rectangle.grd`. Load this file:

```
File
  Open -> rectangle.grd
```

You should obtain your mesh and may check that it consists of 3,036 bilinear elements.

The possibility to divide and unify edges simplifies the case definition in the future.

```
Choose (left wall + right wall (Ctrl down)) -> unify edge
```

You may use the second-order BDF for time integration with 200 steps and a step size of two seconds.

```
Model
  Setup
    Simulation Type = Transient
    Steady state max. iter = 20
    Time Stepping Method = bdf
    BDF Order = 2
    Time Step Intervals = 200
    Time Step Sizes = 2.0
    Gravity = ...
```

In the `Equation` section, you can choose the relevant equations and parameters related to their solution. In this test case, we use the model "Natural Convection," which consists of the heat equation and the incompressible Navier–Stokes equations.

When it comes to the definition of equations and materials, it is possible to assign them to bodies (domains) immediately, or use mouse selection to assign them later. In this test case, we have just one domain and, therefore, it is easier to select `Equation` and `Material` directly. Furthermore, it is important to include the convective term which is responsible for the two-way coupling of the Navier–Stokes system with the heat equation.

The system may include nonlinear iterations for each equation and steady-state iterations to obtain convergence of the coupled system. It is often a good idea to keep the number of nonlinear iterations low in coupled problems. Here, you may select just one nonlinear iteration for both equations. For the linear system solvers, you can use the defaults or try out different preconditioners (ILU1,...) or the direct UMFPACK solver, for example.

```
Model
  Equation
    Name = Natural Convection
    Apply to Bodies = 1
    Heat Equation
      Active = on
      Convection = Computed
      Edit Solver Setting
        Nonlinear System
          Max. iterations = 1
    Navier--Stokes
      Active = on
      Edit Solver Setting
```

```
        Nonlinear System
          Max. iterations = 1
    Add
    OK
```

In the `Material` section, we choose water at room temperature from the material library. You may click through the material parameters of the various solvers to ensure that the properties are indeed as they should be. Any consistent set of units may be used in *Elmer*. One natural choice is of course to perform the computations in SI units.

Apart from the properties from the material database, we define the reference temperature for the Boussinesq approximation.

```
Model
  Material
    Material library
      Water (room temperature)
    General
      Reference Temperature = 293
    Apply to Bodies = 1
    Add
    OK
```

The `Body Force` represents the right-hand side of the Navier–Stokes equations. In this test case, its physical meaning is the buoyancy caused by temperature gradients.

```
Model
  Body Force
    Name = Buoyancy
    Apply to Bodies = 1
    Navier--Stokes
      Boussinesq = on
    Add
    OK
```

Initial conditions should be imposed for transient problems. In this test case, you can choose a constant temperature and a small initial velocity that triggers the symmetry break.

```
Model
  Initial Condition
    Name = Initial Guess
    Heat Equation
      Temperature = 293
    Navier--Stokes
      Velocity 1 = 1.0e-9
      Velocity 2 = 0.0
```

On the boundaries, we prescribe zero velocities and known temperature values. The side walls are assumed to be adiabatic.

```
Model
  BoundaryCondition
    Name = Bottom
```

```
Heat Equation
  Temperature = 293.5
Navier--Stokes
  Velocity 1 = 0.0
  Velocity 2 = 0.0
Add
New

Name = Top
Heat Equation
  Temperature = 293
Navier--Stokes
  Velocity 1 = 0.0
  Velocity 2 = 0.0
Add
New

Name = Sides
Navier--Stokes
  Velocity 1 = 0.0
  Velocity 2 = 0.0
Add
```

As usual, you can check your command file

```
Sif
  Generate
  Edit -> look how your command file came out
```

and save the project files in a directory.

```
File
  Save Project
```

Then you can start the solver

```
Run
  Start solver
```

and, finally, view your results by starting the postprocessor.

```
Run
  Start postprocessor
```

In Figs. 5.31 and 5.32, the obtained temperature distribution and the velocity vectors are presented. The maximum velocity should be about 0.516 mm/s in this simulation.

You may try to solve this problem with different parameters. Changing the temperature difference is one way of affecting the instability of the system. As the temperature difference decreases, the system eventually attains a steady state and the vortices vanish. An increase of the temperature difference may increase the number of vortices and eventually lead to a fully chaotic behavior of the system.

The tutorials in this chapter have introduced you to *Elmer* by exemplifying the setup of a mathematical model and numerical solution procedure. A variety of other models

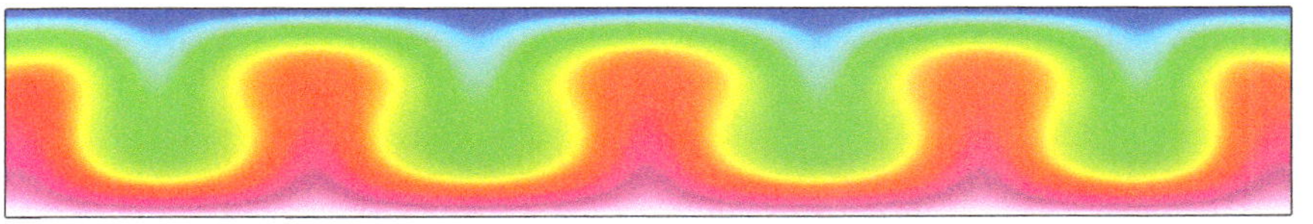

Figure 5.31. *Temperature distribution at 260 s.*

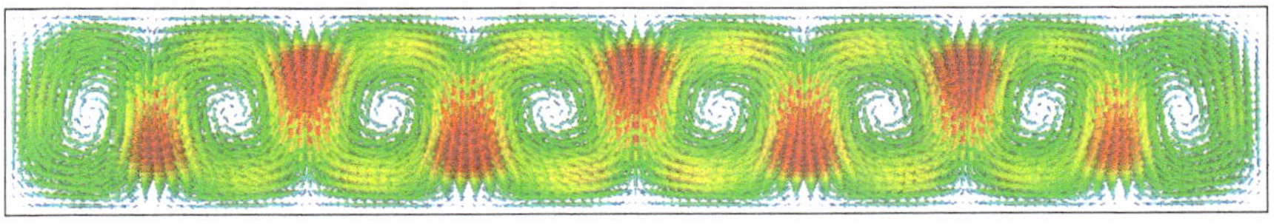

Figure 5.32. *Velocity vectors at 260 s.*

and solution methods can be found in other general-purpose CFD codes. A more detailed description of the physical problems and mathematical models implemented in *Elmer* can be found in the "ElmerModels Manual" on the web pages [62]. In fact, *Elmer* is a multiphysics solver, but we have presented only fluid flow and heat transfer tutorials in this chapter.

Chapter 6

Stabilization Techniques in One Dimension

The Galerkin finite element discretization of the pure diffusion equation enjoys the best approximation property, but the presence of convective effects may give rise to spurious oscillations and numerical instabilities. The lack of stability may result in a complete failure of the simulation run or degrade the rate of convergence to the exact solution. In this chapter, we discuss the numerical treatment of convective terms in the context of 1D model problems discretized using linear finite elements on uniform meshes. Multidimensional extensions and generalizations are presented in the next chapter.

If a finite element scheme turns out to be unstable or produces nonphysical side effects, it can be repaired by modifying the underlying variational formulation or the fully discrete problem. In particular, we will consider artificial diffusion methods, Petrov–Galerkin methods, Taylor–Galerkin methods, and nonlinear high-resolution schemes equipped with flux limiters. As we will see, a proper choice of discretization techniques makes it possible to guarantee positivity preservation and enforce the discrete maximum principle.

Our excursion into the realm of stabilized finite element methods begins in one space dimension. The reasons for this are twofold. First, the design philosophy behind the stabilization techniques to be presented is easier to explain in the 1D case. Second, 1D finite element approximations have the same structure as their finite difference and finite volume counterparts. We will exploit this similarity when it comes to the analysis and design of nonoscillatory FEM for convection-dominated transport problems.

6.1 ▪ Steady Transport Equations

In this section, we consider some finite element approximations and variational stabilization techniques for steady convection-diffusion equations in one dimension. The design of finite element schemes for time-dependent transport equations is addressed in the next section.

We begin with a classical textbook example which illustrates the properties of numerical approximations to the 1D two-point boundary value problem

$$v\frac{\mathrm{d}u}{\mathrm{d}x}-\epsilon\frac{\mathrm{d}^2u}{\mathrm{d}x^2}=0 \qquad \text{in } (0,1), \tag{6.1}$$

$$u(0)=0, \quad u(1)=1, \tag{6.2}$$

where $v>0$ is a constant velocity and $\epsilon>0$ is a constant diffusion coefficient. We use the total derivative notation since (6.1) is, in fact, an ODE.

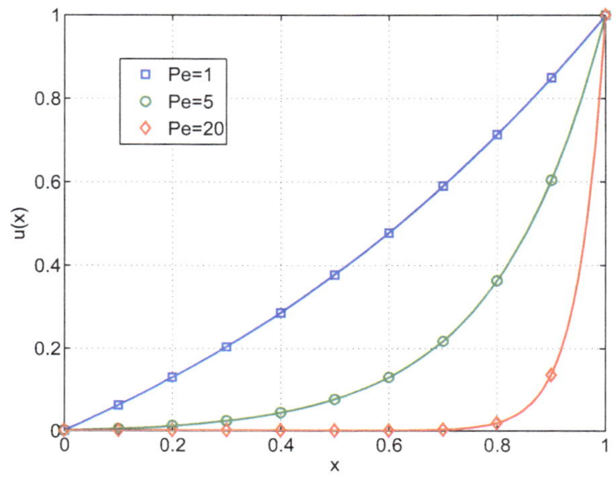

Figure 6.1. *Steady convection-diffusion: exact solution.*

The analytical solution to the above Dirichlet problem is given by the formula

$$u(x) = \frac{e^{\mathrm{Pe}\,x} - 1}{e^{\mathrm{Pe}} - 1}, \tag{6.3}$$

where

$$\mathrm{Pe} = \frac{v}{\epsilon} \tag{6.4}$$

is the Péclet number that corresponds to the characteristic length $L = 1$ (the width of the computational domain). The solution profiles for $\mathrm{Pe} = 1, 5, 20$ are depicted in Fig. 6.1.

In accordance with the maximum principle formulated in Chapter 2, the exact solution is bounded by the prescribed Dirichlet boundary values

$$0 \le u(x) \le 1 \qquad \forall x \in [0,1]. \tag{6.5}$$

The gradient of (6.3) depends on the Péclet number as follows:

$$\frac{\mathrm{d}u}{\mathrm{d}x}(x) = \frac{\mathrm{Pe}\,e^{\mathrm{Pe}\,x}}{e^{\mathrm{Pe}} - 1}. \tag{6.6}$$

In the case $\mathrm{Pe} \gg 1$, a boundary layer develops near the endpoint $x = 1$. We have

$$\frac{\mathrm{d}u}{\mathrm{d}x}(1) = \frac{\mathrm{Pe}\,e^{\mathrm{Pe}}}{e^{\mathrm{Pe}-1}}.$$

Thus, the gradient grows with the Péclet number and becomes infinitely large as the diffusion coefficient goes to zero. This happens because the solution to the reduced problem

$$v\frac{\mathrm{d}u}{\mathrm{d}x} = 0 \qquad \text{in } (0,1), \qquad u(0) = 0 \tag{6.7}$$

is given by $u \equiv 0$ and does not satisfy the boundary condition $u(1) = 1$ imposed in (6.2).

6.1.1 ▪ Galerkin Discretization

Let w be a sufficiently smooth test function satisfying the homogeneous Dirichlet boundary conditions $w(0) = w(1) = 0$. Multiplying (6.1) by w and integrating by parts over $\Omega = (0,1)$, we obtain a variational formulation of our 1D test problem:

$$\int_0^1 \left[vw\frac{du}{dx} + \epsilon \frac{dw}{dx}\frac{du}{dx} \right] dx = 0. \tag{6.8}$$

In this chapter, we restrict ourselves to linear finite elements and uniform meshes. On a mesh that consists of N elements, the global basis function φ_j is defined by

$$\varphi_j(x) = \begin{cases} \frac{x-x_{j-1}}{h} & \text{if } x \in [x_{j-1}, x_j], \\ \frac{x_{j+1}-x}{h} & \text{if } x \in [x_j, x_{j+1}], \\ 0 & \text{otherwise,} \end{cases} \tag{6.9}$$

where $h = \frac{1}{N}$ is the constant mesh size and $x_j = jh$. The approximate solution

$$u_h(x) = \sum_{j=0}^{N} u_j \varphi_j(x), \qquad x \in [0,1], \tag{6.10}$$

is uniquely determined by the $N+1$ nodal values $u_j \approx u(x_j)$. Substituting u_h into

$$\int_0^1 \left[vw_h\frac{du_h}{dx} + \epsilon \frac{dw_h}{dx}\frac{du_h}{dx} \right] dx = 0 \tag{6.11}$$

with $w_h \in \{\varphi_1, \ldots, \varphi_{N-1}\}$, one obtains a system of $N-1$ linear algebraic equations

$$\sum_{j=0}^{N} u_j \left[v \int_0^1 \varphi_i \frac{d\varphi_j}{dx}\, dx + \epsilon \int_0^1 \frac{d\varphi_i}{dx}\frac{d\varphi_j}{dx}\, dx \right] = 0, \tag{6.12}$$

$$i = 1, \ldots, N-1.$$

Two additional equations are provided by the Dirichlet boundary conditions

$$u_0 = 0, \qquad u_N = 1. \tag{6.13}$$

The integrals that appear in (6.12) are calculated element by element using paper and pencil or numerical quadrature. For φ_j given by (6.9), the ith equation becomes

$$\frac{v}{2}(u_{i+1} - u_{i-1}) - \frac{\epsilon}{h}(u_{i-1} - 2u_i + u_{i+1}) = 0. \tag{6.14}$$

Dividing this equation by h, we end up with the central difference discretization

$$v\frac{u_{i+1} - u_{i-1}}{2h} - \epsilon\frac{u_{i-1} - 2u_i + u_{i+1}}{h^2} = 0, \tag{6.15}$$

which is known to be a second-order approximation to (6.1) on uniform meshes.

Fig. 6.2 displays the numerical results for $\mathrm{Pe} = 1, 5, 20$ obtained with 4 linear elements ($h = \frac{1}{4}$). The solutions for $\mathrm{Pe} = 1$ and $\mathrm{Pe} = 5$ satisfy the discrete maximum principle, but the one for $\mathrm{Pe} = 20$ exhibits spurious oscillations and attains negative values at all grid points with odd numbers. As the mesh is refined, the undershoots become smaller and eventually disappear; see Fig. 6.3. The solution calculated on the mesh with $h = \frac{1}{16}$ is nonoscillatory and very close to the analytical solution shown by the dashed line.

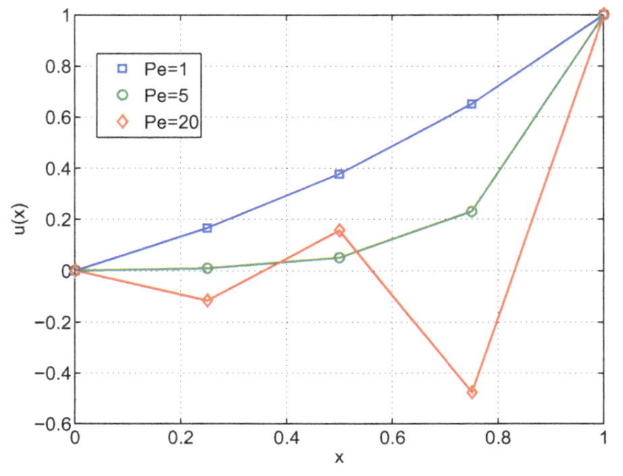

Figure 6.2. *Steady convection-diffusion: $h = \frac{1}{4}$, central difference.*

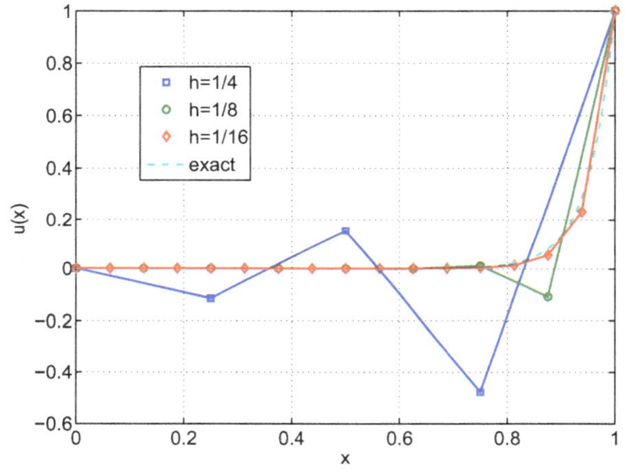

Figure 6.3. *Steady convection-diffusion:* Pe $= 20$, *central difference.*

6.1.2 ▪ Analysis of the Discrete Problem

To see what is wrong with the Galerkin/central difference scheme and find an effective remedy, let us analyze the properties of a generic three-point approximation,

$$a_- u_{i-1} + a_0 u_i + a_+ u_{i+1} = 0, \qquad i = 1,\dots,N-1, \tag{6.16}$$

with constant coefficients a_-, a_0, a_+ satisfying

$$a_- + a_0 + a_+ = 0.$$

Using the fact that $a_0 = -(a_- + a_+)$, we obtain the equivalent representation

$$a_-(u_{i-1} - u_i) + a_+(u_{i+1} - u_i) = 0, \qquad i = 1,\dots,N-1. \tag{6.17}$$

This underdetermined linear system is closed by imposing the boundary conditions

$$u_0 = 0, \qquad u_N = 1. \tag{6.18}$$

In the case $a_- = 0$, $a_+ \neq 0$, equation (6.17) reduces to $u_i = u_{i+1}$. This yields

$$u_0 = 0, \qquad u_1 = \cdots = u_N = 1.$$

The one-sided approximation $a_- \neq 0$, $a_+ = 0$ corresponds to $u_i = u_{i-1}$, whence

$$u_0 = \cdots = u_{N-1} = 0, \qquad u_N = 1.$$

In either case, the solution is bounded by the boundary values 0 and 1, as required by the discrete maximum principle. However, a numerical solution which has the same value at each interior grid point is hardly a good approximation to (6.1).

The solution to (6.17)–(6.18) with $a_- = a_+ \neq 0$ is $u_i = x_i$, $i = 0,\ldots,N$. The general solution to (6.17) with $a_\pm \neq 0$, $a_+ \neq a_-$ is given by [44]

$$u_i = c_1 + c_2\left(\frac{a_-}{a_+}\right)^i, \qquad i = 0,\ldots,N. \tag{6.19}$$

The constants c_1 and c_2 are determined using the Dirichlet boundary conditions

$$c_1 + c_2 = u_0, \qquad c_1 + c_2\left(\frac{a_-}{a_+}\right)^N = u_N.$$

In particular, the boundary values $u_0 = 0$ and $u_N = 1$ correspond to

$$c_1 = -\frac{1}{1-\left(\frac{a_-}{a_+}\right)^N}, \qquad c_2 = \frac{1}{1-\left(\frac{a_-}{a_+}\right)^N}.$$

Substituting these constants into (6.19), we obtain the particular solution

$$u_i = \frac{1-\left(\frac{a_-}{a_+}\right)^i}{1-\left(\frac{a_-}{a_+}\right)^N}, \qquad i = 0,\ldots,N. \tag{6.20}$$

The coefficients of the piecewise-linear Galerkin discretization (6.14) are given by

$$a_- = -\frac{v}{2} - \frac{\epsilon}{h} = -\frac{2+\mathrm{Pe}_h}{2h},$$

$$a_+ = \frac{v}{2} - \frac{\epsilon}{h} = -\frac{2-\mathrm{Pe}_h}{2h},$$

where Pe_h is the mesh Péclet number

$$\mathrm{Pe}_h = \frac{vh}{\epsilon} = \mathrm{Pe}\, h. \tag{6.21}$$

Thus, the sign and magnitude of the ratio $\frac{a_-}{a_+}$ depend on Pe_h. For $|\mathrm{Pe}_h| < 2$, we have

$$a_- < 0, \qquad a_+ < 0$$

with $\frac{a_-}{a_+} \geq 1$ for $0 \leq \mathrm{Pe}_h < 2$ and $0 < \frac{a_-}{a_+} < 1$ for $-2 < \mathrm{Pe}_h < 0$. In either case, the nodal values given by (6.20) satisfy the global discrete maximum principle

$$0 \leq u_i \leq 1, \qquad i = 0,\ldots,N. \tag{6.22}$$

Moreover, the solution values that appear in (6.17) satisfy the local maximum principle

$$\min\{u_{i-1},u_{i+1}\} \le u_i = \frac{a_- u_{i-1}+a_+ u_{i+1}}{a_-+a_+} \le \max\{u_{i-1},u_{i+1}\}.$$

In the case $\mathrm{Pe}_h > 2$, we have $\frac{a_-}{a_+} < -1$ so that the numerator of (6.20) changes its sign at each grid point. This gives rise to spurious oscillations that we have observed in our numerical experiment for $\mathrm{Pe} = 20$ and $h = \frac{1}{4}, \frac{1}{8}$. In the case $\mathrm{Pe}_h < -2$, we have $-1 < \frac{a_-}{a_+} < 0$ so that $u_i \ge 0$ for $i = 0,\dots,N$ but $u_i > 1$ at every other grid point.

It follows that (6.14) satisfies the discrete maximum principle if and only if

$$|\mathrm{Pe}_h| \le 2.$$

This indicates that spurious oscillations can be avoided by using a mesh satisfying

$$h \le \frac{2\epsilon}{|v|}$$

or artificially increasing the value of ϵ in order to decrease the magnitude of Pe_h.

6.1.3 ▪ Artificial Diffusion Method

Many stabilization techniques for finite elements and finite difference approximations boil down to adding a certain amount of artificial diffusion to the governing equation or to the corresponding discrete problem. Consider the modified convection-diffusion equation

$$v\frac{\mathrm{d}u}{\mathrm{d}x} - (\epsilon+\delta)\frac{\mathrm{d}^2u}{\mathrm{d}x^2} = 0, \tag{6.23}$$

where $\delta \ge 0$ is a numerical diffusion coefficient such that $\delta = 0$ in the case of pure diffusion ($v = 0$) and $\delta \to 0$ in the limit $h \to 0$. In what follows, we will use

$$\delta = \alpha\frac{vh}{2}. \tag{6.24}$$

The free parameter α must have the same sign as the velocity v and may depend on Pe_h.

The standard Galerkin discretization of the above problem is given by

$$\frac{v}{2}(u_{i+1}-u_{i-1}) - \left(\frac{\epsilon}{h}+\alpha\frac{v}{2}\right)(u_{i-1}-2u_i+u_{i+1}) = 0, \tag{6.25}$$

which corresponds to the three-point finite difference approximation

$$\frac{v}{2}\left[(1+\alpha)\frac{u_i-u_{i-1}}{h} + (1-\alpha)\frac{u_{i+1}-u_i}{h}\right] - \epsilon\frac{u_{i-1}-2u_i+u_{i+1}}{h^2} = 0. \tag{6.26}$$

In the case $\alpha = 1$, the artificial diffusion term converts a centered approximation of the convective term into the first-order backward difference. This yields

$$v\frac{u_i-u_{i-1}}{h} - \epsilon\frac{u_{i-1}-2u_i+u_{i+1}}{h^2} = 0. \tag{6.27}$$

In the case $\alpha = -1$, the convective part is approximated by the first-order forward difference

$$v\frac{u_{i+1}-u_i}{h}-\epsilon\frac{u_{i-1}-2u_i+u_{i+1}}{h^2}=0. \tag{6.28}$$

The parameter setting $\alpha = \text{sign}(v)$ corresponds to the first-order upwind difference approximation of the convective term. Again, it can be written in the form (6.17) with

$$a_- = -\frac{v+|v|}{2}-\frac{\epsilon}{h}, \qquad a_+ = \frac{v-|v|}{2}-\frac{\epsilon}{h}.$$

In contrast to the central difference scheme ($\alpha = 0$), the solution to this problem satisfies the discrete maximum principle (6.22) regardless of v and h. Indeed, we have

$$a_- < 0, \qquad a_+ < 0.$$

It follows that the numerator and denominator of (6.20) have the same sign (positive for $0 < \frac{a_-}{a_+} < 1$, negative for $\frac{a_-}{a_+} > 1$) and their ratio is bounded above by 1.

The numerical solutions shown in Fig. 6.4 confirm that first-order upwind difference approximation of the convective term ($\alpha = 1$ for $v > 0$) satisfies the discrete maximum principle unconditionally. However, the added artificial diffusion results in a strong smearing of the numerical solution on coarse meshes. Therefore, the use of fine meshes is a must.

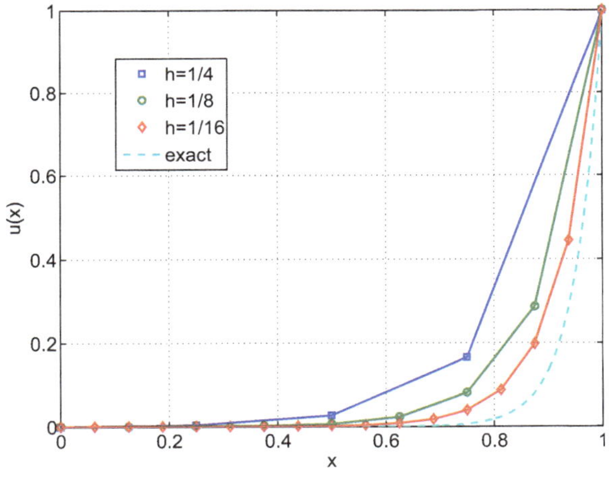

Figure 6.4. *Steady convection-diffusion:* Pe = 20, *upwind difference.*

In order to avoid excess smearing, the value of the mesh Péclet number Pe_h must be taken into account in the formula for the stabilization parameter α. Interestingly enough, the solution to the 1D convection-diffusion equation is nodally exact for

$$\alpha = \coth\left(\frac{\text{Pe}_h}{2}\right)-\frac{2}{\text{Pe}_h}. \tag{6.29}$$

The proof is as follows. Invoking (6.3), we find that the exact nodal values are given by

$$u(x_i) = \frac{e^{i\text{Pe}_h}-1}{e^{N\text{Pe}_h}-1} = u_i.$$

It follows that

$$u_i - u_{i-1} = e^{i\mathrm{Pe}_h}\frac{1-e^{-\mathrm{Pe}_h}}{e^{N\mathrm{Pe}_h}-1},$$

$$u_{i+1} - u_i = e^{i\mathrm{Pe}_h}\frac{e^{\mathrm{Pe}_h}-1}{e^{N\mathrm{Pe}_h}-1}.$$

Substitution into (6.26) reveals that the optimal value of α must satisfy

$$\frac{1+\alpha}{2}(1-e^{-\mathrm{Pe}_h}) + \frac{1-\alpha}{2}(e^{\mathrm{Pe}_h}-1) - \frac{e^{-\mathrm{Pe}_h}-2+e^{\mathrm{Pe}_h}}{\mathrm{Pe}_h} = 0.$$

After some rearrangements, one obtains the following representation of this equation:

$$\underbrace{\frac{e^{\mathrm{Pe}_h}-e^{-\mathrm{Pe}_h}}{2}}_{\sinh \mathrm{Pe}_h} + \left(\alpha + \frac{2}{\mathrm{Pe}_h}\right)\left(1 - \underbrace{\frac{e^{\mathrm{Pe}_h}+e^{-\mathrm{Pe}_h}}{2}}_{\cosh \mathrm{Pe}_h}\right) = 0,$$

which proves that the optimal value of the free parameter α is given by (6.29).

The above result is quite striking at first glance. However, it should be borne in mind that the ability of the "magic formula" (6.29) to produce nodally exact solutions relies on the fact that the coefficients and the mesh size are constant. Nevertheless, this formula is widely used to determine the value of α in more general settings, with varying success.

We remark that (6.29) is commonly approximated by the doubly asymptotic expansion

$$\alpha = \begin{cases} \frac{\mathrm{Pe}_h}{6}, & |\mathrm{Pe}_h| \le 6 \\ \mathrm{sign}(\mathrm{Pe}_h), & |\mathrm{Pe}_h| > 6 \end{cases} = \min\left\{1, \frac{|\mathrm{Pe}_h|}{6}\right\}\mathrm{sign}(\mathrm{Pe}_h), \tag{6.30}$$

as shown in Fig. 6.5. By definition (6.24), the corresponding artificial diffusion coefficient

$$\delta = \min\left\{1, \frac{|v|h}{6\epsilon}\right\}\frac{|v|h}{2} \tag{6.31}$$

is $\mathcal{O}(h^2)$ for $|\mathrm{Pe}_h| \le 6$ and $\mathcal{O}(h)$ for $|\mathrm{Pe}_h| > 6$. Hence, the resultant discretization is second-order accurate for small Pe_h and first-order accurate for large Pe_h.

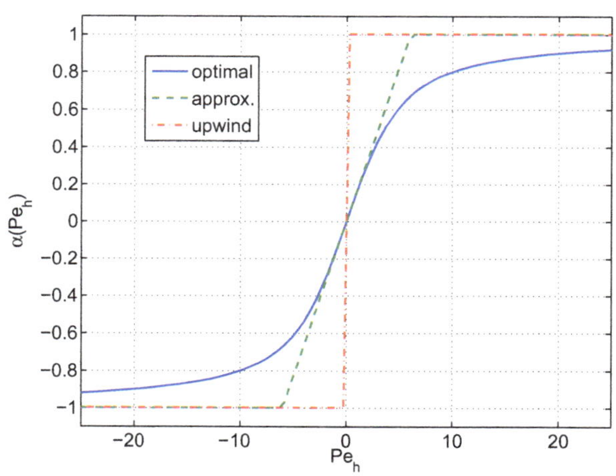

Figure 6.5. *Artificial diffusion parameter:* α *vs.* Pe_h.

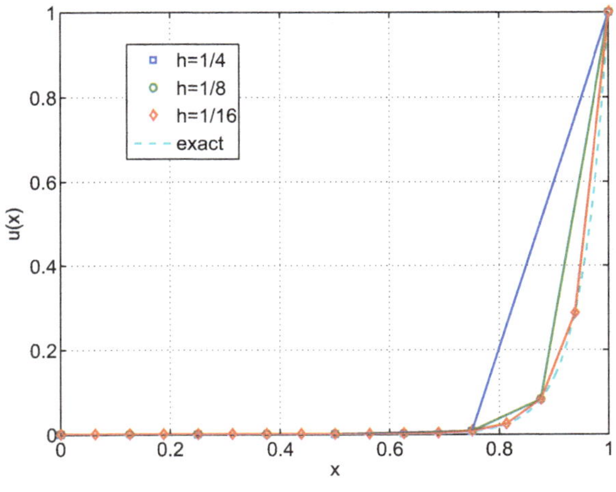

Figure 6.6. *Steady convection-diffusion:* Pe = 20, *optimal* α.

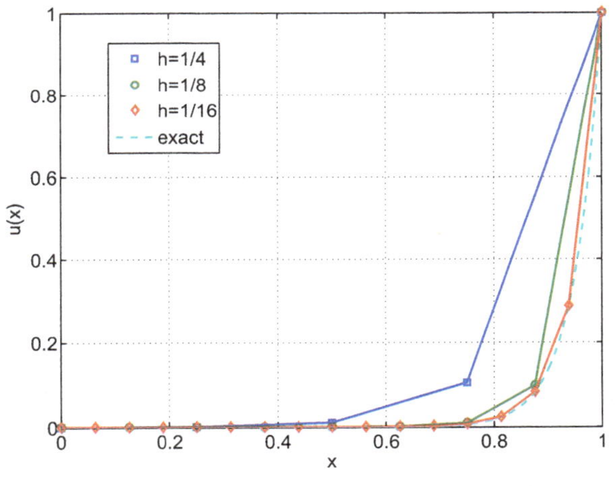

Figure 6.7. *Steady convection-diffusion:* Pe = 20, *approximate* α.

The numerical solutions for Pe = 20 obtained using the Galerkin method stabilized by artificial diffusion based on (6.29) and (6.30) are presented in Figs. 6.6 and 6.7, respectively. As predicted by theory, the optimal value of α delivers exact nodal values, whereas the accuracy of the doubly asymptotic approximation depends on the magnitude of Pe_h.

6.1.4 ▪ Petrov–Galerkin Methods

The variational problem associated with the artificial diffusion method is given by

$$\int_0^1 \left[v w_h \frac{\mathrm{d}u_h}{\mathrm{d}x} + \left(\epsilon + \alpha \frac{v h}{2} \right) \frac{\mathrm{d}w_h}{\mathrm{d}x} \frac{\mathrm{d}u_h}{\mathrm{d}x} \right] \mathrm{d}x = 0. \tag{6.32}$$

Combining the terms that depend on v, we can write (6.32) in the equivalent form

$$\int_0^1 \left[v \tilde{w}_h \frac{\mathrm{d}u_h}{\mathrm{d}x} + \epsilon \frac{\mathrm{d}w_h}{\mathrm{d}x} \frac{\mathrm{d}u_h}{\mathrm{d}x} \right] \mathrm{d}x = 0, \tag{6.33}$$

where $\tilde{w}_h$ is a modified test function incorporating the added artificial diffusion

$$\tilde{w}_h := w_h + \alpha \frac{h}{2} \frac{\mathrm{d}w_h}{\mathrm{d}x}. \tag{6.34}$$

If $w_h = \varphi_i$ is the continuous piecewise-linear test function defined by (6.9), then $\tilde{w}_h = \tilde{\varphi}_i$ is the discontinuous piecewise-linear function depicted in Fig. 6.8. In the case of variable velocities and nonuniform meshes, the stabilization parameter α and the local mesh size h may be defined individually for each element. That is, the amount of artificial diffusion may be adjusted in accordance with the variations of the local mesh Péclet number.

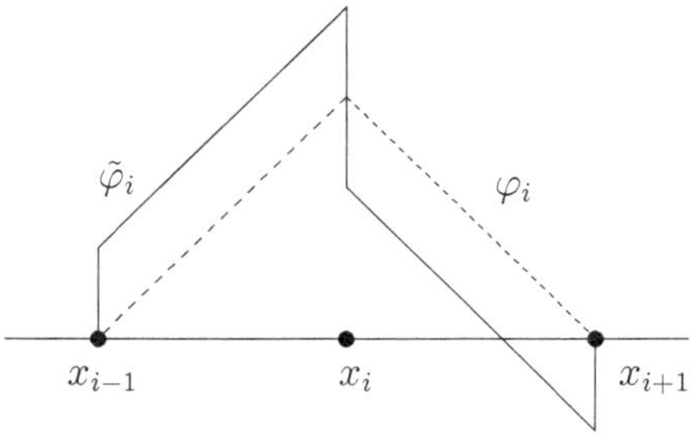

Figure 6.8. *Modified test function,* $\alpha > 0$.

In consistent *Petrov–Galerkin methods* for convection-dominated transport problems, the modified test function $\tilde{w}_h$ is used in all components of the weighted residual

$$\int_0^1 \left[v\tilde{w}_h \frac{\mathrm{d}u_h}{\mathrm{d}x} + \epsilon \frac{\mathrm{d}\tilde{w}_h}{\mathrm{d}x}\frac{\mathrm{d}u_h}{\mathrm{d}x} \right] \mathrm{d}x = 0. \tag{6.35}$$

The so-defined variational problem has the same structure as the standard Galerkin approximation, but the space of test functions is modified so as to achieve the upwinding effect. We will discuss this class of stabilization techniques in more detail in the next chapter.

For linear finite elements, the artificial diffusion method (6.32) and its Petrov–Galerkin counterpart (6.35) are equivalent since $\nabla\tilde{\varphi}_i = \nabla\varphi_i$. The application of these methods to quadratic elements in one dimension is discussed, e.g., in the book by Donea and Huerta [50].

6.2 ▪ Unsteady Transport Equations

Let us now consider the time-dependent convection-diffusion equation which describes the evolution of a scalar quantity $u(x,t)$. The 1D Cauchy problem reads

$$\frac{\partial u}{\partial t} + v\frac{\partial u}{\partial x} - \epsilon \frac{\partial^2 u}{\partial x^2} = 0 \qquad \text{in } \mathbb{R}, \tag{6.36}$$

$$u(x,0) = u_0(x), \qquad x \in \mathbb{R}. \tag{6.37}$$

For the time being, we assume that the velocity v and diffusion coefficient ϵ are constant.

The above model problem can be solved by changing to the Lagrangian coordinates

$$\xi(x,t) := x - vt.$$

Differentiating the solution $u(x,t) = U(\xi(x,t),t)$ by the chain rule, we obtain

$$\frac{\partial u(x,t)}{\partial t} = \frac{\partial U}{\partial t}(\xi(x,t),t) + \frac{\partial U}{\partial \xi}(\xi(x,t),t)\underbrace{\frac{\partial \xi(x,t)}{\partial t}}_{-v},$$

$$\frac{\partial u(x,t)}{\partial x} = \frac{\partial U}{\partial \xi}(\xi(x,t),t), \qquad \frac{\partial^2 u(x,t)}{\partial x^2} = \frac{\partial^2 U}{\partial \xi^2}(\xi(x,t),t).$$

It follows that $U(\xi,t) = u(\xi + vt, t)$ is the solution to the transformed problem

$$\frac{\partial U}{\partial t} - \epsilon \frac{\partial^2 U}{\partial \xi^2} = 0, \qquad U(\xi,0) = u_0(\xi). \tag{6.38}$$

If the diffusion coefficient ϵ is strictly positive, the solution to this problem is

$$U(\xi,t) = \frac{1}{\sqrt{4\pi\epsilon t}} \int_{-\infty}^{\infty} u_0(y) e^{-\frac{(\xi-y)^2}{4\epsilon t}} dy, \qquad t > 0,$$

which corresponds to

$$u(x,t) = \frac{1}{\sqrt{4\pi\epsilon t}} \int_{-\infty}^{\infty} u_0(y) e^{-\frac{(x-vt-y)^2}{4\epsilon t}} dy, \qquad t > 0.$$

If the entire initial mass $m_0 = \int_{-\infty}^{\infty} u_0(x)dx$ is concentrated at a single point $x_0 \in \mathbb{R}$, then $u_0(x) = m_0 \delta(x - x_0)$, where δ is the Dirac delta function. In this case, we have

$$u(x,t) = \frac{m_0}{\sqrt{4\pi\epsilon t}} e^{-\frac{(x-vt-x_0)^2}{4\epsilon t}}, \qquad t > 0,$$

i.e., the exact solution at $t > 0$ is a Gaussian distribution centered at the point $x_0 + vt$.

In the limit of pure convection ($\epsilon = 0$), the Cauchy problem (6.38) reduces to

$$\frac{\partial U}{\partial t} = 0, \qquad U(\xi,0) = u_0(\xi),$$

which implies that the (possibly discontinuous) exact solution

$$u(x,t) := u_0(x - vt) \tag{6.39}$$

remains constant along the characteristic lines given by the formula

$$x = x_0 + vt, \qquad x_0 \in \mathbb{R}.$$

The structure of the solution to the linear convection equation is sketched in Fig. 6.9. For a detailed presentation of this useful model problem, we refer the reader to LeVeque [156].

6.2.1 ▪ Discretization in Space and Time

The variational form of the unsteady convection-diffusion equation to be solved in $\Omega = (0,1)$ is obtained by adding the time derivative to the steady-state problem (6.8). This gives

$$\int_0^1 \left[w\frac{\partial u}{\partial t} + vw\frac{\partial u}{\partial x} + \epsilon \frac{\partial w}{\partial x}\frac{\partial u}{\partial x} \right] dx = 0. \tag{6.40}$$

As before, the space derivatives are discretized using a finite element space spanned by the

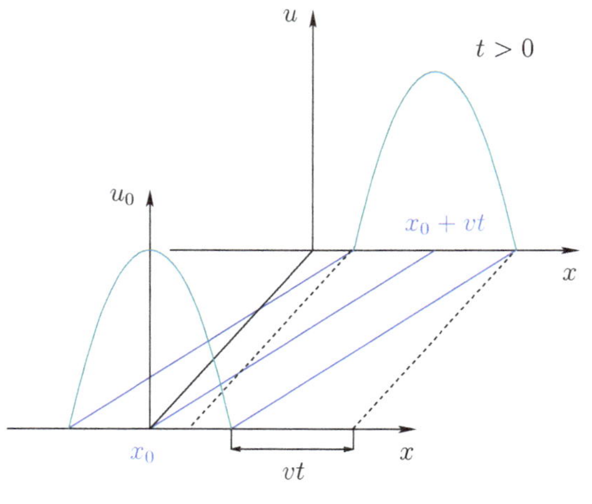

Figure 6.9. *Characteristics of the pure convection equation.*

basis functions $\varphi_0,\ldots,\varphi_N$. The result is a semidiscrete problem of the form

$$\sum_{j=0}^{N} m_{ij}\frac{\mathrm{d}u_j}{\mathrm{d}t}+\sum_{j=0}^{N}\left[vc_{ij}+\epsilon s_{ij}\right]u_j=0,\qquad i=1,\ldots,N-1, \tag{6.41}$$

where

$$m_{ij}=\int_0^1 \varphi_i\varphi_j\,\mathrm{d}x, \tag{6.42}$$

$$c_{ij}=\int_0^1 \varphi_i\frac{\partial\varphi_j}{\partial x}\,\mathrm{d}x, \tag{6.43}$$

$$s_{ij}=\int_0^1 \frac{\partial\varphi_i}{\partial x}\frac{\partial\varphi_j}{\partial x}\,\mathrm{d}x \tag{6.44}$$

are the coefficients of the consistent mass matrix $M_C=\{m_{ij}\}$, the discrete gradient operator $C=\{c_{ij}\}$, and the discrete Laplacian operator $S=\{s_{ij}\}$.

Using a two-level approximation of the time derivative, we obtain a linear system that relates the nodal values of the approximate solution at the time levels t^n and t^{n+1}:

$$Au^{n+1}=Bu^n,\qquad n=0,1,2,\ldots. \tag{6.45}$$

The entries of the matrices $A=\{a_{ij}\}$ and $B=\{b_{ij}\}$ depend on the choice of the time-stepping scheme. For example, the standard two-level θ-scheme corresponds to

$$a_{ij}=\frac{m_{ij}}{\Delta t}+\theta[vc_{ij}+\epsilon s_{ij}], \tag{6.46}$$

$$b_{ij}=\frac{m_{ij}}{\Delta t}-(1-\theta)[vc_{ij}+\epsilon s_{ij}], \tag{6.47}$$

where $\Delta t=t^{n+1}-t^n$ is the time step and $0\le\theta\le 1$. In practice, the implicitness parameter θ is usually used to select one of the following time-stepping schemes:

forward Euler	$\theta=0$,	explicit, first-order,
Crank–Nicolson	$\theta=\frac{1}{2}$,	implicit, second-order,
backward Euler	$\theta=1$,	explicit, first-order.

In the case of a piecewise-linear approximation on a uniform mesh, the basis functions are given by (6.9). The ith equation of the linear system for $\theta = 0$ can be written as

$$\mathcal{M} u_i^{n+1} = \mathcal{M} u_i^n - \Delta t \left[v \frac{u_{i+1}^n - u_{i-1}^n}{2h} - \epsilon \frac{u_{i-1}^n - 2u_i^n + u_{i+1}^n}{h^2} \right], \tag{6.48}$$

$$\mathcal{M} u_i := \frac{u_{i-1} + 4u_i + u_{i+1}}{6}. \tag{6.49}$$

As in the case of the steady convection-diffusion equation, the Galerkin approximation of the convective and diffusive terms is equivalent to the central difference method. However, the time derivative is approximated using the weighted averages $\mathcal{M} u_i$ instead of pointwise values. Hence, the equations of the linear system are coupled even for $\theta = 0$. On the positive side, the use of the consistent mass matrix leads to superconvergent approximations of the convective term (fourth-order spatial accuracy on uniform meshes; see [50, p. 96]).

Replacing $\mathcal{M} u_i$ by u_i, one obtains the fully explicit finite difference scheme

$$u_i^{n+1} = u_i^n - \Delta t \left[v \frac{u_{i+1}^n - u_{i-1}^n}{2h} - \epsilon \frac{u_{i-1}^n - 2u_i^n + u_{i+1}^n}{h^2} \right]. \tag{6.50}$$

The use of u_i in lieu of $\mathcal{M} u_i$ corresponds to approximating M_C by a diagonal matrix M_L which has the same row sums. This trick is known as *row-sum mass lumping* and can be interpreted as inexact calculation of integrals (6.42) by the trapezoidal rule

$$\int_{x_{i-1}}^{x_{i+1}} f(x)\,\mathrm{d}x \approx h \frac{f(x_{i-1}) + f(x_i)}{2} + h \frac{f(x_i) + f(x_{i+1})}{2},$$

where $f(x) = \varphi_i(x)\varphi_j(x)$ so that $f(x_i) = 1$ and $f(x_{i\pm 1}) = 0$ by definition (6.9).

In the first numerical example of this section, we apply the explicit Galerkin method (6.48) and its finite difference counterpart (6.50) to the unsteady convection-diffusion equation (6.37) with $v = 1$. The initial data and exact solution are given by the formula

$$u(x,t) = \frac{0.1}{\sqrt{4\pi(\epsilon t + 0.001)}} \exp\left[-\frac{(x - vt - 0.2)^2}{4(\epsilon t + 0.001)} \right]. \tag{6.51}$$

We restrict ourselves to the domain $\Omega = (0,1)$ and prescribe the boundary conditions

$$u(0) = 0, \qquad \frac{\partial u}{\partial x}(1) = 0. \tag{6.52}$$

The results for $\epsilon = 10^{-3}$ and $t = 0.5$ are shown in Fig. 6.10. To balance the discretization errors due to the first-order accurate forward Euler time stepping and second-order accurate space discretization, we chose $\Delta t = h^2$ in this example. We observe that both the consistent Galerkin method and the central difference scheme produce spurious oscillations on coarse meshes. The consistent-mass approximation is seen to be more accurate on any mesh.

As in the case of the steady convection-diffusion equation, undershoots and overshoots can be eliminated by using a finer mesh. The time step also needs to be refined since an explicit scheme is doomed to become unstable if the time step exceeds a certain upper bound that depends on the mesh size. This means that numerical errors may be amplified until a floating point overflow occurs. The unstable solutions calculated with

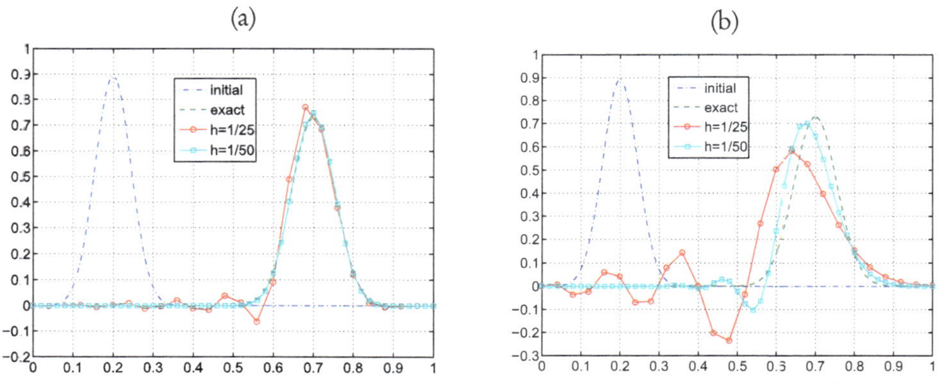

Figure 6.10. *Unsteady convection-diffusion:* (a) *explicit Galerkin method,* (b) *explicit central difference scheme,* $\mathrm{Pe} = 10^3$, $\Delta t = h^2$, $t = 0.5$.

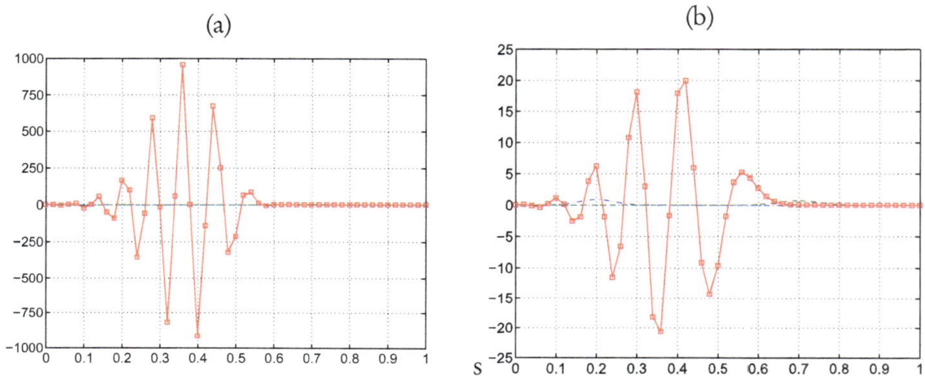

Figure 6.11. *Unsteady convection-diffusion:* (a) *explicit Galerkin method,* (b) *explicit central difference scheme,* $\mathrm{Pe} = 10^3$, $\Delta t = h = 1/50$, $t = 0.5$.

$\Delta t = h = \frac{1}{50}$ are shown in Fig. 6.11. The stability properties of the methods under investigation also depend on the Péclet number. In the case $\epsilon = 0$, the explicit Galerkin/central difference discretization of the convective term is numerically unstable regardless of the time step.

6.2.2 ▪ Stabilized Galerkin Methods

In light of the above, the use of stabilization techniques is a must when it comes to solving unsteady convection problems using linear (and higher-order) finite elements. Our experience with the steady convection-diffusion equation indicates that stability can be enhanced by adding some artificial diffusion or using modified test functions of the form

$$\tilde{w}_h = w_h + \alpha \frac{h}{2} \frac{\partial w}{\partial x}. \tag{6.53}$$

In the classical artificial diffusion method, $\tilde{w}_h$ is used in the convective term only (see Section 6.1.4). Using $\alpha = \mathrm{sign}(v)$ and row-sum mass lumping, we obtain

$$\frac{u_i^{n+1} - u_i^n}{\Delta t} + \left[v \frac{u_{i+1}^n - u_{i-1}^n}{2h} - \left(\epsilon + \frac{|v| h}{2} \right) \frac{u_{i-1}^n - 2u_i^n + u_{i+1}^n}{h^2} \right] = 0, \tag{6.54}$$

which corresponds to the classical upwind difference approximation of the convective term

$$v\frac{u_{i+1}^n - u_{i-1}^n}{2h} + |v|\frac{u_{i-1}^n - 2u_i^n + u_{i+1}^n}{2h} = \begin{cases} v\frac{u_i - u_{i-1}}{h} & \text{if } v \geq 0, \\ v\frac{u_{i+1} - u_i}{h} & \text{if } v \leq 0. \end{cases}$$

In the consistent Petrov–Galerkin method, the modified test function $\tilde{w}_h$ is used in all terms of the variational formulation. The resulting three-point stencil scheme is given by

$$\tilde{\mathcal{M}}\frac{u_i^{n+1} - u_i^n}{\Delta t} + \left[v\frac{u_{i+1}^n - u_{i-1}^n}{2h} - \left(\epsilon + \frac{|v|h}{2}\right)\frac{u_{i-1}^n - 2u_i^n + u_{i+1}^n}{h^2}\right] = 0, \tag{6.55}$$

$$\tilde{\mathcal{M}}u_i := \mathcal{M}u_i - \frac{\alpha}{4}(u_{i+1} - u_{i-1}), \tag{6.56}$$

where $\alpha = \text{sign}(v)$ and $\mathcal{M}$ is the symmetric weighted average operator defined by (6.49).

To give a numerical example which illustrates the accuracy and stability properties of the upwind method, we consider the pure convection equation

$$\frac{\partial u}{\partial t} + \frac{\partial u}{\partial x} = 0 \qquad \text{in } \Omega = (0,1). \tag{6.57}$$

By virtue of (6.39), the exact (weak) solution $u(x,t) = u_0(x - t, t)$ has the same shape as the initial profile. In this example, we will convect a smooth cosine hill,

$$u_1(x,t) = \begin{cases} \frac{1}{2}\left[1 + \cos\left(\pi\frac{x-t-0.2}{0.12}\right)\right] & \text{if } |x - t - 0.2| < 0.12, \\ 0 & \text{otherwise,} \end{cases} \tag{6.58}$$

and a discontinuous step function,

$$u_2(x,t) = \begin{cases} 1 & \text{if } |x - t - 0.2| < 0.1, \\ 0 & \text{otherwise.} \end{cases} \tag{6.59}$$

The inflow boundary condition for both test problems is given by

$$u(0,t) = 0, \qquad t \geq 0.$$

Due to the hyperbolicity, no outflow boundary condition is prescribed at the point $x = 1$.

Fig. 6.12 displays the initial profiles, as well as the exact and numerical solutions at $t = 0.5$. The results presented in Fig. 6.12 were obtained using the lumped-mass upwind approximation (6.54) with $v = 1$, $\epsilon = 0$, and $\Delta t = 0.5h$. There are no spurious oscillations, but the solution for $h = 1/50$ is strongly smeared by numerical diffusion, and even the one for $h = 1/500$ leaves a lot to be desired. Since the numerical diffusion coefficient $\delta = h/2$ is proportional to the mesh size, the rate of convergence to the exact solution is linear, and impractically fine meshes must be used to resolve a smooth peak or a steep gradient. The results produced by the consistent Petrov–Galerkin method are similar (not shown here).

As in the case of unsteady convection-diffusion, the use of sufficiently small steps is a prerequisite for stability of an explicit scheme. If the time step is chosen too large, unbounded oscillations arise, leading to solutions like the ones shown in Fig. 6.11.

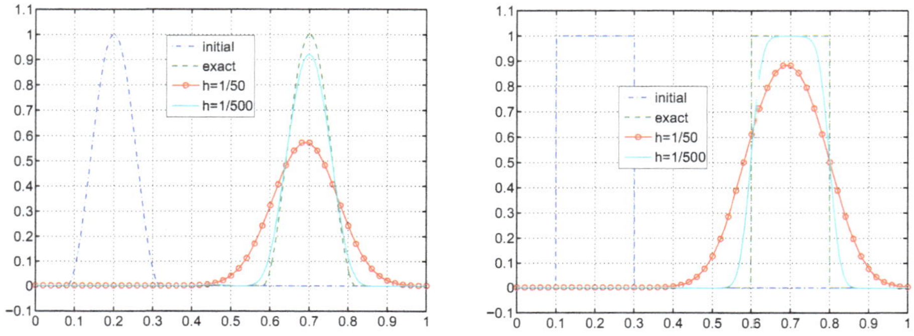

Figure 6.12. *Unsteady convection: lumped mass matrix,* $\delta = h/2$, $\Delta t = \text{C}.5h$, $t = 0.5$.

6.2.3 ▪ Analysis of the Discrete Problem

The above examples demonstrate that stability properties of a numerical scheme and its ability to produce nonoscillatory solutions generally depend on the mesh size, time step, and Péclet number. It is clearly impractical to determine the right values of discretization and stabilization parameters by trial and error. In this section, we present some useful tools for a priori stability analysis, including sufficient conditions of positivity preservation.

Positivity and Maximum Principles

On a uniform mesh, any three-point discretization of the 1D convection-diffusion equation (6.36) with constant coefficients can be written in the generic form

$$a_- u_{i-1}^{n+1} + a_0 u_i^{n+1} + a_+ u_{i+1}^{n+1} = b_- u_{i-1}^n + b_0 u_i^n + b_+ u_{i+1}^n, \qquad (6.60)$$
$$i = 1, \ldots, N-1.$$

We close this system by imposing the homogeneous Dirichlet boundary conditions

$$u_1^{n+1} = u_N^{n+1} = 0. \qquad (6.61)$$

Suppose that the coefficients of our scheme satisfy the following sign conditions:

$$a_- \le 0, \qquad a_0 > 0, \qquad a_+ \le 0, \qquad (6.62)$$

$$b_- \ge 0, \qquad b_0 \ge 0, \qquad b_+ \ge 0, \qquad (6.63)$$

$$a_- + a_0 + a_+ > 0. \qquad (6.64)$$

Then the solution of (6.60)–(6.61) complies with the principle of positivity preservation:

$$u_i^n \ge 0, \quad i = 0, \ldots, N \qquad \Rightarrow \qquad u_i^{n+1} \ge 0, \quad i = 0, \ldots, N. \qquad (6.65)$$

We will prove this fact by contradiction. Suppose that the nodal values u_i^n are all nonnegative but $u_i^{n+1} < 0$ for some i. Let k be the number of a grid point at which the negative global minimum $u_k^{n+1} = \min_i u_i^{n+1}$ is attained. Rearranging (6.60), we obtain

$$a_-(u_{k-1}^{n+1} - u_k^{n+1}) + (a_- + a_0 + a_+) u_k^{n+1} + a_+(u_{k-1}^{n+1} - u_k^{n+1}) = f_i^n,$$

where

$$f_i^n = b_- u_{i-1}^n + b_0 u_i^n + b_+ u_{i+1}^n \ge 0.$$

Since u_k^{n+1} is a global minimum, we have $u_{k\pm1}^{n+1} - u_k^{n+1} \geq 0$. It follows that

$$u_k^{n+1} = \frac{f_i^n - a_-(u_{k-1}^{n+1} - u_k^{n+1}) - a_+(u_{k-1}^{n+1} - u_k^{n+1})}{a_- + a_0 + a_+} \geq 0, \tag{6.66}$$

which contradicts the assumption that the nodal value u_k^{n+1} is strictly negative.

Let us now assume that the coefficients of (6.62) satisfy conditions (6.62)–(6.63) and

$$a_- + a_0 + a_+ = b_- + b_0 + b_+. \tag{6.67}$$

Under this assumption, we will prove the following discrete maximum principle:

$$\min_i u_i^n \leq u_k^{n+1} \leq \max_i u_i^n \qquad \forall k = 0,\dots,N, \tag{6.68}$$

which implies positivity preservation in the case when $u_i^n \geq 0$ for all $i = 0,\dots,N$.

In the case $a_- + a_0 + a_+ = b_- + b_0 + b_+ = 0$, equation (6.60) reduces to

$$u_i^{n+1} = \frac{a_- u_{i-1}^{n+1} + a_+ u_{i+1}^{n+1}}{a_- + a_+}, \qquad i = 1,\dots,N-1.$$

It follows that u_i^{n+1} is a convex combination of u_{i-1}^{n+1} and u_{i+1}^{n+1}. Invoking the boundary conditions (6.61), we obtain the trivial solution $u_i^{n+1} = 0$ for $i = 0,\dots,N$.

In the case $a_- + a_0 + a_+ = b_- + b_0 + b_+ > 0$, we consider the auxiliary grid function $w_i^{n+1} := u_i^{n+1} - \mu^n$, where $\mu^n := \min_i u_i^n$. For any $k = 1,\dots,N-1$, we have

$$\begin{aligned} a_- w_{k-1}^{n-1} + a_0 w_k^{n+1} + a_+ w_{k+1}^{n+1} &= b_- u_{k-1}^n + b_0 u_k^n + b_+ u_{k+1}^n - (a_- + a_0 + a_+)\mu^n \\ &\geq (b_- + b_0 + b_+)\mu^n - (a_- + a_0 + a_+)\mu^n = 0. \end{aligned}$$

If $w_k^{n+1} = \min_i w_i^{n+1}$ is the global minimum value, then $a_\pm(w_{k\pm1}^{n+1} - w_k^{n+1}) \leq 0$, whence

$$\begin{aligned} a_- w_{k-1}^{n+1} + a_0 w_k^{n+1} + a_+ w_{k+1}^{n+1} &= a_-(w_{k-1}^{n+1} - w_k^{n+1}) + a_+(w_{k+1}^{n+1} - w_k^{n+1}) \\ &\quad + (a_- + a_0 + a_+) w_k^{n+1} \leq (a_- + a_0 + a_+) w_k^{n+1}, \end{aligned}$$

where $a_- + a_0 + a_+ > 0$ by assumption. It follows that $u_k^{n+1} - \mu^n = w_k^{n+1} \geq 0$, which proves the lower bound in estimate (6.68). The proof of the upper bound is similar.

Let us verify conditions (6.62)–(6.63) and (6.67) for the Galerkin discretization given by (6.46)–(6.47). On a uniform mesh of linear elements, it reduces to (6.60) with

$$a_0 = \frac{2h}{3\Delta t} + \theta\frac{2\epsilon}{h}, \qquad a_\pm = \frac{h}{6\Delta t} + \theta\left[\pm\frac{v}{2} - \frac{\epsilon}{h}\right],$$

$$b_0 = \frac{2h}{3\Delta t} - (1-\theta)\frac{2\epsilon}{h}, \qquad b_\pm = \frac{h}{6\Delta t} - (1-\theta)\left[\pm\frac{v}{2} - \frac{\epsilon}{h}\right].$$

The values of these coefficients depend on $\mathrm{Pe}_h = \frac{vh}{\epsilon}$ and on the *Courant number*

$$\nu = \frac{v\Delta t}{h}. \tag{6.69}$$

Assuming that $v \neq 0$, we can write the sufficient conditions of the discrete maximum principle as follows:

$$\frac{1}{6}+\theta\nu\left[\pm\frac{1}{2}-\frac{1}{\mathrm{Pe}_h}\right]\leq 0, \qquad \frac{1}{6}-(1-\theta)\nu\left[\pm\frac{1}{2}-\frac{1}{\mathrm{Pe}_h}\right]\geq 0,$$

$$\frac{1}{3}-(1-\theta)\frac{\nu}{\mathrm{Pe}_h}\geq 0.$$

Obviously, the first condition can only be satisfied with $\theta > 0$ and $|\mathrm{Pe}_h| < 2$. The last condition is satisfied automatically for $\theta = 1$ (backward Euler). For any $\theta \in (0,1)$ and $\epsilon > 0$, this condition imposes a stringent upper bound on admissible time steps:

$$\Delta t \leq \frac{1}{1-\theta}\frac{h^2}{3\epsilon}.$$

The coefficients of the explicit lumped-mass Galerkin discretization are given by

$$a_0 = \frac{h}{\Delta t}, \qquad a_\pm = 0, \qquad b_0 = \frac{h}{\Delta t}-\frac{2\epsilon}{h}, \qquad b_\pm = \frac{\epsilon}{h}\pm\frac{v}{2}.$$

The discrete maximum principle holds if $b_0 \geq 0$ and $b_\pm \geq 0$, which corresponds to

$$|\mathrm{Pe}_h| \leq 2, \qquad \frac{\epsilon\Delta t}{h^2}\leq\frac{1}{2}.$$

The right-hand side of the upwind-biased discretization (6.54) is defined by

$$b_0 = \frac{h}{\Delta t}-\frac{2\epsilon}{h}-|v|, \qquad b_\pm = \frac{\epsilon}{h}\pm\frac{v+|v|}{2}.$$

Thus, the discrete maximum principle holds for time steps satisfying

$$\Delta t\left(|v|+\frac{2\epsilon}{h}\right)\leq h.$$

In the limit of vanishing diffusion ($\epsilon = 0$), this constraint reduces to

$$|\nu|\leq 1,$$

the well-known *Courant–Friedrichs–Lewy* (CFL) condition, which is necessary for an explicit discretization of the pure convection equation to be stable [156].

The validity of discrete maximum principle conditions for other explicit and implicit schemes of the form (6.60) can be established in the same way. A numerical method satisfying these conditions is called *monotone* [156]. We remark that monotonicity is sufficient but not necessary for stability. By the Godunov theorem [82], monotonicity constraints impose a first-order accuracy barrier on linear discretizations of the convective term. A monotone linear approximation of the diffusive term can be at most second-order accurate (see [114, pp. 118–120]).

Von Neumann Stability Analysis

If bounded undershoots and overshoots are acceptable, the restrictive monotonicity constraints may be replaced by weaker stability conditions. One of the most popular approaches to analyzing stability of numerical schemes is the *von Neumann method*, which is based on the Fourier series decomposition. To give a brief introduction, we consider the pure Cauchy problem (6.36)–(6.37) with an initial condition u_0 which is assumed to be square integrable and periodic with period 2π. The Fourier series representation of u_0 is given by

$$u_0(x) = \sum_{k=-\infty}^{\infty} c_k e^{ikx}, \qquad x \in \mathbb{R},$$

where i is the imaginary unit, k is the wave number, and c_k is the Fourier coefficient

$$c_k = \frac{1}{2\pi} \int_{-\pi}^{\pi} u_0(x) e^{-ikx}\, dx.$$

The piecewise-linear finite element approximation on a uniform mesh with spacing h is also periodic, and its nodal values are defined by the truncated Fourier series

$$u_j^0 = \sum_{k=-\frac{N}{2}}^{\frac{N}{2}-1} c_k e^{ikx_j} = \sum_{k=-\frac{N}{2}}^{\frac{N}{2}-1} c_k e^{i\theta_k j}, \qquad j \in \mathbb{Z},$$

where

$$x_j = jh, \qquad \theta_k = kh, \qquad N = \frac{2\pi}{h}.$$

By linearity, it is sufficient to perform stability analysis for initial conditions of the form $u_j^0 = e^{i\theta_k j}$. Given a finite difference or finite element approximation of the form (6.60), the *amplification factor* G_k is defined as a complex number such that

$$u_j^0 = e^{i\theta_k j} \quad \forall j \qquad \Rightarrow \qquad u_j^{n+1} = G_k u_j^n = \cdots = G_k^n u_j^0 \quad \forall j.$$

The scheme is stable if it always produces bounded solutions. This is clearly the case if

$$|G_k| \le 1, \qquad k \in \left\{-\frac{N}{2}, \ldots, \frac{N}{2} - 1\right\}.$$

The same argument can be used to prove that errors remain bounded if $|G_k| \le 1$ for all k.

It is easy to verify that the exact solution to (6.36)–(6.37) with $u_0 = e^{ikx}$ is given by

$$u(x,t) = e^{-\epsilon k^2 t} e^{ik(x+vt)}.$$

Hence, the exact amplification factor of the differential equation is given by

$$G_{\text{ex}} = \frac{u(x,t^{n+1})}{u(x,t^n)} = e^{-\epsilon k^2 \Delta t} e^{ikv\Delta t} = e^{-(\delta + i\omega)},$$

where $\delta = \epsilon k^2 \Delta t$ and $\omega = kv\Delta t$ are the exact damping and frequency, respectively.

To quantify the amount of numerical diffusion and dispersion produced by a numerical scheme, the amplification factor $G_{\text{num}} = G_k$ of the Fourier mode $e^{i\theta_k j}$ is written as

$$G_{\text{num}} = e^{-(\delta_{\text{num}} + i\omega_{\text{num}})},$$

where $\delta_{\text{num}} = -\ln|G_{\text{num}}|$ and $\omega_{\text{num}} = -\arg|G_{\text{num}}|$ are the numerical damping and frequency, respectively. The numerical speed of wave propagation is given by

$$\tilde{v} = \frac{\omega_{\text{num}}}{k\Delta t}.$$

The above representation of G_{num} makes it possible to assess the amplitude error

$$E_\delta := \frac{|G_{\text{num}}|}{|G_{\text{ex}}|} = |G_{\text{num}}|e^{\delta}$$

and the phase error

$$E_\omega := \frac{\arg(G_{\text{num}})}{\arg(G_{\text{ex}})} = \frac{\tilde{v}}{v}.$$

If $E_\delta < 1$ for some Fourier component, then the amplitude of the corresponding wave is too small due to excess numerical diffusion. In the case $E_\delta > 1$, the wave amplitude is too large, and numerical instabilities may occur. The value of E_ω measures the error in the speed of wave propagation. The wave is too fast for $E_\omega > 1$ and too slow for $E_\omega < 1$. Spurious oscillations produced by stable schemes are caused by phase errors.

Let us determine the amplification factor G_{num} for the pure convection equation

$$\frac{\partial u}{\partial t} + v\frac{\partial u}{\partial x} = 0 \tag{6.70}$$

discretized using an explicit three-point approximation of the form

$$\frac{u_j^{n+1} - u_j^n}{\Delta t} + \frac{v}{2}\left[(1+\alpha)\frac{u_j^n - u_{j-1}^n}{h} + (1-\alpha)\frac{u_{j+1}^n - u_j^n}{h}\right] = 0.$$

Substituting the trial solution $u_j^n = G_{\text{num}}^n e^{i\theta j}$ into this formula, we obtain

$$(G_{\text{num}}^{n+1} - G_{\text{num}}^n)e^{i\theta j} + \frac{\nu}{2}G_{\text{num}}^n[(1+\alpha)(e^{i\theta j} - e^{i\theta(j-1)}) + (1-\alpha)(e^{i\theta(j+1)} - e^{i\theta j})] = 0,$$

where $\nu = v\frac{\Delta t}{h}$ is the Courant number. Division by $G_{\text{num}}^n e^{i\theta j}$ yields

$$G_{\text{num}} = 1 - \alpha\nu + \frac{\nu}{2}[(1+\alpha)e^{-i\theta} - (1-\alpha)e^{i\theta}].$$

The choice $\alpha = 1$ corresponds to the backward difference approximation with

$$G_{\text{num}} = 1 - \nu + \nu e^{-i\theta} = 1 - \nu + \nu(\cos\theta - i\sin\theta).$$

A sufficient condition of stability is $|G_{\text{num}}| \le 1$ for all phase angles θ. In the case $v > 0$ it reduces to the CFL condition $0 \le \nu \le 1$, which is also necessary for stability. Similarly, the von Neumann stability condition for the forward difference scheme ($\alpha = -1$) is given by $-1 \le \nu \le 0$. The explicit upwind difference scheme ($\alpha = \operatorname{sign}\nu$) is stable for $|\nu| \le 1$, whereas its downwind counterpart ($\alpha = -\operatorname{sign}\nu$) is unstable.

The amplification factor of the central difference scheme ($\alpha = 0$) is

$$G_{\text{num}} = 1 + \frac{\nu}{2}(e^{-i\theta} - e^{i\theta}) = 1 - i\nu\sin\theta.$$

This scheme is unconditionally unstable since $|G_{\text{num}}|^2 = 1 + \nu^2\sin^2\theta > 1$ for $\theta \ne 0$.

Let us now discretize (6.70) using the fully implicit three-point approximation

$$\frac{u_j^{n+1} - u_j^n}{\Delta t} + \frac{v}{2}\left[(1+\alpha)\frac{u_j^{n+1} - u_{j-1}^{n+1}}{h} + (1-\alpha)\frac{u_{j+1}^{n+1} - u_j^{n+1}}{h}\right] = 0.$$

Substitution of $u_j^n = G_{\rm num}^n e^{i\theta j}$ leads to the following equation for the amplification factor:

$$G_{\rm num} = \frac{1}{1+\alpha\nu - \frac{\nu}{2}[(1+\alpha)e^{-i\theta} - (1-\alpha)e^{i\theta}]}. \tag{6.71}$$

The implicit central difference approximation ($\alpha = 0$) is unconditionally stable. We have

$$G_{\rm num} = \frac{1}{1+i\nu\sin\theta}, \qquad |G_{\rm num}|^2 = \frac{1}{1+\nu^2\sin^2\theta} \le 1.$$

The amplification factor of the implicit upwind scheme ($\alpha = \operatorname{sign}\nu$) is given by

$$G_{\rm num} = \frac{1}{1+|\nu|-|\nu|e^{-i\theta}}.$$

Note that $1/G_{\rm num}(\theta)$ is the equation of a circle with center $1+|\nu|$ and radius $|\nu|$. This circle lies outside the unit circle for all ν. It follows that the amplification factor $G_{\rm num}(\theta)$ lies inside the unit circle, which proves unconditional stability [228].

The implicit downwind scheme ($\alpha = -\operatorname{sign}\nu$) has the amplification factor

$$G_{\rm num} = \frac{1}{1-|\nu|+|\nu|e^{i\theta}}.$$

Now $1/G_{\rm num}(\theta)$ is the equation of a circle with center $1-|\nu|$ and radius $|\nu|$. Thus, this scheme is stable for $|\nu| > 1$. However, it is not convergent for convection problems in bounded domains since it does not use the inflow boundary data [228].

Von Neumann stability analysis can only be performed for linear convection-diffusion equations with constant coefficients and periodic boundary conditions. Nevertheless, it represents a very useful tool for the analysis and design of numerical approximations.

The Modified Equation Method

Qualitative properties of numerical schemes can also be investigated using the *modified equation method* [3, 50, 240]. As a model problem, consider the 1D convection equation (6.70) and a numerical scheme which leads to a system of algebraic equations for approximate nodal values $\{u_j^n\}$. Using Taylor series expansions in space and time, one can show that these values coincide with the exact solution of the modified equation

$$\frac{\partial u}{\partial t} + v\frac{\partial u}{\partial x} = \tau_h, \tag{6.72}$$

where the truncation error τ_h depends on spatial and temporal derivatives. Time derivatives of higher than first order can be eliminated by differentiating (6.72), which leads to an equivalent representation of τ_h in terms of spatial derivatives only. Suppose that

$$\tau_h = \sum_{k=1}^{\infty} \mu_{2k}\frac{\partial^{2k}u}{\partial x^{2k}} + \sum_{k=1}^{\infty}\mu_{2k+1}\frac{\partial^{2k+1}u}{\partial x^{2k+1}}, \tag{6.73}$$

where $\mu_k = \mu_k(v,h,\Delta t)$. For (6.72) to be well-posed, the coefficient μ_{2k} of the first nonvanishing even-order derivative must satisfy the necessary stability condition

$$\operatorname{sign}(\mu_{2k}) = (-1)^{k+1}.$$

In particular, the numerical diffusion coefficient μ_2 is required to be nonnegative.

The truncated version of the modified equation gives an insight into the nature of both dissipative and dispersive errors [240]. The even- and odd-order derivatives are associated with amplitude and phase errors, respectively. If the modified equation does not reveal any deficiency, a detailed analysis of stability and accuracy may be performed using the von Neumann method. The derivation of modified equations is a straightforward but tedious exercise. For details and examples, the interested reader is referred to Anderson [3]. In this book, we will present modified equations without derivation.

The modified equation of the explicit consistent-mass Galerkin approximation reads

$$\frac{\partial u}{\partial t}+v\frac{\partial u}{\partial x}=-\frac{v^2 h}{2}\frac{\partial^2 u}{\partial x^2}-\frac{vh^2}{3}\nu^2\frac{\partial^3 u}{\partial x^3}+\cdots.$$

The one for the explicit lumped-mass Galerkin/central difference scheme is similar:

$$\frac{\partial u}{\partial t}+v\frac{\partial u}{\partial x}=-\frac{v^2 h}{2}\frac{\partial^2 u}{\partial x^2}-\frac{vh^2}{6}(1+2\nu^2)\frac{\partial^3 u}{\partial x^3}+\cdots.$$

Both schemes are unstable since the numerical diffusion coefficient $\mu_2=-\frac{v^2 h}{2}$ is negative.

In the case $v>0$, the modified equation of the explicit upwind difference scheme is given by

$$\frac{\partial u}{\partial t}+v\frac{\partial u}{\partial x}=\frac{vh}{2}(1-\nu)\frac{\partial^2 u}{\partial x^2}+\frac{vh^2}{6}(3\nu-2\nu^2-1)\frac{\partial^3 u}{\partial x^3}+\cdots.$$

The numerical diffusion coefficient $\mu_2=\frac{vh}{2}(1-\nu)$ is nonnegative under the CFL condition $0\le\nu\le 1$. Interestingly enough, the truncation error vanishes and the nodally exact solution is obtained in the case $\nu=1$. This phenomenon is known as the *unit CFL property*.

6.2.4 ▪ Lax–Wendroff Method

It turns out that the instability of the explicit Galerkin/central difference scheme can be cured by using a higher-order time discretization. For simplicity, we consider the pure convection equation (6.70) and discretize it in time using the Taylor series expansion

$$u^{n+1}=u^n+\Delta t\, u_t^n+\frac{(\Delta t)^2}{2}u_{tt}^n+\mathcal{O}[(\Delta t)^3]. \tag{6.74}$$

The first time derivative is defined by the governing equation. By (6.70), we have

$$u_t=-v u_x.$$

Differentiating this formula w.r.t. t and interchanging the mixed partials, we obtain

$$u_{tt}=(u_t)_t=(-v u_x)_t=-v(u_t)_x=v^2 u_{xx}.$$

Substituting the time derivatives into (6.74) and neglecting the remainder, we end up with a second-order accurate explicit time discretization known as the *Lax–Wendroff method*:

$$u^{n+1}\approx u^n-v\Delta t\, u_x^n+\frac{(v\Delta t)^2}{2}u_{xx}^n. \tag{6.75}$$

Note that this equation corresponds to the forward Euler time discretization of the artificial diffusion method with the numerical diffusion coefficient $\delta=\frac{v^2\Delta t}{2}$.

The semidiscrete problem (6.75) can be discretized in space using a finite difference or finite element method. The central difference approximation yields

$$\frac{u_i^{n+1}-u_i^n}{\Delta t}+v\frac{u_{i+1}^n-u_{i-1}^n}{2h}=\frac{v^2\Delta t}{2}\frac{u_{i-1}^n-2u_i^n+u_{i+1}^n}{h^2},$$

or, equivalently,

$$u_i^{n+1}=u_i^n-\frac{\nu}{2}(u_{i+1}^n-u_{i-1}^n)+\frac{\nu^2}{2}(u_{i-1}^n-2u_i^n+u_{i+1}^n),$$

where $\nu=v\frac{\Delta t}{h}$ is the Courant number. This scheme is second-order accurate in space and time. It possesses the unit CFL property, and the modified equation is given by [52]

$$\frac{\partial u}{\partial t}+v\frac{\partial u}{\partial x}=-\frac{vh^2}{6}(1-\nu^2)\frac{\partial^3 u}{\partial x^3}-\frac{vh^3}{8}\nu(1-\nu^2)\frac{\partial^4 u}{\partial x^4}+\cdots.$$

It can readily be seen that the CFL condition $\nu^2\leq 1$ is necessary for stability. Using the von Neumann stability analysis, one can verify that it is also sufficient.

The finite element Lax–Wendroff method differs from its finite difference counterpart in the approximation of the time derivative. The fully discrete problem can be written as

$$\mathcal{M}\frac{u_i^{n+1}-u_i^n}{\Delta t}+v\frac{u_{i+1}^n-u_{i-1}^n}{2h}=\frac{v^2\Delta t}{2}\frac{u_{i-1}^n-2u_i^n+u_{i+1}^n}{h^2},$$

where $\mathcal{M}u_i$ is the weighted average defined by (6.49). The modified equation becomes

$$\frac{\partial u}{\partial t}+v\frac{\partial u}{\partial x}=\frac{vh^2}{6}\nu^2\frac{\partial^3 u}{\partial x^3}-\frac{vh^3}{24}\nu(1-3\nu^2)\frac{\partial^4 u}{\partial x^4}+\cdots.$$

In contrast to the finite difference version, this scheme does not possess the unit CFL property and is stable for $\nu^2\leq\frac{1}{3}$ in the 1D case. Moreover, the von Neumann stability condition becomes more restrictive in multidimensions ($\nu^2\leq\frac{1}{24}$ in two dimensions, $\nu^2\leq\frac{1}{81}$ in three dimensions).

6.2.5 ▪ Taylor–Galerkin Methods

The combination of the Galerkin space discretization with higher-order time stepping makes it possible to achieve a further gain of accuracy in the finite element context. The *Taylor-Galerkin* method introduced by Donea [48] is based on the third-order Taylor expansion

$$u^{n+1}=u^n+\Delta t\,u_t^n+\frac{(\Delta t)^2}{2}u_{tt}^n+\frac{(\Delta t)^3}{6}u_{ttt}^n+\mathcal{O}[(\Delta t)^4]. \tag{6.76}$$

Since the presence of third space derivatives would prevent the use of linear finite elements, the third-order term is evaluated using the finite difference approximation

$$u_{ttt}^n=(u_{tt})_t^n=\frac{u_{tt}^{n+1}-u_{tt}^n}{\Delta t}+\mathcal{O}(\Delta t)=v^2\frac{u_{xx}^{n+1}-u_{xx}^n}{\Delta t}+\mathcal{O}(\Delta t).$$

The use of this approximation in the above Taylor expansion leads to the following third-order accurate generalization of the classical Lax–Wendroff method:

$$u^{n+1}\approx u^n-v\Delta t\,u_x^n+\frac{(v\Delta t)^2}{2}u_{xx}^n+\frac{(v\Delta t)^2}{6}(u_{xx}^{n+1}-u_{xx}^n). \tag{6.77}$$

Using linear finite elements to discretize this problem in space, one obtains

$$\tilde{\mathscr{M}}\frac{u_i^{n+1}-u_i^n}{\Delta t}+v\frac{u_{i+1}^n-u_{i-1}^n}{2h}=\frac{v^2\Delta t}{2}\frac{u_{i-1}^n-2u_i^n+u_{i+1}^n}{h^2},$$

where

$$\tilde{\mathscr{M}}u_i=\frac{u_{i+1}+4u_i+u_{i-1}}{6}-\frac{\nu^2}{6}(u_{i-1}-2u_i+u_{i+1}).$$

Thus, the contribution of u_{ttt}^n gives rise to a modification of the consistent mass matrix. All other terms are the same as in the second-order Lax-Wendroff method. Hence, the two methods are equivalent in the steady-state limit, i.e., when $u^{n+1}=u^n$.

The modified equation of the third-order Taylor–Galerkin method reads

$$\frac{\partial u}{\partial t}+v\frac{\partial u}{\partial x}=-\frac{vh^3}{24}\nu(1-\nu^2)\frac{\partial^4 u}{\partial x^4}+\frac{vh^4}{180}(1-5\nu^2+4\nu^4)\frac{\partial^5 u}{\partial x^5}+\cdots.$$

The increased order of the time discretization results in further shrinkage of the local truncation error. The third-order scheme enjoys the unit CFL property, and the von Neumann stability condition $\nu^2\le 1$ is optimal in one dimension. However, the multidimensional stability conditions are still rather restrictive ($\nu^2\le\frac{1}{8}$ in two dimensions, $\nu^2\le\frac{1}{27}$ in three dimensions).

The predictor-corrector approach makes it possible to derive multistep Taylor–Galerkin methods with improved accuracy and stability properties. The two-step algorithm [212]

$$\tilde{u}^n=u^n+\frac{\Delta t}{3}u_t^n+\alpha(\Delta t)^2u_{tt}^n, \tag{6.78}$$

$$u^{n+1}=u^n+\Delta t\,u_t^n+\frac{(\Delta t)^2}{2}\tilde{u}_{tt}^n \tag{6.79}$$

is fourth-order accurate for $\alpha=\frac{1}{12}$ and third-order accurate for any value of the free parameter α. As shown by Quartapelle and Selmin [193], the fourth-order scheme is stable for $\nu^2\le 1$, also in two and three dimensions. The third-order scheme defined by $\alpha=\frac{1}{9}$ has the same phase response as the single-step Taylor–Galerkin method and is stable for $\nu^2\le\frac{3}{4}$.

Many other multistage time-stepping methods can be derived using various factorizations of the Taylor series or the concept of Padé approximations [50, 53]. Unconditional stability can be achieved with an implicit treatment of second-order derivatives. Safjan and Oden [202, 203] introduced a family of semi-implicit multistep Taylor–Galerkin schemes which are of order $2s$ for s stages. A detailed review of the state of the art is beyond the scope of this text. We refer the reader to [50, 53, 192] for a more detailed introduction to the field.

6.2.6 ▪ Lapidus Viscosity

Let us apply the finite difference Lax–Wendroff method (6.75) and the third-order Taylor–Galerkin scheme (6.77) to equation (6.57) with initial conditions given by (6.58) and (6.59). The numerical solutions for $h=\frac{1}{50}$, $\Delta t=\frac{1}{100}$, and $t=0.5$ are presented in Fig. 6.13. It can be seen that the phase accuracy of Taylor–Galerkin is better than that of Lax–Wendroff. A detailed analysis of phase errors can be performed using the von Neumann method [52]. In the case of the smooth cosine hill, the Lax–Wendroff solution has a small undershoot, while the Taylor–Galerkin solution is almost perfect. In the case of the

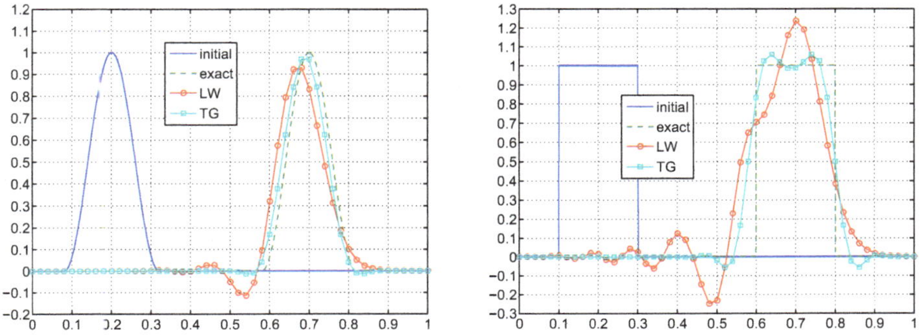

Figure 6.13. *Unsteady convection: Lax-Wendroff (LW) and Taylor-Galerkin (TG) methods without discontinuity capturing;* $h = \frac{1}{50}$, $\Delta t = \frac{1}{100}$, $t = 0.5$.

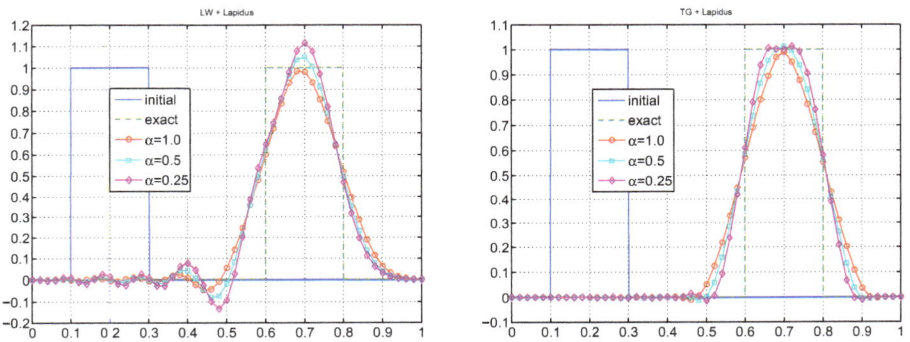

Figure 6.14. *Unsteady convection: Lax-Wendroff and Taylor-Galerkin methods with discontinuity capturing;* $h = \frac{1}{50}$, $\Delta t = \frac{1}{500}$, $t = 0.5$.

discontinuous profile, Taylor–Galerkin is still more accurate than Lax–Wendroff, but both methods produce bounded undershoots and overshoots around the discontinuities.

In contrast to the first-order upwind difference scheme, which is guaranteed to produce nonoscillatory solutions, high-order methods like Lax–Wendroff and Taylor–Galerkin violate the sufficient conditions of positivity preservation, as formulated in Section 6.2.3. In order to suppress spurious oscillations, such schemes must incorporate a certain mechanism for increasing the level of numerical diffusion in the neighborhood of steep fronts. This nonlinear feedback mechanism is called *shock capturing* or *discontinuity capturing*. In the 1D case, the desired effect can be achieved by using artificial diffusion coefficients which depend on the gradients of the transported quantity. A classical example is the *Lapidus viscosity* [150]

$$\delta_L = \alpha_L h^2 \left| \frac{\partial u}{\partial x} \right|, \tag{6.80}$$

where α_L is a user-defined positive constant. In the case of a piecewise-linear finite element approximation, the element matrix associated with the Lapidus viscosity is given by

$$L_j^{(e)} = \alpha_L |u_j - u_{j-1}| \begin{bmatrix} 1 & -1 \\ -1 & 1 \end{bmatrix}.$$

The numerical solutions presented in Fig. 6.14 were obtained using the Lax–Wendroff and Taylor–Galerkin methods with added Lapidus viscosity. Varying the value of α_L, one can reduce the magnitude of spurious oscillations around discontinuities. Unfortunately, this improvement comes at the expense of increasing the dissipative errors. The numerical results for $\alpha_L = 1.0, 0.5$, and 0.25 show the influence of the Lapidus viscosity on the tradeoff between the numerical dispersion and dissipation. We see that no value of α_L guarantees that the discrete maximum principle will hold. Hence, a more sophisticated form of discontinuity capturing is required to eliminate undershoots/overshoots without generating an inordinate amount of artificial diffusion. We will discuss advanced discontinuity capturing techniques in Section 6.3 and extend them to the multidimensional case in the next chapter.

6.2.7 ▪ Operator Splitting Methods

In our introduction to the Lax–Wendroff and Taylor–Galerkin methods, we used the pure convection equation as a model problem. The extension to the convection-diffusion equation

$$\frac{\partial u}{\partial t} + v\frac{\partial u}{\partial x} - \epsilon\frac{\partial^2 u}{\partial x^2} = 0 \tag{6.81}$$

is fairly straightforward. For this second-order PDE, the formula for u_{tt} is given by

$$u_{tt} = (\epsilon u_{xx} - v u_x)_t = \epsilon(u_t)_{xx} - v(u_t)_x,$$

where

$$u_t = \epsilon u_{xx} - v u_x.$$

Hence, the contribution of u_{tt} depends on fourth-order spatial derivatives which call for the use of quadratic finite elements when it comes to the discretization in space.

Another approach to solving (6.81) is based on the use of *operator splitting*. In the simplest case, the solution is advanced in time by solving the pure convection problem

$$\begin{cases} \dfrac{\partial u}{\partial t} + v\dfrac{\partial u}{\partial x} = 0, & t \in (t^n, t^{n+1}), \\ u(t^n) := u^n, & \tilde{u}^n := u(t^{n+1}) \end{cases} \tag{6.82}$$

followed by the pure diffusion problem

$$\begin{cases} \dfrac{\partial u}{\partial t} - \epsilon\dfrac{\partial^2 u}{\partial x^2} = 0, & t \in (t^n, t^{n+1}), \\ u(t^n) := \tilde{u}^n, & u^{n+1} := u(t^{n+1}). \end{cases} \tag{6.83}$$

This particular decomposition is known as the *Marchuk–Yanenko* method, and the splitting error is $\mathcal{O}(\Delta t)$. Since the splitting is performed at the differential level, each subproblem must be endowed with boundary conditions that ensure its well-posedness.

The pure convection equation is hyperbolic. Hence, the solution values or fluxes should be prescribed on the inflow boundary. The resulting initial-boundary value problem can be solved, e.g., using the Lax–Wendroff or Taylor–Galerkin method. The pure diffusion equation is parabolic, so a boundary condition is required at each point on the boundary of the domain. The combination of the standard Galerkin approximation with the implicit Crank–Nicolson or backward Euler time stepping is unconditionally stable for this subproblem.

A second-order operator splitting scheme can be designed using *Strang symmetrization*. Its application to (6.81) leads to the following three-step algorithm:

$$\begin{cases} \dfrac{\partial u}{\partial t} - \epsilon \dfrac{\partial^2 u}{\partial x^2} = 0, & t \in (t^n, t^{n+1/2}), \\ u(t^n) := u^n, & \tilde{u}^{n+1/2} := u(t^{n+1/2}). \end{cases} \tag{6.84}$$

$$\begin{cases} \dfrac{\partial u}{\partial t} + v \dfrac{\partial u}{\partial x} = 0, & t \in (t^n, t^{n+1}), \\ u(t^n) := \tilde{u}^{n+1/2}, & u^{n+1/2} := u(t^{n+1}). \end{cases} \tag{6.85}$$

$$\begin{cases} \dfrac{\partial u}{\partial t} - \epsilon \dfrac{\partial^2 u}{\partial x^2} = 0, & t \in (t^{n+1/2}, t^{n+1}), \\ u(t^{n+1/2}) := u^{n+1/2}, & u^{n+1} := u(t^{n+1}). \end{cases} \tag{6.86}$$

Again, the choice of boundary conditions for each subproblem depends on the PDE type, and custom-made discretization techniques may be employed for each subproblem.

Source terms can also be included using operator splitting techniques. The segregated treatment of different physical effects makes it possible to develop simulation software in a modular way using optimal methods and discretization parameters for each subproblem. We remark that fractional-step algorithms based on operator splitting are generally less robust than monolithic discretizations and may require the use of small time steps. For a detailed introduction to the field and analysis of existing schemes, we refer the reader to [80, 114, 195].

6.3 ▪ High-Resolution Schemes

In many practical applications, even small undershoots and overshoots may result in a crash of the code or give rise to physically unrealistic solutions. In this section, we will be concerned with the design of nonlinear high-resolution schemes which are guaranteed to satisfy the discrete maximum principle and have a low level of numerical diffusion. The underlying design philosophy is quite simple: the difference between a high-order scheme and its nonoscillatory low-order counterpart is decomposed into numerical fluxes, and the magnitude of these fluxes is adjusted so as to construct an optimal combination of the two methods. The concept of *flux correction* was introduced in the context of finite difference and finite volume approximations. As we will see, it can be extended to finite elements and constitutes an attractive alternative to parameter-dependent discontinuity capturing techniques.

6.3.1 ▪ Flux-Based Approximations

We consider, for a 1D model problem, the scalar conservation law

$$\frac{\partial u}{\partial t} + \frac{\partial f(u)}{\partial x} = 0, \quad x \in \mathbb{R}, \quad t > 0, \tag{6.87}$$

$$u(x,0) = u_0(x), \tag{6.88}$$

where $f(u)$ is a given *flux function*. The linear convection model corresponds to

$$f(u) = vu.$$

In general, the dependence of $f(u)$ on u may be nonlinear. As a consequence, the weak solution to (6.87) may become discontinuous even if the initial data u_0 is smooth [156].

Multiplying (6.87) by a test function w and integrating over $\Omega = \mathbb{R}$, we obtain

$$\int_{\mathbb{R}} \left[w \frac{\partial u}{\partial t} - f(u) \frac{\partial w}{\partial x} \right] \mathrm{d}x = 0, \tag{6.89}$$

where we have shifted the space derivative from $f(u)$ to w using integration by parts.

Let us discretize in space using linear finite elements. As before, the space of numerical solutions is spanned by a set of basis functions $\{\varphi_j\}$ defined by (6.9). We have

$$u_h(x,t) = \sum_{j=0}^{N} u_j(t)\varphi_j(x). \tag{6.90}$$

In the standard FEM, the flux $f(u)$ is approximated by $f(u_h)$. Alternatively, one may use a mixed finite element formulation in which the numerical approximation $f_h(u_h, x)$ is defined as a linear combination of piecewise-constant basis functions

$$f_h(u_h, x) = \sum_{j=0}^{N-1} f_{j+1/2}(u_h)\psi_{j+1/2}(x), \tag{6.91}$$

$$\psi_{j+1/2}(x) := \begin{cases} 1 & \text{if } x \in (x_j, x_{j+1}), \\ 0 & \text{otherwise.} \end{cases} \tag{6.92}$$

The result is a semidiscrete scheme which can be written in the form

$$\mathcal{M} \frac{\mathrm{d}u_i}{\mathrm{d}t} + \frac{f_{i+1/2} - f_{i-1/2}}{h} = 0, \tag{6.93}$$

where $\mathcal{M}$ is defined by (6.49). The corresponding lumped-mass approximation

$$\frac{\mathrm{d}u_i}{\mathrm{d}t} + \frac{f_{i+1/2} - f_{i-1/2}}{h} = 0 \tag{6.94}$$

exhibits the same structure as a conservative finite difference/finite volume scheme.

To obtain a closed system in which the number of equations equals the number of unknowns, we must define the value of $f_{i+1/2}$ in terms of u_h. The natural choice

$$f_{i+1/2}^{CD} = \frac{f(u_{i+1}) + f(u_i)}{2} \tag{6.95}$$

corresponds to the second-order central difference approximation. The fluxes

$$f_{i+1/2}^{FD} = f(u_{i+1}), \tag{6.96}$$

$$f_{i+1/2}^{BD} = f(u_i) \tag{6.97}$$

are associated with the first-order forward and backward difference, respectively.

The characteristic velocity of wave propagation $v(u) = f'(u)$ is approximated by

$$v_{i+1/2} = \begin{cases} \frac{f(u_{i+1}) - f(u_i)}{u_{i+1} - u_i} & \text{if } u_{i+1} \neq u_i, \\ f'(u_i) & \text{if } u_{i+1} = u_i. \end{cases} \tag{6.98}$$

The numerical flux of the first-order upwind approximation is given by

$$f_{i+1/2}^{UD} = \begin{cases} f_{i+1/2}^{BD} & \text{if } v_{i+1/2} \geq 0, \\ f_{i+1/2}^{FD} & \text{if } v_{i+1/2} \leq 0 \end{cases} \tag{6.99}$$

and admits the equivalent representation

$$f_{i+1/2}^{UD} = f_{i+1/2}^{CD} - \frac{|v_{i+1/2}|}{2}(u_{i+1} - u_i). \tag{6.100}$$

Hence, the upwinding effect can be achieved by adding artificial diffusion to $f_{i+1/2}^{CD}$.

In a similar vein, the generalized Lax–Wendroff method is defined by the flux

$$f_{i+1/2}^{LW} = \frac{f(u_{i+1}) + f(u_i)}{2} - |\nu_{i+1/2}|\frac{|v_{i+1/2}|}{2}(u_{i+1} - u_i), \tag{6.101}$$

where

$$\nu_{i+1/2} = v_{i+1/2}\frac{\Delta t}{h}$$

is the Courant number based on the local characteristic velocity. Invoking (6.100), we find that the Lax–Wendroff flux can be obtained by adding *antidiffusion* to the upwind flux:

$$f_{i+1/2}^{LW} = f_{i+1/2}^{UD} + (1 - |\nu_{i+1/2}|)\frac{|v_{i+1/2}|}{2}(u_{i+1} - u_i). \tag{6.102}$$

The flux-based representation of the third-order Taylor–Galerkin method is defined by

$$f_{i+1/2}^{TG} = f_{i+1/2}^{LW} + \frac{1}{6}(1 - \nu_{i+1/2}^2)\left[\frac{u_{i+1}^{n+1} - u_{i+1}^n}{\Delta t} - \frac{u_i^{n+1} - u_i^n}{\Delta t}\right]. \tag{6.103}$$

We leave it as an exercise for the reader to verify that the methods analyzed in the previous sections are recovered in the case of the linear convection equation ($f(u) = vu$).

6.3.2 ▪ Flux-Corrected Transport

In modern high-resolution schemes, monotonicity constraints are enforced by blending high- and low-order approximations in an adaptive manner. This methodology traces its origins to the flux-corrected transport (FCT) algorithm of Boris and Book [18, 17, 19], generalized by Zalesak [245, 246, 247]. The classical version of FCT is a predictor-corrector algorithm in which a *transported and diffused* solution is corrected by adding limited antidiffusive fluxes. The limiting procedure is designed to minimize the level of numerical diffusion while keeping the solution free of undershoots/overshoots.

The main ingredients of an explicit FCT scheme are a high-order approximation,

$$u_i^H = u_i^n - \frac{\Delta t}{h}[f_{i+1/2}^H - f_{i-1/2}^H], \tag{6.104}$$

and a nonoscillatory low-order approximation,

$$u_i^L = u_i^n - \frac{\Delta t}{h}[f_{i+1/2}^L - f_{i-1/2}^L]. \tag{6.105}$$

The difference between the high- and low-order solutions depends on the difference between the fluxes $f^H_{i+1/2}$ and $f^L_{i+1/2}$. Subtracting (6.105) from (6.104), we obtain

$$u^H_i = u^L_i - \frac{\Delta t}{h}[f^{AD}_{i+1/2} - f^{AD}_{i-1/2}], \tag{6.106}$$

where

$$f^{AD}_{i+1/2} = f^H_{i+1/2} - f^L_{i+1/2}$$

is the *antidiffusive* flux which must be added to $f^L_{i+1/2}$ to obtain $f^H_{i+1/2}$. For example, the pair of approximations $f^H_{i+1/2} := f^{LW}_{i+1/2}$ and $f^L_{i+1/2} := f^{UD}_{i+1/2}$ corresponds to

$$f^{AD}_{i+1/2} = (1 - |\nu_{i+1/2}|)\frac{|v_{i+1/2}|}{2}(u^n_{i+1} - u^n_i).$$

If we replace $f^{LW}_{i+1/2}$ by $f^{TG}_{i+1/2}$, the antidiffusive flux $f^{AD}_{i+1/2}$ will include the contribution of the modified mass matrix, which should be evaluated using u^H in place of u^{n+1}.

In order to prevent the birth and growth of spurious oscillations, each raw antidiffusive flux is multiplied by a correction factor $\alpha_{i+1/2} \in [0, 1]$. The final solution is given by

$$u^{n+1}_i = u^L_i - \frac{\Delta t}{h}[\alpha_{i+1/2} f^{AD}_{i+1/2} - \alpha_{i-1/2} f^{AD}_{i-1/2}]. \tag{6.107}$$

By definition of $f^{AD}_{i+1/2}$, the so-defined FCT scheme can be written in the form

$$u^{n+1}_i = u^n_i - \frac{\Delta t}{h}[f^{FCT}_{i+1/2} - f^{FCT}_{i-1/2}], \tag{6.108}$$

where

$$f^{FCT}_{i+1/2} = \alpha_{i+1/2} f^H_{i+1/2} + (1 - \alpha_{i+1/2}) f^L_{i+1/2}. \tag{6.109}$$

The choice $\alpha_{i+1/2} = 1$ corresponds to the high-order approximation, while $\alpha_{i+1/2} = 0$ corresponds to the low-order approximation. Hence, the value of $\alpha_{i+1/2}$ should be chosen as close to 1 as possible without generating an undershoot/overshoot at node i or $i + 1$.

The algorithm for calculating the correction factors $\alpha_{i+1/2}$ is called a *flux limiter*. The FCT limiting strategy is designed to guarantee that the antidiffusive fluxes cannot conspire to create a new local maximum or minimum. That is, the definition of $\alpha_{i+1/2}$ should guarantee that the solution value u^{n+1}_i given by (6.107) satisfies the inequality constraints

$$u^{\min}_i \le u^{n+1}_i \le u^{\max}_i, \tag{6.110}$$

where $u^{\min}_i$ and $u^{\max}_i$ represent the local extrema of u^L. In the 1D case, we have

$$u^{\max}_i = \max\{u^L_{i-1}, u^L_i, u^L_{i+1}\},$$

$$u^{\min}_i = \min\{u^L_{i-1}, u^L_i, u^L_{i+1}\}.$$

In the worst-case scenario, the fluxes $f^{AD}_{i+1/2}$ and $-f^{AD}_{i-1/2}$ have the same sign. The maximum and minimum increments to the low-order solution value u^L_i are given by

$$P^+_i = \frac{\Delta t}{h}[\max\{0, -f^{AD}_{i+1/2}\} + \max\{0, f^{AD}_{i-1/2}\}],$$

$$P^-_i = \frac{\Delta t}{h}[\min\{0, -f^{AD}_{i+1/2}\} + \min\{0, f^{AD}_{i-1/2}\}].$$

In accordance with (6.110), the bounds for positive and negative increments are defined by

$$Q_i^+ = u_i^{\max} - u_i^L, \qquad Q_i^- = u_i^{\min} - u_i^L.$$

The unconstrained high-order approximation satisfies (6.110) if $P_i^+ \leq Q_i^+$ and $P_i^- \geq Q_i^-$.

In order to control the magnitude of positive and/or negative increments, we introduce

$$R_i^+ = \min\left\{1, \frac{Q_i^+}{P_i^+}\right\}, \qquad R_i^- = \min\left\{1, \frac{Q_i^-}{P_i^-}\right\}$$

and define the correction factor $\alpha_{i+1/2}$ using Zalesak's [245, 247] symmetric formula

$$\alpha_{i+1/2} = \begin{cases} \min\{R_{i+1}^+, R_i^-\} & \text{if } f_{i+1/2}^{AD} \geq 0, \\ \min\{R_{i+1}^-, R_i^+\} & \text{if } f_{i+1/2}^{AD} < 0. \end{cases}$$

This limiting strategy guarantees that the flux-corrected solution (6.107) satisfies (6.110).

In summary, the predictor-corrector FCT algorithm involves the following steps:

- Calculate the low-order solution u^L and determine its local extrema $u_i^{\max}_{\min}$.
- Calculate the antidiffusive fluxes $f_{i+1/2}^{AD}$ and the correction factors $\alpha_{i+1/2}$.
- Add the sum of limited antidiffusive fluxes to the low-order predictor u^L.

To show how FCT works, we apply it to our test problem (6.57)–(6.59). Fig. 6.15 displays the numerical solutions calculated using the Lax–Wendroff and Taylor–Galerkin methods as high-order schemes for FCT. In both cases, the first-order upwind approximation was used to calculate the low-order predictor u^L. The flux-corrected solutions are free of spurious oscillations, and the level of numerical diffusion is remarkably low. Since the unconstrained Taylor–Galerkin method produces smaller ripples than Lax–Wendroff, the overall accuracy of the Taylor–Galerkin FCT approximation is higher than that of Lax–Wendroff FCT. The results for the smooth cosine hill illustrate a well-known drawback of FCT: the *peak clipping* phenomenon [16]. Since the flux limiter is designed to prevent formation of new maxima and minima, it always switches to the low-order approximation at a local extremum.

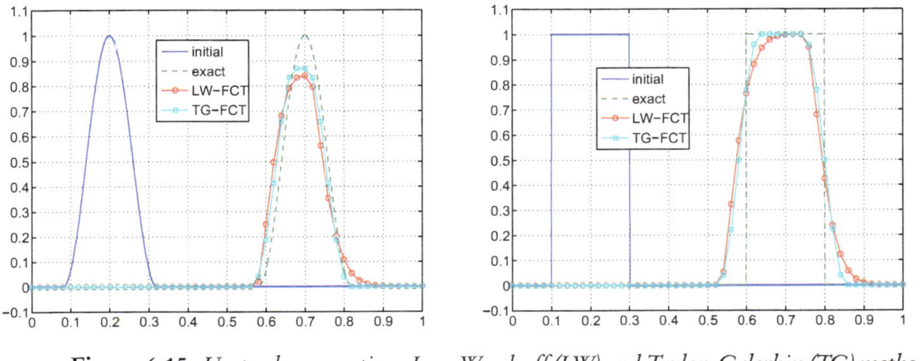

Figure 6.15. *Unsteady convection: Lax–Wendroff (LW) and Taylor–Galerkin (TG) methods with FCT;* $h = \frac{1}{50}$, $\Delta t = \frac{1}{100}$, $t = 0.5$.

6.3.3 ▪ Total Variation Diminishing Methods

In the classical FCT algorithm, the bounds are defined as the local extrema of the low-order predictor. Another approach to the design of nonlinear high-resolution schemes is based on the concept of *total variation stability* [98, 99, 219]. Consider a semidiscrete scheme of the form (6.94). The total variation of the numerical solution is defined by

$$TV(u_h(t)) = \sum_{i=-\infty}^{\infty} |u_i(t) - u_{i-1}(t)|.$$

The space discretization technique is called *total variation diminishing* (TVD) if

$$\frac{\mathrm{d}}{\mathrm{d}t} TV(u_h(t)) \leq 0, \qquad t \geq 0.$$

At the fully discrete level, the scheme is said to possess the TVD property if

$$TV(u_h^{n+1}) \leq TV(u_h^n), \qquad n = 0, 1, 2, \ldots.$$

It is easy to show that an explicit scheme is TVD if it can be written as [98]

$$u_i^{n+1} = u_i^n + c_{i-1/2}(u_{i-1}^n - u_i^n) + c_{i+1/2}(u_{i+1}^n - u_i^n), \tag{6.111}$$

$$c_{i+1/2} \geq 0, \qquad c_{i-1/2} \geq 0, \qquad c_{i+1/2} + c_{i-1/2} \leq 1. \tag{6.112}$$

Note that the same conditions are sufficient for the scheme to be positivity-preserving and satisfy the discrete maximum principle formulated in Section 6.2.3.

The first-order upwind discretization of the linear convection equation is TVD under the usual CFL condition. By the Godunov theorem [82], no linear approximation of second or higher order can satisfy the above sign conditions. Hence, the coefficients of a higher-order TVD scheme must depend on the solution. As in the case of FCT, the corresponding flux function is constructed by blending high- and low-order approximations:

$$f_{i+1/2}^{TVD} = \alpha_{i+1/2} f_{i+1/2}^{H} + (1 - \alpha_{i+1/2}) f_{i+1/2}^{L}. \tag{6.113}$$

The correction factor $\alpha_{i+1/2} \in [0, 2]$ is defined so that the resulting scheme

$$u_i^{n+1} = u_i^n - \frac{\Delta t}{h}[f_{i+1/2}^{TVD} - f_{i-1/2}^{TVD}] \tag{6.114}$$

admits representation (6.111) with solution-dependent coefficients $c_{i\pm 1/2}$ satisfying (6.112).

We will illustrate the design of TVD schemes using the linear convection equation

$$\frac{\partial u}{\partial t} + v\frac{\partial u}{\partial x} = 0, \qquad v > 0, \tag{6.115}$$

as a model problem. An extension to general conservation laws will be presented later.

Let us use the first-order upwind method as the low-order scheme and the Lax–Wendroff method as the high-order scheme. The corresponding fluxes are given by

$$f_{i+1/2}^{L} = v u_i^n, \qquad f_{i+1/2}^{H} = v u_i^n + (1 - \nu)\frac{v}{2}(u_{i+1}^n - u_i^n).$$

Using these fluxes to define $f_{i+1/2}^{TVD}$ in equations (6.113) and (6.114), we obtain

$$u_i^{n+1} = u_i^n - \nu(u_i^n - u_{i-1}^n) - (1 - \nu)\frac{\nu}{2}[\alpha_{i+1/2}(u_{i+1}^n - u_i^n) - \alpha_{i-1/2}(u_i^n - u_{i-1}^n)]. \tag{6.116}$$

This scheme reduces to the Lax–Wendroff method for $\alpha_{i\pm 1/2} = 1$ and to the upwind method for $\alpha_{i\pm 1/2} = 0$. The optimal choice of $\alpha_{i\pm 1/2}$ depends on the local smoothness of the solution, which can be measured using the ratio of two consecutive gradients:

$$r_i = \frac{u_i - u_{i-1}}{u_{i+1} - u_i}.$$

The *slope ratio* r_i is negative at a local extremum, where $\alpha_{i\pm 1/2} = 0$ should be used for safety reasons. In the case $r_i = 1$, the solution is linear and it is safe to use $\alpha_{i\pm 1/2} = 1$.

Using the blending factor $\alpha_{i+1/2} = r_i$, one obtains the second-order accurate *Beam–Warming method*. The corresponding flux function $f^{BW}_{i+1/2}$ is given by

$$f^{BW}_{i+1/2} = v u_i^n + (1-\nu)\frac{v}{2}(u_i^n - u_{i-1}^n).$$

Any linear combination of $f^{LW}_{i+1/2}$ and $f^{BW}_{i+1/2}$ also leads to a second-order accurate scheme.

A nonlinear TVD scheme discretization of (6.115) can be constructed using

$$\alpha_{i+1/2} = \Phi(r_i),$$

where Φ is a suitable limiter function which guarantees the equivalence of (6.116) to (6.111) with solution-dependent coefficients $c_{i+1/2}$ satisfying the TVD constraints (6.112).

It is natural to require that $\Phi(r_i) = 0$ for $r_i \le 0$. This criterion implies that the low-order flux will be used at a local extremum. For $r_i > 0$, we can write (6.116) thus:

$$u_i^{n+1} = u_i^n - \nu(u_i^n - u_{i-1}^n) - (1-\nu)\frac{\nu}{2}\left[\frac{\Phi(r_i^n)}{r_i^n} - \Phi(r_{i-1}^n)\right](u_i^n - u_{i-1}^n),$$

which corresponds to (6.111) with

$$c_{i+1/2} = 0, \qquad c_{i-1/2} = \nu + (1-\nu)\frac{\nu}{2}\left[\frac{\Phi(r_i^n)}{r_i^n} - \Phi(r_{i-1}^n)\right].$$

The scheme is TVD if $0 \le c_{i-1/2} \le 1$. To ensure this, $\Phi(r)$ is required to satisfy [156]

$$0 \le \Phi(r) \le 2, \qquad 0 \le \frac{\Phi(r)}{r} \le 2 \qquad \forall r > 0.$$

In summary, the criterion for the design of a TVD limiter function Φ is as follows:

$$0 \le \Phi(r) \le \max\{0, \min\{2, 2r\}\}.$$

Thus, a linear second-order scheme defined by $\hat{\Phi}(r)$ can be constrained by using

$$\Phi(r) = \max\{0, \min\{2, \hat{\Phi}(r), 2r\}\}.$$

Some of the most popular TVD limiter functions are listed in Table 6.1. All of them lie in the *second-order TVD region*, which is bounded above by the graph of the superbee limiter and below by the graph of the minmod limiter; see Fig. 6.16.

Table 6.1. *Standard TVD limiters* [156, 219].

	$\hat{\Phi}(r)$	$\Phi(r)$
minmod	$\min\{1,r\}$	$\max\{0,\min\{1,r\}\}$
van Leer	$\frac{2r}{1+r}$	$\frac{r+\lvert r\rvert}{1+\lvert r\rvert}$
monotonized centered (MC)	$\frac{1+r}{2}$	$\max\{0,\min\{2,\frac{1+r}{2},2r\}$
superbee	$\max\{1,r\}$	$\max\{0,\min\{1,2r\},\min\{2,r\}\}$

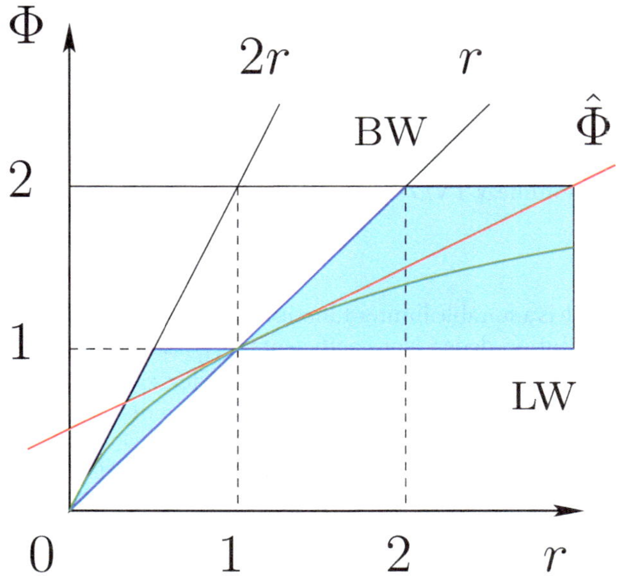

Figure 6.16. *Second-order TVD region* [156, 219].

As shown by Jameson [116], standard TVD limiter functions can be interpreted and implemented as *limited average operators* $\mathscr{L}(\cdot,\cdot)$ such that $\Phi(r)=\mathscr{L}(1,r)$. These two-parameter functions are characterized by the following common properties:

$$\text{(P1)}\qquad \mathscr{L}(a,b)=\mathscr{L}(b,a).$$

$$\text{(P2)}\qquad \mathscr{L}(ca,cb)=c\mathscr{L}(a,b).$$

$$\text{(P3)}\qquad \mathscr{L}(a,a)=a.$$

$$\text{(P4)}\qquad \mathscr{L}(a,b)=0 \quad \text{if } ab\leq 0.$$

In particular, the last requirement ensures that $\Phi(r)=0$ if $r\leq 0$. Thus, the accuracy of a TVD discretization inevitably degrades to first order at a local maximum or minimum.

Another important implication is the symmetry of the flux limiting procedure

$$\Phi(r)=\mathscr{L}(1,r)=r\mathscr{L}(1/r,1)=r\Phi(1/r), \tag{6.117}$$

which follows from (P1) and (P2). In essence, this property implies that an antidiffusive flux from node $i+1$ into node i is proportional to the limited average of two slopes and has the same effect as a diffusive flux from node $i-1$ since $\Phi(1/r_i)$ is nonnegative:

$$\Phi(r_i)(u_{i+1}-u_i)=\mathscr{L}(u_{i+1}-u_i,u_i-u_{i-1})=\Phi(1/r_i)(u_i-u_{i-1}). \qquad (6.118)$$

The numerical flux (6.113) for a TVD scheme defined by $\mathscr{L}(\cdot,\cdot)$ is given by

$$f_{i+1/2}^{TVD}=vu_i^n+(1-\nu)\frac{v}{2}\mathscr{L}(u_{i+1}-u_i,u_i-u_{i-1}).$$

The two-parameter representation of standard TVD limiters is as follows [116]:

minmod: $\mathscr{L}(a,b)=\mathscr{S}(a,b)\min\{|a|,|b|\}$,

van Leer: $\mathscr{L}(a,b)=\mathscr{S}(a,b)\dfrac{2|a||b|}{|a|+|b|}$,

MC: $\mathscr{L}(a,b)=\mathscr{S}(a,b)\min\left\{\dfrac{|a+b|}{2},2|a|,2|b|\right\}$,

superbee: $\mathscr{L}(a,b)=\mathscr{S}(a,b)\max\{\min\{2|a|,|b|\},\min\{|a|,2|b|\}\}$,

where

$$\mathscr{S}(a,b)=\frac{\operatorname{sign}(a)+\operatorname{sign}(b)}{2}=\begin{cases}1 & \text{if } a>0 \wedge b>0,\\ -1 & \text{if } a<0 \wedge b<0,\\ 0 & \text{otherwise.}\end{cases}$$

The results for our test problem (6.57)–(6.59) are displayed in Fig. 6.17. The most diffusive numerical solution is produced by the first-order upwind scheme, which corresponds to $\Phi(r)\equiv 0$. In nonlinear TVD schemes, the level of numerical diffusion depends on the choice of limiter. Since minmod is too diffusive and superbee is too antidiffusive, limiters like MC and van Leer are usually employed in practical applications.

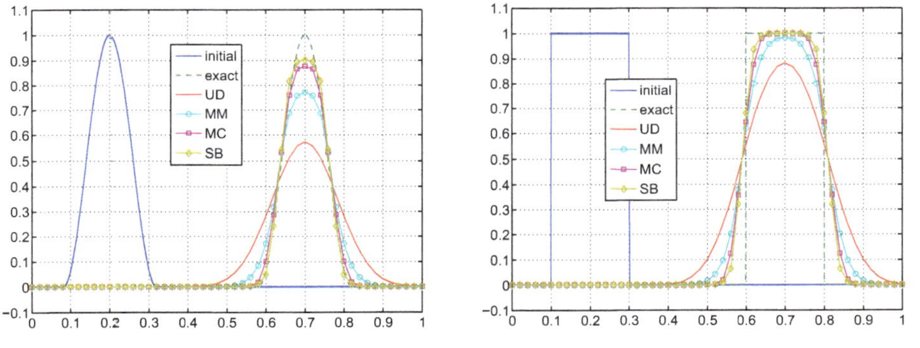

Figure 6.17. *Unsteady convection: Upwind vs. TVD;* $h=\frac{1}{50}$, $\Delta t=\frac{1}{100}$, $t=0.5$.

For a general conservation law of the form (6.87), the TVD flux is defined by [156]

$$f_{i+1/2}^{TVD}=f_{i+1/2}^{UD}+(1-|\nu_{i+1/2}|)\frac{|v_{i+1/2}|}{2}\mathscr{L}(u_{i+1}-u_i,u_j-u_{j-1}),$$

where $|v_{i+1/2}|$ is the local characteristic velocity defined by (6.98), $f_{i+1/2}^{UD}$ is the generalized upwind flux defined by (6.100), and j is the number of the upwind node:

$$j=i-\operatorname{sign}(v_{i+1/2}).$$

Let us apply the generalized TVD scheme to the inviscid Burgers equation

$$\frac{\partial u}{\partial t} + \frac{\partial f(u)}{\partial x} = 0, \qquad f(u) = \frac{u^2}{2}, \tag{6.119}$$

with the discontinuous initial condition

$$u_0(x) = \begin{cases} 1 & \text{if } x \in [0.1, 0.6], \\ 0 & \text{otherwise.} \end{cases} \tag{6.120}$$

The exact solution to this nonlinear problem can be determined as explained in [156]. It consists of a smooth rarefaction wave and a shock propagating at speed $s = 0.5$.

The numerical solutions shown in Fig. 6.18 were calculated using the first-order upwind scheme and the TVD scheme equipped with the MC limiter. In both cases, the results are nonoscillatory. The upwind solution is strongly smeared by numerical diffusion. The TVD scheme resolves the rarefaction wave much better because the MC limiter corresponds to the second-order accurate approximation $\hat{\Phi}(r) = \frac{1+r}{2}$ in smooth regions.

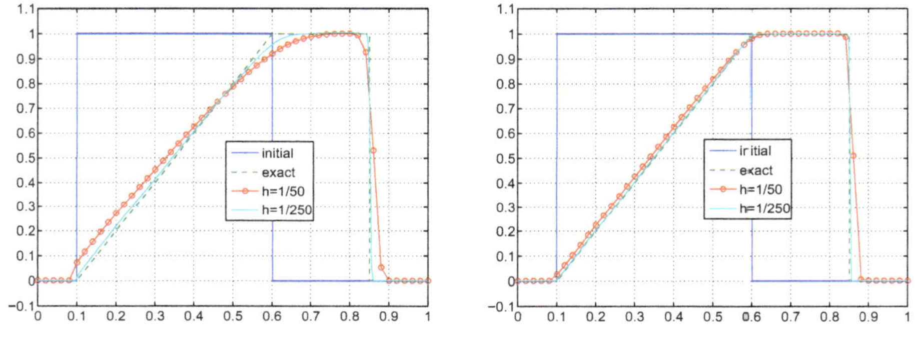

Figure 6.18. *Burgers equation: Upwind vs. TVD;* $\Delta t = 0.5h$, $t = 0.5$.

6.4 ▪ Summary

In this chapter, we discussed the design of finite element approximations to 1D convection-dominated transport problems. To simplify the presentation of basic concepts, we restricted ourselves to uniform meshes, linear finite elements, and explicit time discretizations. Moreover, we wrote the resulting schemes in the finite difference notation to take advantage of existing similarities and analyze the properties of the discrete problem. The reader is encouraged to consult further texts [104, 114, 156, 157, 228, 241] for additional insights into the development of advanced numerical methods for conservation laws.

Many additional issues arise when it comes to the generalization of the presented algorithms to multidimensional problems, unstructured meshes, high-order approximations, and implicit time-stepping schemes. We will address these issues in the next chapter.

Chapter 7

Stabilization Techniques in Multidimensions

The numerical treatment of convective terms becomes more involved when it comes to solving multidimensional transport equations on unstructured meshes. In this chapter, we review various extensions and generalizations of the 1D algorithms presented so far. As in Chapter 6, we consider both classical stabilization techniques for continuous finite elements and advanced discontinuity capturing techniques, including a family of nonlinear high-resolution schemes based on the concept of algebraic flux correction. Furthermore, we introduce the group finite element approach to discretization of flux functions in general conservation laws and present edge-based data structures which offer a number of significant advantages compared to traditional element-based implementations.

7.1 ▪ Variational Formulation

Let $\Omega \subset \mathbb{R}^d$ be a bounded domain with a piecewise-smooth boundary $\Gamma = \partial\Omega$. The most general model problem to be considered in this chapter is the conservation law

$$\frac{\partial u}{\partial t} + \nabla \cdot \mathbf{f} = s \qquad \text{in } \Omega, \tag{7.1}$$

where u is a conserved scalar quantity, s is a source term, and $\mathbf{f}$ is a given flux function.

A typical set of initial and boundary conditions for problem (7.1) is as follows:

$$u = u_0 \qquad \text{at } t = 0, \tag{7.2}$$

$$u = g \qquad \text{on } \Gamma_D \subseteq \Gamma, \tag{7.3}$$

$$\mathbf{f} \cdot \mathbf{n} = h \qquad \text{on } \Gamma_N \subseteq \Gamma, \tag{7.4}$$

where $\mathbf{n}$ denotes the unit outward normal vector. If equation (7.1) is of hyperbolic type, then $\Gamma_- = \Gamma_D \cup \Gamma_N$ should be the inflow boundary of Ω. In the presence of diffusive fluxes, a boundary condition must be prescribed at each point on $\Gamma = \Gamma_D \cup \Gamma_N$.

The weak form of the above problem is given by the weighted residual formulation

$$\int_\Omega w \left[\frac{\partial u}{\partial t} + \nabla \cdot \mathbf{f} \right] d\mathbf{x} = \int_\Omega w s \, d\mathbf{x}. \tag{7.5}$$

This relation must hold for any test function w vanishing on Γ_D. In the case $\Gamma_N \neq \emptyset$, we integrate by parts and substitute the flux boundary condition (7.4). This yields

$$\int_\Omega \left[w \frac{\partial u}{\partial t} - \nabla w \cdot \mathbf{f} \right] d\mathbf{x} = \int_\Omega w s \, d\mathbf{x} - \int_{\Gamma_N} w h \, ds. \tag{7.6}$$

To begin with, we consider a linear convection-diffusion equation,

$$\frac{\partial u}{\partial t}+\nabla\cdot(\mathbf{v}u-\epsilon\nabla u)=s \qquad \text{in } \Omega, \tag{7.7}$$

where $\mathbf{f}=\mathbf{v}u-\epsilon\nabla u$ is the linear flux function derived in Chapter 1. The velocity field $\mathbf{v}$ and the constant diffusion coefficient $\epsilon>0$ are given.

The flux boundary conditions for our linear model problem are defined as follows:

$$(\mathbf{v}u-\epsilon\nabla u)\cdot\mathbf{n}=(\mathbf{v}g)\cdot\mathbf{n} \qquad \text{on } \Gamma_-, \tag{7.8}$$

$$\epsilon\nabla u\cdot\mathbf{n}=0 \qquad \text{on } \Gamma\backslash\Gamma_-. \tag{7.9}$$

This definition corresponds to setting the diffusive flux equal to zero on Γ and calculating the convective flux across $\Gamma_-:=\{\mathbf{x}\in\Gamma\,|\,\mathbf{v}\cdot\mathbf{n}<0\}$ using the boundary value $u=g$.

Using (7.8)–(7.9) to define the normal flux $h=\mathbf{f}\cdot\mathbf{n}$ in equation (7.6), we obtain

$$\int_\Omega\left[w\frac{\partial u}{\partial t}-\nabla w\cdot(\mathbf{v}u-\epsilon\nabla u)\right]d\mathbf{x}+\int_{\Gamma\backslash\Gamma_-}w(\mathbf{v}u)\cdot\mathbf{n}\,ds$$
$$=\int_\Omega ws\,d\mathbf{x}-\int_{\Gamma_-}w(\mathbf{v}g)\cdot\mathbf{n}\,ds, \tag{7.10}$$

where the surface integrals represent the contribution of the weakly imposed flux boundary conditions. The choice $w\equiv 1$ corresponds to the integral form of the conservation law.

Integrating by parts in the convective term, one obtains the equivalent representation

$$\int_\Omega\left[w\frac{\partial u}{\partial t}+w\nabla\cdot(\mathbf{v}u)+\epsilon\nabla w\cdot\nabla u\right]d\mathbf{x}-\int_{\Gamma_-}w(\mathbf{v}u)\cdot\mathbf{n}\,ds$$
$$=\int_\Omega ws\,d\mathbf{x}-\int_{\Gamma_-}w(\mathbf{v}g)\cdot\mathbf{n}\,ds.$$

Thus, the variational form of our initial-boundary value problem can be written as

$$\left(w,\frac{\partial u}{\partial t}\right)+a(u,w)=b(w) \qquad \forall w\in\mathcal{W} \tag{7.11}$$

where $\mathcal{W}$ is the space of admissible test functions and $(\cdot,\cdot)$ is the L^2 scalar product, i.e.,

$$\left(w,\frac{\partial u}{\partial t}\right)=\int_\Omega w\frac{\partial u}{\partial t}\,d\mathbf{x}.$$

The bilinear form $a(\cdot,\cdot)$ and linear functional $b(\cdot)$ are defined by

$$a(w,u)=\int_\Omega w\nabla\cdot(\mathbf{v}u)\,d\mathbf{x}+\epsilon\int_\Omega\nabla w\cdot\nabla u\,d\mathbf{x}-\int_{\Gamma_-}w(\mathbf{v}u)\cdot\mathbf{n}\,ds, \tag{7.12}$$

$$b(w)=\int_\Omega ws\,d\mathbf{x}-\int_{\Gamma_-}w(\mathbf{v}g)\cdot\mathbf{n}\,ds. \tag{7.13}$$

Variational problems for other boundary conditions may be derived in a similar way.

7.2 ▪ Galerkin Discretization

Let us discretize (7.11) in space using finite elements. Let $\{\varphi_i\}$ be a set of basis functions spanning the finite-dimensional space $\mathcal{V}_h$. The numerical solution $u_h \in \mathcal{V}_h$ is defined by

$$u_h = \sum_j u_j \varphi_j. \tag{7.14}$$

When it comes to differentiating u_h, time derivatives are applied to u_j, whereas space derivatives are applied to φ_j. For example, the gradient of u_h is given by

$$\nabla u_h = \sum_j u_j \nabla \varphi_j. \tag{7.15}$$

The straightforward approach to discretization of the convective flux $\mathbf{v}u$ involves the multiplication of u_h by the finite element interpolant of the velocity

$$\mathbf{v}_h = \sum_j \mathbf{v}_j \varphi_j. \tag{7.16}$$

For the time being, the nodal values $\mathbf{v}_j$ are assumed to be known. In real-life applications, they are usually calculated by solving a momentum balance equation.

Instead of interpolating each variable individually, it is possible to interpolate the nodal values of a flux function that depends on a group of variables. This approach is known as the *group finite element formulation* [71, 72, 73]. It leads to simpler discretized equations and offers a significant reduction of the computational cost, especially in applications to nonlinear problems. In particular, this unconventional approximation technique provides a natural treatment of inviscid fluxes in conservation laws. It turns out that the approximation of the product is often more accurate than the product of approximations [72, 73].

The group finite element approximation of the convective flux is given by

$$(\mathbf{v}u)_h = \sum_j (\mathbf{v}_j u_j)\varphi_j. \tag{7.17}$$

Obviously, this function is easier to differentiate than the product $\mathbf{v}_h u_h$. For example, the divergence of the convective flux that appears in (7.12) is approximated by

$$\nabla \cdot (\mathbf{v}u)_h = \sum_j u_j(\mathbf{v}_j \cdot \nabla \varphi_j). \tag{7.18}$$

The resultant semidiscrete problem can be written in the generic form

$$\left(w_h, \frac{\partial u_h}{\partial t}\right) + a(u_h, w_h) = b(w_h) \qquad \forall w_h \in \mathcal{W}_h, \tag{7.19}$$

where $\mathcal{W}_h$ is the finite-dimensional space of test functions and

$$a(w_h, u_h) = \int_\Omega w_h \nabla \cdot (\mathbf{v}u)_h \,\mathrm{d}\mathbf{x} + \epsilon \int_\Omega \nabla w_h \cdot \nabla u_h \,\mathrm{d}\mathbf{x} - \int_{\Gamma_-} w_h (\mathbf{v}u)_h \cdot \mathbf{n} \,\mathrm{d}s$$

is the bilinear form associated with the group finite element approximation (7.17).

Using $w_h \in \{\varphi_i\}$, one obtains a sparse system of differential-algebraic equations,

$$M_C \frac{\mathrm{d}u}{\mathrm{d}t} = Ku + r, \tag{7.20}$$

where $M_C = \{m_{ij}\}$ is the consistent mass matrix, $K = \{k_{ij}\}$ is the discrete transport operator, and $r = \{r_i\}$ is the vector of discretized source terms. By (7.13) we have

$$r_i = b(\varphi_i) = \int_\Omega \varphi_i s \, d\mathbf{x} - \int_{\Gamma_-} \varphi_i(\mathbf{v}g)\cdot\mathbf{n}\, ds. \tag{7.21}$$

By (7.14), (7.15), and (7.18), the entries of the matrices M_C and K are given by

$$m_{ij} = (\varphi_i, \varphi_j) = \int_\Omega \varphi_i \varphi_j \, d\mathbf{x}, \tag{7.22}$$

$$k_{ij} = -a(\varphi_i, \varphi_j) = \int_{\Gamma_-} \varphi_i \varphi_j \mathbf{v}\cdot\mathbf{n}\, ds - \mathbf{c}_{ij}\cdot\mathbf{v}_j - \varepsilon s_{ij}, \tag{7.23}$$

where the nodal velocities $\mathbf{v}_j$ may be time-dependent, and the diffusion coefficient ε is constant,

$$\mathbf{c}_{ij} = \int_\Omega \varphi_i \nabla\varphi_j \, d\mathbf{x}, \tag{7.24}$$

$$s_{ij} = \int_\Omega \nabla\varphi_i \cdot \nabla\varphi_j \, d\mathbf{x}. \tag{7.25}$$

If the mesh does not change, the mass matrix M_C and the coefficient matrices $\mathbf{C} = \{\mathbf{c}_{ij}\}$, $S = \{s_{ij}\}$ need to be assembled just once at the beginning of the simulation run. The entries of K must be updated using (7.23) every time the nodal velocities change.

Integration in time by the two-level θ-scheme yields a system of algebraic equations

$$\begin{aligned}[M_C + \theta\Delta t K]u = {} & [M_C - (1-\theta)\Delta t K]u^n \\ & + \Delta t[\theta r(t^{n+1}) + (1-\theta) r(t^n)],\end{aligned}$$

where $\theta \in [0,1]$ is the implicitness parameter and $\Delta t = t^{n+1} - t^n$ is the time step.

7.2.1 ▪ Properties of Discrete Operators

Before dwelling on the design of stabilization techniques for multidimensional convection-dominated transport equations, it is instructive to examine the properties of the matrices M_C, $\mathbf{C}$, and S that result from the standard Galerkin discretization. This exercise will give us a better understanding of the (semi)discrete problem and equip us with tools that we will need when it comes to constructing stabilized numerical approximations.

Consistency requires that (7.14) and (7.15) be exact for constant functions. If the coefficients u_j represent the values of u_h at some points, the basis functions must satisfy

$$\sum_j \varphi_j \equiv 1. \tag{7.26}$$

It follows that the gradients of the basis functions must sum to zero at every point:

$$\sum_j \nabla\varphi_j \equiv 0. \tag{7.27}$$

By (7.26) the ith row sum of the consistent mass matrix M_C is given by

$$\sum_j m_{ij} = \int_\Omega \varphi_i \sum_j \varphi_j \, d\mathbf{x} = \int_\Omega \varphi_i \, d\mathbf{x}.$$

Furthermore, summation over all indices yields the volume (area) of the domain Ω:

$$\sum_i \sum_j m_{ij} = \int_\Omega \sum_i \varphi_i \, d\mathbf{x} = |\Omega|. \tag{7.28}$$

The discrete Laplacian $S = \{s_{ij}\}$ is symmetric with zero row and column sums:

$$\sum_j s_{ij} = \int_\Omega \nabla \varphi_i \cdot \sum_j \nabla \varphi_j \, d\mathbf{x} = 0, \tag{7.29}$$

$$\sum_i s_{ij} = \int_\Omega \sum_i \nabla \varphi_i \cdot \nabla \varphi_j \, d\mathbf{x} = 0. \tag{7.30}$$

The discrete gradient/divergence operator $\mathbf{C} = \{\mathbf{c}_{ij}\}$ is nonsymmetric. We have

$$\sum_j \mathbf{c}_{ij} = \int_\Omega \varphi_i \sum_j \nabla \varphi_j \, d\mathbf{x} = 0, \tag{7.31}$$

$$\sum_i \mathbf{c}_{ij} = \int_\Omega \sum_i \varphi_i \nabla \varphi_j \, d\mathbf{x} = \int_\Omega \nabla \varphi_j \, d\mathbf{x}. \tag{7.32}$$

The zero sum property implies that the application of $\mathbf{C}$ and S to a constant vector produces a zero vector in accordance with the fact that the derivative of a constant is zero.

Using the continuity of φ_i and integration by parts, one can easily show that

$$\sum_i \mathbf{c}_{ij} = \int_\Gamma \varphi_j \, ds, \qquad \mathbf{c}_{ij} + \mathbf{c}_{ji} = \int_\Gamma \varphi_i \varphi_j \, \mathbf{n} \, ds. \tag{7.33}$$

In the case of linear and bilinear finite elements, the latter relation implies that $\mathbf{c}_{ji} = -\mathbf{c}_{ij}$ unless both nodes lie on the outer boundary Γ of the domain Ω. For any interior node i, we have $\mathbf{c}_{ii} = 0$ since φ_i vanishes on the boundary of the integration region. The symmetric vector-valued weights $\mathbf{b}_{ij} := \mathbf{c}_{ij} + \mathbf{c}_{ji}$ distribute the surface area between pairs of nodes in much the same way as the coefficients m_{ij} distribute the volume of the domain Ω.

7.2.2 ▪ Conservation and Mass Lumping

The global conservation property can be easily inferred from the above properties of coefficient matrices. The total "mass" of the numerical solution u_h is defined as the integral

$$m = \int_\Omega u_h \, d\mathbf{x} = \sum_i m_i u_i, \tag{7.34}$$

where the row sum m_i represents the share of the domain Ω associated with node i:

$$m_i = \int_\Omega \varphi_i \, d\mathbf{x} = \sum_j m_{ij}. \tag{7.35}$$

Summing the semidiscrete equations over i, one obtains the global mass balance equation

$$\frac{dm}{dt} = \sum_i \sum_j k_{ij} u_j + \sum_i r_i.$$

If the coefficients of $K=\{k_{ij}\}$ are defined by (7.23), the first sum reduces to

$$\sum_i \sum_j k_{ij} u_j = -a(1,u_h) = -\int_{\Gamma\setminus\Gamma_-} (\mathbf{v}u)_h \cdot \mathbf{n}\, ds.$$

Thus, only source/sink terms and fluxes across the boundary may change the total mass.

The imposition of Dirichlet boundary conditions in the strong sense may invalidate the proof of global conservation. The elimination of the corresponding equations means that $w_h \equiv 1$ is no longer an admissible test function, and the column sums of the resultant coefficient matrices are different. This problem can be rectified by including a set of equations for the boundary fluxes weighted by the missing basis functions [108].

As we mentioned in Chapter 6, it is common practice to approximate the consistent mass matrix M_C by a diagonal matrix M_L in order to update the solution in a fully explicit way or make an implicit algorithm more robust and efficient. This trick is particularly useful in steady-state computations since the approximation of the time derivative has no influence on the final solution. In the case of time-dependent problems, mass lumping is usually undesirable since it may adversely affect the phase accuracy of the finite element scheme. Moreover, it may result in a loss of the discrete conservation property.

In general, a diagonal mass matrix M_L is a conservative approximation to M_C if

$$\sum_i (M_L u)_i = \sum_i (M_C u)_i \tag{7.36}$$

holds for any vector u. In particular, the sum of matrix entries should be equal to the volume/area of the computational domain, as required by (7.28) and (7.36) with $u \equiv 1$. These requirements are clearly satisfied by the *row-sum* mass lumping technique,

$$M_L = \mathrm{diag}\{m_i\}, \qquad m_i = \sum_j m_{ij}. \tag{7.37}$$

By definition, the sum of all nodal values multiplied by the diagonal coefficients of the so-defined lumped mass matrix M_L is equal to the total mass (7.34).

Mass lumping can also be performed using diagonal scaling or inexact evaluation of the coefficients m_{ij} by a nodal quadrature rule that produces a diagonal matrix. In fact, the integrals given by (7.35) can be interpreted as weights of a Newton–Cotes numerical integration scheme. This observation makes it possible to quantify the loss of accuracy due to mass lumping using the available estimates of the quadrature error [96].

The quadrature-based approach is attractive from the viewpoint of analysis, but only the row-sum lumping (7.35) conserves mass in the above sense [96]. Suppose that there exists another diagonal mass matrix $\tilde{M}_L = \mathrm{diag}\{\tilde{m}_i\}$ satisfying (7.36). Then we have

$$\sum_i (M_L u)_i - \sum_i (\tilde{M}_L u)_i = \sum_i (m_i - \tilde{m}_i) u_i = 0.$$

Since u is arbitrary, it follows that $m_i = \tilde{m}_i$ for all i. This proves the uniqueness of the lumped mass matrix M_L, which enjoys the global conservation property.

Linear and bilinear finite element approximations give rise to mass matrices satisfying

$$m_{ij} \geq 0, \quad m_i > 0 \qquad \forall i,j$$

since the basis functions are nonnegative everywhere. In higher-order approximations, the off-diagonal entries of M_C are of variable sign, so row-sum lumping may produce M_L with zero or negative diagonal entries. For example, the lumped mass matrix for the six-node quadratic finite element approximation on a triangular mesh assigns zero masses to all vertex-based degrees of freedom [40]. Such a lumped-mass Galerkin discretization can be interpreted as a finite volume method with edge-centered degrees of freedom.

7.3 ▪ Stabilized Approximations

At high Péclet numbers, the discrete transport operator K is dominated by the nonsymmetric convective part, which has very unfavorable properties. Our 1D experiments in the previous chapter have shown that the Galerkin discretization of convective terms is similar to a central difference scheme and tends to produce spurious oscillations. Moreover, an iterative algorithm or an explicit time integration scheme may become unstable.

The lack of robustness can be compensated for by adding some artificial diffusion or using modified test functions in the variational formulation. In either case, K is replaced by its stabilized counterpart $\tilde{K}$, which may or may not be associated with a continuous bilinear form. At the discrete level, the difference between the two approximations is a sparse matrix $D = \{d_{ij}\}$. The modified transport operator admits the following decomposition:

$$\tilde{K} = K + D, \qquad \tilde{k}_{ij} = k_{ij} + d_{ij}. \tag{7.38}$$

It is essential to ensure that the modified Galerkin scheme remains consistent and conservative. Furthermore, the added perturbation should not make it too diffusive.

In this section, we review some ways to localize the wiggles and keep them bounded. The topics to be discussed include first-order upwinding, linear stabilization, and residual-based shock capturing for problems with steep fronts. A review of discrete maximum principles and nonlinear high-resolution schemes will follow in the next sections.

7.3.1 ▪ First-Order Upwinding

In an unstructured mesh FEM, a nonoscillatory upwind-biased approximation of the convective term can be constructed using the equivalence between linear finite elements and vertex-centered finite volumes [7, 213, 214]. In *edge-based* finite element methods [163, 170, 178], the contribution of the Galerkin transport operator K is expressed in terms of numerical fluxes. The flux-based representation makes it easy to extend 1D finite difference/volume schemes to finite elements and unstructured meshes. It is commonly used in conjunction with edge-based data structures, which we discuss in Section 7.6.10.

Consider a piecewise-linear group finite element approximation of the form

$$\mathbf{f}_h = \sum_j \mathbf{f}_j \varphi_j.$$

In the linear convection model, $\mathbf{f}_j = \mathbf{v}_j u_j$ and the coefficients of the discrete transport operator are given by (7.23). By the zero row-sum property (7.31), we have

$$\sum_j \mathbf{c}_{ij} \cdot \mathbf{f}_j = \sum_j \mathbf{c}_{ij} \cdot \mathbf{f}_j + \mathbf{f}_i \cdot \sum_j \mathbf{c}_{ij} = \sum_j \mathbf{c}_{ij} \cdot (\mathbf{f}_j + \mathbf{f}_i). \tag{7.39}$$

Recall that $\mathbf{c}_{ji} = -\mathbf{c}_{ij}$ and $\mathbf{c}_{ii} = 0$ unless the pair of nodes i and j is associated with a boundary edge. It follows that the discretized convective term can be written as

$$\sum_j \mathbf{c}_{ij} \cdot \mathbf{f}_j = \sum_{j \neq i} f_{ij},$$

where

$$f_{ij} = \mathbf{c}_{ij} \cdot (\mathbf{f}_j + \mathbf{f}_i)$$

is the numerical flux from node j into node i. Due to $\mathbf{c}_{ji} = -\mathbf{c}_{ij}$, the fluxes satisfy

$$f_{ji} = -f_{ij}.$$

To achieve the upwinding effect, the centered Galerkin flux f_{ij} is replaced by [213]

$$\tilde{f}_{ij} = \mathbf{c}_{ij} \cdot (\mathbf{f}_j + \mathbf{f}_i) - |\mathbf{c}_{ij} \cdot \mathbf{v}_{ij}|(u_j - u_i) \qquad \forall j \neq i, \tag{7.40}$$

where $\mathbf{v}_{ij}$ denotes the edge-averaged velocity

$$\mathbf{v}_{ij} = \begin{cases} \frac{\mathbf{f}_j - \mathbf{f}_i}{u_j - u_i} & \text{if } u_j \neq u_i, \\ \mathbf{f}'(u_i) & \text{if } u_j = u_i. \end{cases} \tag{7.41}$$

The above definition of the upwind flux $\tilde{f}_{ij}$ remains valid for nonlinear conservation laws.

In fact, the implementation of first-order upwinding in a finite element code does not require the use of numerical fluxes or edge-based data structures. The desired effect can be achieved directly by adding a discrete diffusion operator $D = \{d_{ij}\}$ defined by

$$d_{ii} = -\sum_{j \neq i} d_{ij}, \qquad d_{ij} = |\mathbf{c}_{ij} \cdot \mathbf{v}_{ij}| \quad \forall j \neq i.$$

Note that the so-defined matrix D is symmetric with zero row and column sums. A general approach to the design of such artificial diffusion operators and the concept of "discrete upwinding" for finite elements on unstructured meshes are presented in Section 7.5.

7.3.2 ▪ Artificial Diffusion

First-order upwinding is rarely used in finite element methods because it produces a lot of numerical diffusion and does not fit into the analytical framework based on the calculus of variations. The mainstream approach to the design of stabilized FEM for convection-dominated transport problems involves a manipulation of the bilinear form $a(\cdot,\cdot)$ which defines the coefficients $k_{ij} = -a(\varphi_i, \varphi_j)$ of the Galerkin transport operator K.

Adding a stabilization term $b(\cdot,\cdot)$ to $a(\cdot,\cdot)$, we obtain the modified bilinear form

$$\tilde{a}(w,u) := a(w,u) + b(w,u). \tag{7.42}$$

For the modified Galerkin scheme to be weakly consistent, it is sufficient to require that

$$\lim_{h \to 0} b(w,u) = 0.$$

If the classical solution of the governing equation satisfies the modified variational formulation, the stabilized finite element scheme is called strongly consistent.

The matrix $\tilde{K}$ with coefficients $\tilde{k}_{ij} = -\tilde{a}(\varphi_i, \varphi_j)$ is given by (7.38), where

$$d_{ij} = -b(\varphi_i, \varphi_j).$$

In a classical artificial diffusion method, the stabilization operator is defined by

$$b(w,u) = \sum_{K \in \mathcal{T}_h} \int_K \nabla w \cdot (\mathcal{D} \nabla u)\, \mathrm{d}\mathbf{x}, \tag{7.43}$$

where $\mathscr{D}$ denotes a tensor diffusivity. The optimal amount of artificial diffusion depends on the local mesh size h and on the magnitude of the velocity vector $\mathbf{v}$.

The simplest way to offset the intrinsic negative diffusion of the Galerkin scheme is to add (7.43) with *isotropic* balancing dissipation of the form $\mathscr{D} = \delta \mathbf{I}$, where

$$\delta = \alpha \frac{|\mathbf{v}|h}{2} \tag{7.44}$$

is a scalar diffusion coefficient and $\mathbf{I}$ is the identity tensor. The free parameter α should be chosen sufficiently large to achieve stability. However, inordinately large values of α should be avoided since the additional discretization error is proportional to α.

Any constant value $\alpha \in [0,1]$ corresponds to a first-order perturbation error and renders the use of higher-order finite elements useless. To avoid a global loss of accuracy, the parameter α is commonly defined as a function of the mesh Péclet number

$$\mathrm{Pe}_h = \frac{|\mathbf{v}|h}{\varepsilon}$$

associated with a given element K of the triangulation $\mathscr{T}_h$. The local mesh size h may represent the longest edge or the diameter of the area/volume equivalent circle/sphere.

The 1D analysis presented in Chapter 6 indicates that for $\mathrm{Pe}_h \leq 2$ the solution is well resolved and no stabilization is required. A larger value of Pe_h indicates that the flow is too fast or the mesh is too coarse. In order to avoid the arising of spurious oscillations, it is necessary to refine the mesh or apply some artificial diffusion.

Due to the lack of alternatives, the value of α is commonly defined using the formula

$$\alpha = \coth\left(\frac{\mathrm{Pe}_h}{2}\right) - \frac{2}{\mathrm{Pe}_h} \tag{7.45}$$

or the corresponding "doubly asymptotic" approximation

$$\alpha = \min\left\{1, \frac{\mathrm{Pe}_h}{6}\right\}. \tag{7.46}$$

Clearly, the results are unlikely to be as good as in the case of the stationary 1D convection-diffusion equation we considered in Chapter 6. Neither (7.45) nor (7.46) is guaranteed to be optimal when it comes to solving multidimensional transport equations with variable coefficients on unstructured meshes. The best choice of α is highly problem-dependent and difficult to guess correctly. In advanced artificial diffusion methods, the definition of α may depend not only on Pe_h but also on the derivatives of certain quantities, residuals of the continuous problem, or other detectors of spurious oscillations [117, 120, 163].

7.3.3 ▪ Streamline Upwinding

Convective fluxes transport the conserved quantity along the streamlines of the velocity field or, generally speaking, along the characteristics of the hyperbolic equation which is obtained in the limit of vanishing physical diffusivity ($\epsilon = 0$). The artificial diffusion method with $\mathscr{D} = \delta \mathscr{I}$ corresponds to the standard Galerkin discretization of the transport equation augmented by $\delta \Delta u$. The isotropic structure of this term gives rise to unnecessary smearing of the solution in directions orthogonal to that of the velocity vector.

Since numerical instabilities are caused by streamline derivatives of the form $\mathbf{v} \cdot \nabla u$, it is sufficient to add *anisotropic balancing dissipation*, which acts along the streamlines of

$\mathbf{v}$ but not transversely [107, 125]. A family of streamline upwind finite element methods is based on (7.43) with anisotropic tensor diffusivity of the form [23, 50, 122, 230]

$$\mathcal{D} = \tau(\mathbf{v} \otimes \mathbf{v}), \tag{7.47}$$

where τ is the so-called *intrinsic time*. The components of $\mathcal{D}$ are given by

$$\mathcal{D}_{ij} = \tau v_i v_j.$$

After some rearrangements, the stabilization term (7.43) can be written in the form

$$b(w,u) = \sum_{K \in \mathcal{T}_h} \int_K \tau(\mathbf{v} \cdot \nabla w)(\mathbf{v} \cdot \nabla u)\,\mathrm{d}\mathbf{x} \tag{7.48}$$

and represents the Galerkin weak form of $\tau(\mathbf{v}\cdot\nabla)^2 u$. Hence, anisotropic tensor diffusivity of this kind is equivalent to adding *streamline diffusion* proportional to τ.

The stabilization parameter τ for the streamline upwind method is defined as

$$\tau = \frac{\delta}{|\mathbf{v}|^2} = \frac{\alpha h}{2|\mathbf{v}|},$$

where δ is given by (7.44) and depends on the choice of α. The default is (7.45) or (7.46). The value $\alpha = 1$ corresponds to first-order upwinding along the streamlines.

The elimination of crosswind artificial diffusion results in a smaller perturbation of the original Galerkin scheme than the use of $\mathcal{D} = \delta\mathcal{I}$. However, streamline diffusion stabilization is insufficient to guarantee the validity of the discrete maximum principle and/or positivity preservation. As a result, spurious undershoots and overshoots are likely to occur in the neighborhood of steep fronts [120]. Traditional remedies to this problem are discussed in Section 7.3.6.

7.3.4 ▪ Petrov–Galerkin Methods

Instead of modifying the Galerkin bilinear form, streamline diffusion can be generated by using a consistent Petrov–Galerkin method with suitably defined test functions. Let

$$\mathcal{L}u = \sigma u + \mathbf{v} \cdot \nabla u - \epsilon \Delta u \tag{7.49}$$

denote the differential operator associated with the conservative ($\sigma = \nabla \cdot \mathbf{v}$) or nonconservative ($\sigma = 0$) form of the transport equation. The corresponding bilinear form is

$$a(w,u) = \sum_{K \in \mathcal{T}_h} \int_K (w\sigma u + w\mathbf{v} \cdot \nabla u + \epsilon \nabla w \cdot \nabla u)\,\mathrm{d}\mathbf{x}. \tag{7.50}$$

The addition of the streamline upwind stabilization term (7.48) transforms $a(w,u)$ into

$$\begin{aligned}\tilde{a}(w,u) &= a(w,u) + \sum_{K \in \mathcal{T}_h} \int_K \tau(\mathbf{v} \cdot \nabla w)(\mathbf{v} \cdot \nabla u)\,\mathrm{d}\mathbf{x} \\ &= \sum_{K \in \mathcal{T}_h} \int_K (w\sigma u + \tilde{w}\mathbf{v} \cdot \nabla u + \epsilon \nabla w \cdot \nabla u)\,\mathrm{d}\mathbf{x},\end{aligned}$$

where the convective term is weighted by the nonconforming test function

$$\tilde{w} = w + \tau \mathbf{v} \cdot \nabla w. \tag{7.51}$$

In the *streamline upwind Petrov–Galerkin* (SUPG) method [23], the modified test function $\tilde{w}$ is applied to all components of $\mathscr{L}u$ and the bilinear form is redefined as

$$\tilde{a}(w,u) := a(u,\tilde{w}). \tag{7.52}$$

The L^2 scalar product associated with the contribution of the time derivative and the linear form that depends on the source term (if any) are also defined using $\tilde{w}$ in place of w. The SUPG approach to stabilization of convective terms preserves strong consistency since the residual of the variational problem vanishes if u is the classical solution. This property makes it possible to prove optimal convergence rates [195, p. 269].

The invention of SUPG was followed by the development of other Petrov–Galerkin methods. The corresponding stabilization terms can be written in the generic form [50]

$$b(w,u) = \sum_{K\in\mathscr{T}_h} \int_K \tau \mathscr{P} w \mathscr{R}(u)\,\mathrm{d}\mathbf{x}, \tag{7.53}$$

where $\mathscr{R}(u)$ is the residual of the transport equation and $\mathscr{P}$ is a differential operator that defines the space of test functions for the stabilized finite element approximation.

In the original SUPG method, $\mathscr{P}$ is taken to be the streamline derivative operator

$$\mathscr{P}w = \mathbf{v}\cdot\nabla w.$$

The resulting stabilization term $b(w,u)$ is nonsymmetric unless $\mathscr{L} = \mathscr{P}$. A symmetric version can be constructed using the *Galerkin least-squares* (GLS) method [110]

$$\mathscr{P}w = \mathscr{L}w.$$

A better weighting of σu is provided by the *subgrid scale* method [50]

$$\mathscr{P}w = -\mathscr{L}^* w,$$

where $\mathscr{L}^*$ is the differential operator adjoint to $\mathscr{L}$. In this Petrov–Galerkin method [50] and its relatives [56, 75], the sign of the symmetric terms is reversed.

All of the above Petrov–Galerkin methods reduce to the classical SUPG method if $\sigma = 0$ and the test function w is linear inside each element. In general, the three schemes are not equivalent but produce solutions of similar quality. Other definitions of $\mathscr{P}$ are unlikely to make a big difference. For a comprehensive review of stabilized finite element methods based on various generalizations of SUPG, we refer the reader to [37, 50, 120, 195] and references therein.

7.3.5 ▪ Taylor–Galerkin Methods

A major drawback of SUPG and other Petrov–Galerkin methods based on (7.53) is the presence of the free parameter τ. Guessing the right value of τ is particularly difficult for time-dependent problems since both amplitude and phase errors must be taken into account. A parameter-free alternative to streamline upwinding and SUPG-like stabilization is the use of inherently stable Taylor–Galerkin methods [48, 50, 53, 192], which we introduced in Chapter 6 using the 1D linear convection equation as a model problem.

In this section, we consider a multidimensional transport equation of the form

$$\frac{\partial u}{\partial t} + \mathscr{L}u = s \qquad \text{in } \Omega, \tag{7.54}$$

where $\mathscr{L}u$ represents the quasi-linear form of $\nabla\cdot\mathbf{f}(u)$ for a given flux function $\mathbf{f}(u)$. The generalization of Taylor–Galerkin methods presented in Chapter 6 to this problem makes it possible to achieve stability in a natural way, perhaps under a certain time step restriction.

Second-Order Taylor–Galerkin Approximation

To illustrate the relationship between Taylor–Galerkin methods and streamline upwind/streamline diffusion stabilization, we consider the second-order approximation

$$\frac{u^{n+1}-u^n}{\Delta t}=\left(\frac{\partial u}{\partial t}\right)^n+\frac{\Delta t}{2}\left(\frac{\partial^2 u}{\partial t^2}\right)^{n+\theta}, \tag{7.55}$$

which corresponds to the explicit ($\theta=0$) or semi-implicit ($\theta=1$) Lax-Wendroff (second-order Taylor-Galerkin (TG2)) method [37, 41]. The stability limit (if any) for the time step Δt depends on the structure of the spatial differential operator $\mathscr{L}$ and on the number of space dimensions [52].

Invoking the governing equation (7.54) and assuming that its coefficients are independent of t, the time derivatives are expressed it terms of space derivatives as follows:

$$\frac{\partial u}{\partial t}=s-\mathscr{L}u, \qquad \frac{\partial^2 u}{\partial t^2}=-\mathscr{L}\frac{\partial u}{\partial t}=\mathscr{L}(\mathscr{L}u-s). \tag{7.56}$$

Using this representation of the time derivatives in (7.55), we obtain

$$\frac{u^{n+1}-u^n}{\Delta t}+\mathscr{L}u^n+\frac{\Delta t}{2}\mathscr{L}(s-\mathscr{L}u^{n+\theta})=s. \tag{7.57}$$

The Galerkin variational form of this semidiscrete problem is given by

$$\left(w,\frac{u^{n+1}-u^n}{\Delta t}\right)+a(u^n,w)+b(u^{n+\theta},w)=(w,s), \tag{7.58}$$

where the contributions of the time derivatives (7.56) are represented by the bilinear forms

$$a(w,u)=(w,\mathscr{L}u-s), \tag{7.59}$$

$$b(w,u)=\frac{\Delta t}{2}(w,\mathscr{L}(s-\mathscr{L}u)). \tag{7.60}$$

The latter represents an explicit or implicit correction to the forward Euler/Galerkin discretization, which makes it more stable and second-order accurate in time.

After integration by parts, the following representation of $b(w,u)$ is obtained for a pure Dirichlet problem such that $w=0$ on the boundary Γ:

$$b(w,u)=\frac{\Delta t}{2}(\mathscr{L}^*w,s-\mathscr{L}u)=\frac{\Delta t}{2}\int_\Omega \mathscr{L}^*w\mathscr{R}(u)\,\mathrm{dx}. \tag{7.61}$$

Thus, the second-order term is identical to (7.53) with $\tau=\Delta t/2$ and $\mathscr{P}=-\mathscr{L}^*$. Like many other stable approximations, (7.58) turns out to be a disguised streamline upwind method [37, 206].

The above relationship between Taylor–Galerkin methods and SUPG-like finite element schemes can be exploited to define the stabilization parameter τ for transient computations, or the local time step Δt for marching the solution to a steady state [12, 41].

Linear Hyperbolic Equations

The simplest model problem of the form (7.54) is the linear convection equation in which

$$\mathscr{L}u=\mathbf{v}\cdot\nabla u$$

so that $\frac{\partial^2}{\partial t^2} = (\mathbf{v}\cdot\nabla)^2$ is the streamline diffusion operator. Integration by parts yields

$$b(w,u) = \frac{\Delta t}{2}\left[\int_\Omega \nabla\cdot(\mathbf{v}w)(\mathbf{v}\cdot\nabla u)\,\mathrm{d}\mathbf{x} - \int_\Gamma w\,\mathbf{v}\cdot\mathbf{n}(\mathbf{v}\cdot\nabla u)\,\mathrm{d}s\right].$$

In incompressible flow problems, the velocity field is divergence-free ($\nabla\cdot\mathbf{v}=0$), so the volume integral reduces to the streamline upwind stabilization term (7.48) with $\tau = \Delta t/2$:

$$b(w,u) = \frac{\Delta t}{2}\left[\int_\Omega (\mathbf{v}\cdot\nabla w)(\mathbf{v}\cdot\nabla u)\,\mathrm{d}\mathbf{x} - \int_\Gamma w\,\mathbf{v}\cdot\mathbf{n}(\mathbf{v}\cdot\nabla u)\,\mathrm{d}s\right].$$

When a Dirichlet boundary condition is imposed at the inlet $\Gamma_- = \{\mathbf{x}\in\Gamma\,|\,\mathbf{v}\cdot\mathbf{n} < 0\}$, as required by the nature of hyperbolic problems, the surface integral reduces to that over the outflow boundary $\Gamma_+ = \{\mathbf{x}\in\Gamma\,|\,\mathbf{v}\cdot\mathbf{n} > 0\}$. Such boundary terms are missing in the original streamline upwind formulation (7.48) but have been incorporated into the GLS method [59]. The failure to include them might cause spurious reflections at outflow boundaries [12, 52].

Nonlinear Hyperbolic Equations

For a nonlinear hyperbolic conservation law of the form (7.1), the operator $\mathscr{L}$ is defined by

$$\mathscr{L}u = \nabla\cdot\mathbf{f}(u),$$

where $\mathbf{f}(u)$ is an inviscid flux vector. Invoking approximation (7.55), one obtains

$$u^{n+1} = u^n + \Delta t(s - \nabla\cdot\mathbf{f}^n) + \frac{(\Delta t)^2}{2}\nabla\cdot\mathbf{f}_t^{n+\theta}, \tag{7.62}$$

where $\mathbf{f}^n = \mathbf{f}(u^n)$ and $\mathbf{f}_t^n$ is the first time derivative. By the chain rule, we have

$$\mathbf{f}_t = \left(\frac{\mathrm{d}\mathbf{f}}{\mathrm{d}u}\right)\frac{\partial u}{\partial t} = -\mathbf{a}\nabla\cdot\mathbf{f}, \tag{7.63}$$

where the vector-valued quantity

$$\mathbf{a} = \left(\frac{\mathrm{d}\mathbf{f}}{\mathrm{d}u}\right)$$

represents the characteristic speed of wave propagation. In extensions to nonlinear hyperbolic systems, $\mathbf{a}$ is replaced by a Jacobian matrix [168].

The variational form of the semidiscrete scheme (7.62)–(7.63) is given by (7.58) with

$$a(w,u) = \int_\Gamma w\,\mathbf{f}(u)\cdot\mathbf{n}\,\mathrm{d}s - \int_\Omega \nabla w\cdot\mathbf{f}(u)\,\mathrm{d}\mathbf{x}, \tag{7.64}$$

$$b(w,u) = \frac{\Delta t}{2}\left[\int_\Omega (\mathbf{a}\cdot\nabla w)\nabla\cdot\mathbf{f}(u)\,\mathrm{d}\mathbf{x} - \int_\Gamma w(\mathbf{a}\cdot\mathbf{n})\nabla\cdot\mathbf{f}(u)\,\mathrm{d}s\right]. \tag{7.65}$$

In an implicit Taylor–Galerkin algorithm, $b(u^{n+\theta}, w)$ may be linearized using the approximation [163]

$$\mathbf{f}^{n+\theta} \approx \mathbf{f}^n + \theta\mathbf{a}^n(u^{n+1} - u^n).$$

For $\theta = 0$, the combined contribution of $a(u^n, w)$ and $b(u^n, w)$ can be written as [54]

$$\tilde{a}(u^n, w) = \int_\Gamma w\,\mathbf{f}^{n+1/2}\cdot\mathbf{n}\,\mathrm{d}s - \int_\Omega \nabla w\cdot\mathbf{f}^{n+1/2}\,\mathrm{d}\mathbf{x}, \tag{7.66}$$

where $\mathbf{f}^{n+1/2}$ represents a second-order accurate approximation to the flux $\mathbf{f}(u^{n+1/2})$:

$$\mathbf{f}^{n+1/2} = \mathbf{f}^n + \frac{\Delta t}{2}\mathbf{f}_t^n = \mathbf{f}^n - \frac{\Delta t}{2}\mathbf{a}^n \nabla \cdot \mathbf{f}^n. \tag{7.67}$$

Note that no derivatives of the (possibly discontinuous) flux function $\mathbf{f}(u)$ appear in (7.64) and (7.66). However, the evaluation of (7.65) and (7.67) may be computationally intensive. For this reason, a fractional-step (predictor-corrector) approach is frequently employed when it comes to solving nonlinear conservation laws using Taylor–Galerkin methods.

Two-Step Implementation

An alternative implementation [6, 169] of the explicit second-order Taylor–Galerkin method for solving (7.1) is based on the two-step Runge–Kutta time-stepping scheme

$$u^{n+1/2} = u^n + \frac{\Delta t}{2}\left(\frac{\partial u}{\partial t}\right)^n = u^n + \frac{\Delta t}{2}(s - \nabla \cdot \mathbf{f}^n), \tag{7.68}$$

$$u^{n+1} = u^n + \Delta t\left(\frac{\partial u}{\partial t}\right)^{n+1/2} = u^n + \Delta t(s - \nabla \cdot \mathbf{f}^{n+1/2}). \tag{7.69}$$

In contrast to (7.67), the flux $\mathbf{f}^{n+1/2}$ is evaluated using the intermediate solution

$$\mathbf{f}^{n+1/2} = \mathbf{f}(u^{n+1/2}). \tag{7.70}$$

The finite difference discretization of (7.68)–(7.69) is known as the *Richtmyer scheme*. In the 1980s, a finite element version of this popular predictor-corrector algorithm was developed [6, 5, 169] and combined with the multidimensional version of the flux-corrected transport (FCT) algorithm to enforce monotonicity on unstructured meshes [166, 167].

The weak form of the midpoint rule corrector (7.69) can be written as follows:

$$\left(w, \frac{u^{n+1} - u^n}{\Delta t}\right) + \tilde{a}(u^n, w) = (w, s), \tag{7.71}$$

where $\tilde{a}(\cdot,\cdot)$ is the Taylor–Galerkin bilinear form defined by (7.66). The only difference compared to (7.58) is the calculation of $\mathbf{f}^{n+1/2}$ using (7.70) rather than (7.67).

Donea and Huerta [50, p. 158] use the strong form of (7.68) and the group finite element approximation to calculate $\mathbf{f}^{n+1/2}$ at the Gauss integration points. The performance of different flux representations is illustrated by the numerical examples in [50].

Alternatively, the Richtmyer scheme can be implemented using a finite element approximation to $u^{n+1/2}$. The variational form of the predictor step (7.68) is given by

$$\left(w, \frac{u^{n+1/2} - u^n}{\Delta t/2}\right) + a(u^n, w) = (w, s), \tag{7.72}$$

where $a(\cdot,\cdot)$ is the bilinear form defined by (7.64). The corrector step (7.71) becomes

$$\left(w, \frac{u^{n+1} - u^n}{\Delta t}\right) + a(u^{n+1/2}, w) = (w, s). \tag{7.73}$$

The use of equal-order interpolations for $u^{n+1/2}$ and u^{n+1} widens the stencil of the discrete scheme and is not to be recommended for steady-state computations since the standard Galerkin approximation $a(w,u) = (w,s)$ is recovered for $u = u^n = u^{n+1/2} = u^{n+1}$.

Therefore, it is common practice to use a piecewise-constant basis for $u^{n+1/2}$ and (multi)linear elements for u^{n+1} [54, 169]. This strategy eliminates the need for extra stabilization at steady state [163]. For the linear convection equation in one dimension, it leads to the same algebraic system as the Taylor–Galerkin scheme based on (7.67). For a detailed presentation of the two-step Taylor–Galerkin method (7.72)–(7.73), the reader is referred to Löhner et al. [163, 169, 166].

7.3.6 ▪ Discontinuity Capturing

As shown in Chapter 6, linear stability of a numerical scheme is insufficient to prevent violations of the discrete maximum principle in the presence of interior and/or boundary layers. Finite element methods that rely on streamline diffusion stabilization of convective terms are known to produce undershoots and overshoots in regions where the solution gradients are steep and not aligned with the flow direction. Small imperfections can often be tolerated, but in many practical applications the use of nonoscillatory schemes is a must.

Maximum principles can be enforced by increasing the amount of numerical dissipation in thin layers where the solution undergoes abrupt changes and linear stabilization is insufficient. Since the location of layers is generally unknown, it must be determined using suitable smoothness sensors such as gradients or residuals. The process of incorporating residual-based nonlinear dissipation into stabilized finite element methods is commonly referred to as *discontinuity capturing* or *shock capturing* [36, 111, 206].

Adding a discontinuity capturing component $c(w,u)$ to the linear stabilization term $b(w,u)$, we define

$$\tilde{b}(w,u) = b(w,u) + c(w,u).$$

The additional dissipation $c(w,u)$ should be designed so as to control the solution gradients not only along the streamlines but also in the crosswind direction. A typical discontinuity capturing term depends on the residual $\mathscr{R}(u)$ of the governing equation as follows [50]:

$$c(w,u) = \sum_{K\in\mathscr{T}_h} \int_K \hat{\tau}(\hat{\mathbf{v}}\cdot\nabla w)\mathscr{R}(u)\,d\mathbf{x}, \tag{7.74}$$

where $\hat{\tau}$ is another free parameter and $\hat{\mathbf{v}}$ is a solution-dependent vector field. Obviously, the addition of $c(w,u)$ does not destroy the strong consistency of a Petrov–Galerkin method.

A constant solution cannot have undershoots or overshoots. Thus, it is appropriate to set $\hat{\mathbf{v}} := 0$ in elements where $|\nabla u| = 0$. In the remaining elements, the discontinuity capturing velocity field $\hat{\mathbf{v}}$ can be defined, e.g., as the orthogonal projection of $\mathbf{v}$ onto the gradient of u [111]:

$$\hat{\mathbf{v}} = \left(\frac{\mathbf{v}\cdot\nabla u}{|\nabla u|^2}\right)\nabla u. \tag{7.75}$$

The resultant vector field $\hat{\mathbf{v}}$ is called the *effective transport velocity* [206] because

$$\hat{\mathbf{v}}\cdot\nabla u = \mathbf{v}\cdot\nabla u. \tag{7.76}$$

Substitution of (7.75) into (7.74) gives rise to the isotropic artificial dissipation term

$$c(w,u) = \sum_{K\in\mathscr{T}_h} \int_K \nu(u)\nabla w\cdot\nabla u\,d\mathbf{x}, \tag{7.77}$$

where the coefficient $\nu(u)$ depends on the gradient and residual of u as follows:

$$\nu(u) = \begin{cases} \hat{\tau}\left(\frac{\mathbf{v}\cdot\nabla u}{|\nabla u|^2}\right)\mathscr{R}(u) & \text{if } |\nabla u| \neq 0, \\ 0 & \text{if } |\nabla u| = 0. \end{cases}$$

An obvious drawback to this approach is the possibility of adding artificial viscosity that may become negative and destabilize the finite element discretization.

To ensure that the sign of $\nu(u)$ is correct, the nonlinear discontinuity capturing velocity $\hat{\mathbf{v}}$ can be redefined using the residual instead of the streamline derivative [77, 97, 123]:

$$\hat{\mathbf{v}} = \left(\frac{\mathscr{R}(u)}{|\nabla u|^2}\right)\nabla u.$$

Substitution into (7.74) yields a dissipative term of the form (7.77), where

$$\nu(u) = \begin{cases} \hat{\tau}\left(\frac{\mathscr{R}(u)}{|\nabla u|}\right)^2 & \text{if } |\nabla u| \neq 0, \\ 0 & \text{if } |\nabla u| = 0. \end{cases}$$

As before, the nonlinearity of $\nu(u)$ stems from the definition of $\hat{\mathbf{v}}$ in terms of ∇u.

In the context of SUPG methods, a reasonable value of $\hat{\tau}$ is even more difficult to find than that of the linear stabilization parameter τ. A typical setting is [112]

$$\hat{\tau} = \max\{0, \tau(\hat{\mathbf{v}}) - \tau(\mathbf{v})\},$$

where the dependence of $\tau(\mathbf{v})$ on $\mathbf{v}$ is the same as in the linear term $b(w,u)$. Thus, no extra numerical diffusion is added in the case when $\hat{\mathbf{v}} = \mathbf{v}$ is parallel to ∇u.

Alternatively, an anisotropic discontinuity capturing operator can be designed to act in the crosswind direction only. The removal of streamline diffusion from (7.77) leads to [36]

$$c(w,u) = \sum_{K\in\mathcal{T}_h}\int_K \nu(u)\left(\nabla w\cdot\nabla u - \frac{(\mathbf{v}\cdot\nabla w)(\mathbf{v}\cdot\nabla u)}{|\mathbf{v}|^2}\right)\mathrm{d}\mathbf{x}.$$

This nonlinear crosswind diffusion term is equivalent to (7.43) with $\mathscr{D}$ given by

$$\mathscr{D} = \nu(u)\left(\mathscr{I} - \frac{\mathbf{v}\otimes\mathbf{v}}{|\mathbf{v}|^2}\right),$$

where $\mathscr{I}$ denotes the unit tensor and it is tacitly assumed that $|\mathbf{v}| \neq 0$. The use of anisotropic artificial diffusion makes it possible to reduce excess numerical damping.

De Sampaio and Coutinho [206] combine streamline diffusion and discontinuity capturing within the framework of Petrov–Galerkin and Taylor–Galerkin methods for unsteady transport equations. Blending $\mathbf{v}$ and $\hat{\mathbf{v}}$, they define the combined velocity

$$\bar{\mathbf{v}} = \gamma\mathbf{v} + (1-\gamma)\hat{\mathbf{v}}, \qquad 0 \leq \gamma \leq 1.$$

To ensure that $\hat{\mathbf{v}}$ is well-defined, the weighting factor $\gamma = 1$ is employed for $|\nabla u| = 0$. The dissipative term $\tilde{b}(w,u)$ differs from its linear counterpart $b(w,u)$ in that the convective derivatives are evaluated using the above *combined velocity* $\bar{\mathbf{v}}$ rather than $\mathbf{v}$. This replacement is consistent since $\bar{\mathbf{v}}\cdot\nabla u = \mathbf{v}\cdot\nabla u$ for the continuous problem [206].

Many alternative definitions of $\tilde{b}(w,u)$ and of the involved parameters can be found in the literature. We refer the reader to John et al. [120, 121] for a comprehensive review of the state of the art and a detailed comparative study of discontinuity capturing techniques.

7.3.7 ▪ Interior Penalty Methods

A promising new approach to the design of linear stabilization and discontinuity capturing operators is the use of *interior penalty* terms to control the jumps of the solution gradient across the boundaries of mesh elements. This methodology was introduced by Douglas and Dupont [55] and recently revived by Burman and Hansbo [24, 26], who called it *edge stabilization* and backed it by rigorous numerical analysis. The potential of such continuous interior penalty (CIP) methods was shown in many recent publications [209, 234].

Edge stabilization methods penalize the jumps of the normal derivative via [26]

$$b(w,u)=\sum_{k\neq l}\int_{\Gamma_{kl}}\gamma h^2[\mathbf{n}\cdot\nabla w][\mathbf{n}\cdot\nabla u]\,ds, \tag{7.78}$$

where γ is a free parameter and $[q]$ stands for the jump of a scalar quantity q across the common boundary Γ_{kl} of two adjacent mesh elements with numbers k and l.

In contrast to Petrov–Galerkin and Taylor–Galerkin methods, the symmetric CIP term (7.78) is independent of the underlying equation and vanishes if $[\mathbf{n}\cdot\nabla u]=0$. The amount of stabilization is determined solely by the smoothness of the numerical solution, and the results are typically not sensitive to the choice of γ. The price to be paid for these advantages is a denser matrix with a rather unpleasant sparsity pattern.

The discontinuity capturing term for a CIP method can be defined as follows [25]:

$$c(w,u)=\sum_{k\neq l}\delta(u)\sum_{E\in\Gamma_{kl}}\int_E\frac{(\mathbf{e}\cdot\nabla w)(\mathbf{e}\cdot\nabla u)}{|\mathbf{e}\cdot\nabla u|}\,de, \tag{7.79}$$

where $\mathbf{e}$ is a vector parallel to an edge E of Γ_{kl}. In two space dimensions, the second sum consists of a single term which represents a line integral over the edge $E=\Gamma_{kl}$.

Burman and Ern [25] employ the following definition of the smoothness sensor

$$\delta(u)=\hat{\gamma}h\int_{\Gamma_{kl}}|[\mathbf{n}\cdot\nabla u]|\,ds \tag{7.80}$$

and prove that no spurious oscillations are generated for sufficiently large values of $\hat{\gamma}$.

The edge/face integrals that appear in (7.78)–(7.80) are commonly evaluated using the midpoint rule, which is exact for piecewise-linear finite element approximations.

7.3.8 ▪ Modulated Dissipation

In the late 1980s, the diffusive nature of row-sum mass lumping for (multi)linear finite elements was exploited to construct explicit Taylor–Galerkin schemes that introduce *modulated dissipation* in proximity to discontinuities and steep fronts. In these methods, the added *mass diffusion* admits a conservative flux decomposition which facilitates extensions of the FCT algorithm and TVD schemes to finite elements [54, 210, 211]. As explained in Chapter 6, the basic idea behind such high-resolution schemes is to suppress dispersive ripples using a combination of (linear) first-order numerical diffusion and (nonlinear) balancing *antidiffusion*. The latter must be controlled by a sensor/limiter capable of detecting discontinuities and steep gradients.

High-Order Scheme

Consider a time-dependent conservation law (7.1) and its weak form associated with a linearly stable (e.g., Taylor–Galerkin or Petrov–Galerkin) finite element scheme

$$\left(w, \frac{u^{n+1}-u^n}{\Delta t}\right) + \tilde{a}(w, u^n) = (w, s).$$

Inserting the finite-dimensional counterparts of u and w, one obtains the linear system

$$M_C u^{n+1} = M_C u^n + \Delta t\, r^n, \tag{7.81}$$

where M_C is the consistent mass matrix and r is the vector of nodal increments

$$m_{ij} = (\varphi_i, \varphi_j), \qquad r_i = \sum_j \tilde{k}_{ij} u_j, \qquad \tilde{k}_{ij} = \tilde{a}(\varphi_i, \varphi_j).$$

The replacement of M_C by its lumped counterpart M_L, as defined in (7.37), is possible, but the use of mass lumping degrades the phase accuracy of the finite element scheme.

Note that the use of M_C on the left-hand side of (7.81) results in an implicit coupling of the degrees of freedom, although the underlying time discretization is fully explicit. However, the symmetry and diagonal dominance of M_C make it possible to solve (7.81) efficiently using the following iterative algorithm proposed by Donea et al. [49]:

$$M_L u^{(m+1)} = M_C u^n + \Delta t\, r^n + (M_L - M_C) u^{(m)}, \qquad m = 0, \ldots, L-1. \tag{7.82}$$

The first and last iterates are given by $u^{(0)} = u^n$ and $u^{n+1} = u^{(L)}$, respectively. The derivation of this preconditioned Richardson scheme is based on an approximate factorization of the consistent mass matrix [49, 192]. In the context of explicit Taylor–Galerkin schemes, the three-pass solver ($L = 3$) was found to offer a good cost-accuracy ratio.

Low-Order Scheme

Let us now consider (7.82) with $L = 1$ and $u^{(0)} = 0$. The resulting explicit approximation

$$M_L u^{n+1} = M_C u^n + \Delta t\, r^n \tag{7.83}$$

corresponds to the original finite element scheme (7.81) with mass lumping on the left-hand side only. For our purposes, it is worthwhile to write (7.83) in the form

$$M_L u^{n+1} = M_L u^n + \Delta t\, r^n + (M_C - M_L) u^n. \tag{7.84}$$

As shown by Hansbo [96], the matrix $M_C - M_L$ represents a discrete diffusion operator. Its contribution renders the solution to problem (7.84) nonoscillatory (at least for sufficiently small time steps) but results in a dramatic loss of accuracy.

For 1D hyperbolic problems, the low-order counterpart (7.84) of the TG2 scheme corresponds to the Lax–Wendroff method with added dissipation. It proves stable and monotone for Courant numbers ν in the range $|\nu| \le \sqrt{2/3}$; see [100, 210] for details.

Selective Mass Lumping

Since $M_C - M_L$ is a symmetric matrix with zero row/column sums, explicit mass diffusion built into the right-hand side of (7.84) can be expressed as a sum of numerical fluxes:

$$(M_C u - M_L u)_i = \sum_j m_{ij} u_j - m_i u_i = \sum_{j \ne i} m_{ij}(u_j - u_i). \tag{7.85}$$

The amount of numerical diffusion can be reduced using *modulation coefficients* $\alpha_{ij} \in [0,1]$ to replace $M_C - M_L$ by another symmetric matrix $D = \{d_{ij}\}$ such that

$$d_{ii} = -\sum_{j \neq i} d_{ij}, \qquad d_{ij} = (1-\alpha_{ij}) m_{ij} \quad \forall j \neq i.$$

The setting $\alpha_{ij} \equiv 0$ corresponds to $D = M_C - M_L$, while the use of $0 < \alpha_{ij} \leq 1$ leads to a less diffusive approximation. The matrix form of the resulting scheme reads

$$M_L u^{n+1} = M_L u^n + \Delta t r^n + D u^n. \tag{7.86}$$

The last term vanishes for $\alpha_{ij} \equiv 1$, which corresponds to standard mass lumping:

$$M_L u^{n+1} = M_L u^n + \Delta t r^n. \tag{7.87}$$

Varying the modulation coefficients α_{ij} between 0 and 1, it is possible to switch between (7.83) and (7.87) in a conservative fashion. This strategy can be interpreted as selective mass lumping on the right-hand side of the low-order scheme (7.83).

Two-Stage Implementation

If the transient nature of the problem at hand dictates the use of small time steps, operator splitting may be used to separate convective transport (phase errors) from added mass diffusion (amplitude errors). The resultant two-step algorithm reads [54, 210, 211]

$$M_C u^H = M_C u^n + \Delta t r^n, \tag{7.88}$$

$$M_L u^{n+1} = (M_L + D) u^H. \tag{7.89}$$

The first stage corresponds to (7.81) and inherits its superb phase accuracy. The superscript H stands for "high-order." At the second stage, the amplitudes are corrected by adding a certain amount of mass diffusion in regions where steep gradients are detected by the smoothness sensor built into the definition of the coefficients α_{ij}.

The high-order solution $u^{n+1} = u^H$ is obtained for $D = 0$. The associated low-order scheme corresponds to $D = M_C - M_L$ and $u^{n+1} = M_L^{-1} M_C u^H$. Importantly, neither first-order nor modulated dissipation introduces any phase errors. That is, the predictor u^H is smeared but not displaced. For truly time-dependent problems, this makes (7.88)–(7.89) more attractive than the one-step implementation (7.86) of selective mass lumping.

Modulation Coefficients

In essence, the use of modulated mass diffusion is a discontinuity capturing technique that operates at the fully discrete level. The overall complexity and performance of the resulting scheme depend on the philosophy behind the computation of the modulation coefficients α_{ij}. First or second derivatives of flow variables can be used to construct empirical smoothness sensors, but the presence of a free parameter undermines the practical utility of such schemes. Alternatively, the values of α_{ij} can be determined using Zalesak's multidimensional FCT algorithm [163, 166, 211] or symmetric TVD limiters [54, 211]. Below we outline the former approach since it is easier to implement in explicit codes.

Explicit FCT algorithms add limited antidiffusion to a nonoscillatory low-order solution u^L. In the present context, this "transported and diffused" solution is computed from (7.86) or (7.89) with $D = M_C - M_L$. It follows that u^{n+1} is given by

$$m_i u_i^{n+1} = m_i u_i^L + \sum_{j \neq i} \alpha_{ij} m_{ij} (u_i - u_j). \tag{7.90}$$

The last term consists of limited antidiffusive fluxes which are evaluated using the solution $u = u^n$ in (7.86) and $u = u^H$ in (7.89). The correction factors α_{ij} are chosen so that u_i^{n+1} is bounded by the local extrema $u_i^{\max}$ and $u_i^{\min}$ defined by

$$u_i^{\max} := \max\{u_i^L, \max_{j \in S_i} u_j^L\}, \tag{7.91}$$

$$u_i^{\min} := \min\{u_i^L, \min_{j \in S_i} u_j^L\}, \tag{7.92}$$

where $\mathcal{S}_i = \{j \neq i \,|\, m_{ij} \neq 0\}$ is the stencil of node i. The definition of α_{ij} for (7.90) is based on Zalesak's multidimensional FCT algorithm [163, 245].

Let $f_{ij} = m_{ij}(u_i - u_j)$ denote the raw antidiffusive flux from node j into node i. By definition, the companion flux f_{ji} has the same magnitude and opposite sign. To prevent an overshoot, an upper bound must be imposed on the sum of positive fluxes. Likewise, a lower bound must be imposed on the sum of positive fluxes to prevent an undershoot. In accordance with the FCT design philosophy, the fluxes must be limited so as to prevent the formation of new local extrema. That is, the nodal values of u^{n+1} must satisfy

$$u_i^{\min} \leq u_i^{n+1} \leq u_i^{\max} \quad \forall i. \tag{7.93}$$

In Zalesak's FCT algorithm [245, 247], the correction factors α_{ij} are calculated as follows:

1. Compute the sums of positive and negative raw antidiffusive fluxes:

$$P_i^+ = \sum_{j \neq i} \max\{0, f_{ij}\}, \qquad P_i^- = \sum_{j \neq i} \min\{0, f_{ij}\}.$$

2. Compute the upper/lower bounds for the sum of limited antidiffusive fluxes:

$$Q_i^+ = m_i(u_i^{\max} - u_i^L), \qquad Q_i^- = m_i(u_i^{\min} - u_i^L).$$

3. Calculate the nodal correction factors for positive and negative fluxes:

$$R_i^+ = \min\left\{1, \frac{Q_i^+}{P_i^+}\right\}, \qquad R_i^- = \min\left\{1, \frac{Q_i^-}{P_i^-}\right\}.$$

4. Limit the fluxes f_{ij} and f_{ji} using the common correction factor:

$$\alpha_{ij} = \begin{cases} \min\{R_i^+, R_j^-\} & \text{if } f_{ij} \geq 0, \\ \min\{R_i^-, R_j^+\} & \text{if } f_{ij} < 0. \end{cases}$$

This parameter-free definition of α_{ij} guarantees that the maximum principle (7.93) holds.

In Section 7.5, we will use Zalesak's algorithm and generalizations thereof to enforce discrete maximum principles in implicit finite element methods. Building on the FCT and TVD algorithms presented in Chapter 6, we will develop a general framework for the design of algebraic flux correction schemes for convection-dominated transport equations.

Constrained Projections

The need for initialization and projection of data onto a given mesh frequently arises in numerical algorithms for systems of conservation laws. If the initial data are prescribed analytically, it is essential to guarantee that the numerical solution has the right total mass, momentum, and energy when the simulation begins. The pointwise definition

$$u_i^0 = u_0(\mathbf{x}_i) \tag{7.94}$$

is generally nonconservative. This may result in significant errors if the computational mesh is too coarse in regions where u_0 is discontinuous. On the other hand, conservative high-order projections tend to produce undershoots and overshoots.

The first use of FCT-like limiting techniques in the context of constrained data projection (initialization, interpolation, remapping) dates back to the work of Smolarkiewicz and Grell [216], who introduced a class of nonconservative monotone interpolation schemes. Conservative projection algorithms based on FCT were developed by Váchal and Liska [236] and Liska et al. [161]. Farrell et al. [67] introduced a bounded L^2 projection operator for globally conservative interpolation between unstructured meshes. In the monograph by Löhner [163, pp. 257–260], the multidimensional FCT limiter is applied to the difference between the consistent and lumped-mass L^2 projections. The latter serves as the low-order method that satisfies the discrete maximum principle for linear finite elements [67].

A general approach to synchronized FCT projections for systems of conserved variables was presented in [141]. Let u denote the initial data or a numerical solution from an arbitrary finite element space. The standard L^2 projection is defined by

$$\int_\Omega w_h u_h^H \, d\mathbf{x} = \int_\Omega w_h u \, d\mathbf{x} \qquad \forall w_h. \tag{7.95}$$

The nodal values of the high-order approximation u_h^H satisfy the linear system

$$M_C u^H = r, \tag{7.96}$$

where $M_C = \{m_{ij}\}$ is the consistent mass matrix and r is the load vector

$$r_i = \int_\Omega \varphi_i u \, d\mathbf{x}. \tag{7.97}$$

If the functions φ_i and u are defined on different meshes, numerical integration can be performed using a *supermesh* that represents the union of the two meshes [67].

The lumped-mass approximation to (7.96) is a linear system with a diagonal matrix

$$M_L u^L = r. \tag{7.98}$$

The so-defined low-order solution u_h^L has the same "mass" as u_h^H but is free of undershoots and overshoots, at least in the case of linear finite elements [67].

The difference between u_i^H and u_i^L admits the conservative flux decomposition

$$u_i^H = u_i^L + \sum_{j \neq i} f_{ij}, \qquad f_{ij} = m_{ij}(u_i^H - u_j^H), \tag{7.99}$$

and the FCT-constrained L^2 projection is defined by

$$u_i^{FCT} = u_i^L + \sum_{j \neq i} \alpha_{ij} f_{ij}, \qquad 0 \leq \alpha_{ij} \leq 1. \tag{7.100}$$

The calculation of the correction factors α_{ij} is based on Zalesak's FCT algorithm.

7.4 ▪ Discrete Maximum Principles

A good numerical scheme must respect the known properties of exact solutions. In this section, we review algebraic constraints which imply a discrete maximum principle and/or ensure positivity preservation. Later we will use these sufficient conditions to constrain the Galerkin discretization of the transport equation. We tacitly assume that one or two terms in this equation may be missing, so that it represents all models listed in Table 2.1. In this section, we briefly review the theory of discrete maximum principles. A practically oriented reader may want to skip the formal mathematical analysis of the discrete problem and proceed to the self-contained presentation of numerical algorithms in Section 7.5.

7.4.1 ▪ Semidiscrete Problem

For simplicity, we consider the linear convection-diffusion equation (7.7) without source terms ($s = 0$) and prescribe the homogeneous boundary condition $g = 0$ at the inlet. As a result, the matrix form (7.20) of the Galerkin space discretization reduces to

$$M_C \frac{\mathrm{d}u}{\mathrm{d}t} = Ku. \tag{7.101}$$

As in the continuous case, the unknown solution values are known to be bounded under certain assumptions. The semidiscrete equation associated with node i reads

$$\sum_j \left(m_{ij} \frac{\mathrm{d}u_j}{\mathrm{d}t} \right) = \sum_j k_{ij} u_j. \tag{7.102}$$

The following theorem provides a set of sufficient conditions which guarantee the validity of a semidiscrete maximum principle and/or positivity preservation.

Theorem 7.1. *Consider a semidiscrete scheme of the form* (7.102). *Suppose that*

$$m_{ii} > 0, \quad m_{ij} = 0, \quad k_{ij} \geq 0 \quad \forall j \neq i. \tag{7.103}$$

Then the following estimates hold:

(i) $\sum_j k_{ij} = 0 \quad \wedge \quad u_i \geq u_j \quad \forall j \neq i \quad \Rightarrow \quad \frac{\mathrm{d}u_i}{\mathrm{d}t} \leq 0$ *(semidiscrete maximum principle).*

(ii) $u_j(0) \geq 0 \quad \forall j \quad \Rightarrow \quad u_i(t) \geq 0 \quad \forall t > 0$ *(positivity preservation).*

Proof. The zero row- sum property of K is a discrete version of the constraint $\nabla \cdot \mathbf{v} = 0$. If $\sum_j k_{ij} = 0$, then $k_{ii} = -\sum_{j \neq i} k_{ij}$ so that (7.102) can be written in the form

$$\frac{\mathrm{d}u_i}{\mathrm{d}t} = \frac{1}{m_{ii}} \sum_{j \neq i} k_{ij}(u_j - u_i). \tag{7.104}$$

Suppose that $u_i = \max_j u_j$ By assumption, we have $m_{ii} > 0$ and $k_{ij}(u_j - u_i) \leq 0$ for all $j \neq i$. Thus $\frac{\mathrm{d}u_i}{\mathrm{d}t} \leq 0$, i.e., a maximum cannot increase. This proves (i).

To prove (ii), suppose that $u_i(t)=0$ and $u_j(t)\geq 0$ for all $j\neq i$. It follows that

$$\frac{\mathrm{d}u_i}{\mathrm{d}t}=\frac{1}{m_{ii}}\sum_{j\neq i}k_{ij}u_j, \tag{7.105}$$

where $k_{ij}u_j\geq 0$ for all $j\neq i$. Thus, the solution value u_i cannot become negative. □

Definition 7.2. *A space discretization of the form* (7.104) *with* $m_{ii}>0$ *and* $k_{ij}\geq 0$ *for all* i *and* $j\neq i$ *is called local extremum diminishing (LED).*

The LED criterion was introduced by Jameson [117, 118] in the context of finite volume methods for unstructured grids. It is consistent with the FCT philosophy introduced in the previous chapter: *no new extrema can form and existing extrema cannot grow.* The word *local* refers to the fact that the coefficient matrices are sparse, so only the nearest neighbors of node i make a nonzero contribution to (7.102) and define the bounds for u_i.

It is easy to prove that an LED scheme is TVD in one dimension. By the Godunov theorem [82], a linear positivity-preserving/LED discretization of a hyperbolic transport equation can be at most first-order accurate. The order barrier for a linear LED approximation of the diffusive term is 2 (see [114, pp. 118–120]). Hence, the sufficient conditions formulated in Theorem 7.1 are very restrictive in the linear case, but they turn out to be a handy tool for the design of *nonlinear* high-resolution schemes [137].

7.4.2 ▪ Fully Discrete Problem

The fully discrete counterpart of problem (7.101) is a sparse linear system of the form

$$\begin{pmatrix} A_\Omega & A_\Gamma \\ 0 & I \end{pmatrix}\begin{pmatrix} u_\Omega \\ u_\Gamma \end{pmatrix}=\begin{pmatrix} B_\Omega & B_\Gamma \\ 0 & I \end{pmatrix}\begin{pmatrix} g_\Omega \\ g_\Gamma \end{pmatrix}, \tag{7.106}$$

where I is the identity matrix and $u_\Gamma=g_\Gamma$ is the vector of Dirichlet boundary values.

For a two-level time-stepping scheme, $u_\Omega=u_\Omega^{n+1}$ is the vector of unknowns and $g_\Omega=u_\Omega^n$ is the vector of solution values from the last time step. For stationary problems $B_\Omega=0$ and $B_\Gamma=0$. The general form of the ith equation reads

$$a_{ii}u_i=b_{ii}g_i+\sum_{j\in S_i}(b_{ij}g_j-a_{ij}u_j), \tag{7.107}$$

where $S_i:=\{j\neq i\mid a_{ij}\neq 0\vee b_{ij}\neq 0\}$ is the set of nearest neighbors of node i.

Definition 7.3. *The solution to* (7.107) *satisfies the* local *discrete maximum principle if*

$$u_i^{\min}\leq u_i\leq u_i^{\max}, \tag{7.108}$$

where

$$u_i^{\max}:=\max\{\max_{j\in S_i\cup\{i\}}g_j,\max_{k\in S_i}u_k\}, \tag{7.109}$$

$$u_i^{\min}:=\min\{\min_{j\in S_i\cup\{i\}}g_j,\min_{k\in S_i}u_k\} \tag{7.110}$$

are the largest and smallest solution values that appear on the right-hand side of (7.107).

The so-defined local discrete maximum principle implies that u_i should not decrease as result of increasing any other nodal value that contributes to the discretized equation for node i [185]. Conversely, u_i should not increase if another nodal value is decreased, all other things being fixed. If the given solution values are all nonnegative, then so is u_i.

Definition 7.4. *The solution to* (7.107) *is said to be* locally *positivity-preserving if*

$$u_i^{\min} \geq 0 \quad \Rightarrow \quad u_i \geq 0. \tag{7.111}$$

The following theorem presents sufficient conditions of local positivity preservation and an additional constraint which guarantees the validity of (7.108) [137].

Theorem 7.5. *Suppose that the coefficients of the discrete problem* (7.107) *satisfy*

$$a_{ii} > 0, \qquad b_{ii} \geq 0, \quad a_{ij} \leq 0, \quad b_{ij} \geq 0 \qquad \forall j \in S_i. \tag{7.112}$$

Then the following a priori estimates hold for the solution value u_i*:*

(i) $\sum_j a_{ij} = \sum_j b_{ij} \quad \Rightarrow \quad u_i^{\min} \leq u_i \leq u_i^{\max}$ *(local discrete maximum principle).*

(ii) $u_i^{\min} \geq 0 \quad \Rightarrow \quad u_i \geq 0$ *(positivity preservation).*

Proof. To prove (i), we define $w_j := u_j - u_i^{\max}$ and $v_j := g_j - u_i^{\max}$ such that

$$w_j \leq 0 \quad \forall j \in S_i, \qquad v_j \leq 0 \quad \forall j \in S_i \cup \{i\}. \tag{7.113}$$

Using the row sum condition $\sum_j a_{ij} = \sum_j b_{ij}$, we can express (7.107) as follows:

$$a_{ii} w_i = b_{ii} v_i + \sum_{j \in S_i} (b_{ij} v_j - a_{ij} w_j). \tag{7.114}$$

By (7.112) and (7.113), the right-hand side of (7.114) is nonpositive. Since $a_{ii} > 0$, we have $w_i \leq 0$ or, equivalently, $u_i \leq u_i^{\max}$. The proof for $u_i \geq u_i^{\min}$ is similar.

To prove (ii), suppose that $u_i^{\min} \geq 0$, i.e., $u_j \geq 0$ for all $j \in S_i$ and $g_j \geq 0$ for all $j \in S_i \cup \{i\}$. By (7.112), the right-hand side of (7.107) is nonnegative, so $a_{ii} > 0 \Rightarrow u_i \geq 0$. □

The theorem implies that u_i is bounded by the solution values in a neighborhood of node i. If the local discrete maximum principle holds for all nodes, then global maxima and minima must occur on the Dirichlet boundary or at the previous time level. Likewise, local positivity preservation for all nodes implies global positivity preservation.

Definition 7.6. *The solution to* (7.106) *satisfies the* global *discrete maximum principle if*

$$\min g \leq u \leq \max g, \tag{7.115}$$

where u denotes the vector of unknowns and g is the vector of given solution values.

Definition 7.7. *The solution to* (7.106) *is said to be* globally *positivity-preserving if*

$$g \geq 0 \quad \Rightarrow \quad u \geq 0. \tag{7.116}$$

A typical proof of (7.115) and (7.116) is based on the theory of monotone matrices [238].

Definition 7.8. *A regular matrix A is called* monotone *if $A^{-1} \geq 0$ or, equivalently, if*

$$u \geq 0 \quad \Rightarrow \quad Au \geq 0.$$

Definition 7.9. *A monotone matrix A with $a_{ij} \leq 0$ for all $j \neq i$ is called an* M-matrix.

Theorem 7.10. *Consider a fully discrete scheme of the form $Au = Bg$. Suppose that the coefficients of $A = \{a_{ij}\}$ and $B = \{b_{ij}\}$ satisfy conditions (7.112) for all i.*

If A is strictly or irreducibly diagonally dominant, then A is an M-matrix and

(i) *the global discrete maximum principle holds if $\sum_j a_{ij} = \sum_j b_{ij}\ \forall i$;*

(ii) *the scheme is globally positivity-preserving.*

Proof. We refer the reader to Varga [238] for a proof of the M-matrix property. To prove the global discrete maximum principle, define the vectors $w := u - \max g$ and $v := g - \max g$. Invoking (7.114), we obtain a linear system of the form $Aw = Bv$, where A is monotone, $B \geq 0$, and $v \leq 0$. Hence $w = A^{-1}Bv \leq 0$, which implies $u \leq \max g$. Similarly, the solution to $Au = Bg$ proves positivity-preserving since $u = A^{-1}Bg \geq 0$ whenever $g \geq 0$. □

7.5 ▪ Algebraic Flux Correction

The fully discrete counterpart of (7.101) is a linear system of the form $Au^{n+1} = Bu^n$. The diagonal entries of the matrices A and B are positive, at least for sufficiently small time steps Δt. However, a violation of the discrete maximum principle conditions (7.112) may be caused by

- positive off-diagonal entries of the consistent mass matrix M_C;
- negative off-diagonal entries of the discrete transport operator K.

In the process of algebraic flux correction, we will constrain the contribution of these troublesome entries in a conservative manner. To this end, we will split the Galerkin scheme

$$M_C \frac{\mathrm{d}u}{\mathrm{d}t} = Ku \tag{7.117}$$

into a positivity-preserving low-order approximation of the form

$$M_L \frac{\mathrm{d}u}{\mathrm{d}t} = Lu \tag{7.118}$$

and the antidiffusive remainder

$$f(u) = (M_L - M_C)\frac{\mathrm{d}u}{\mathrm{d}t} + (K - L)u. \tag{7.119}$$

By definition of $f(u)$, the high-order system (7.117) is equivalent to

$$M_L \frac{\mathrm{d}u}{\mathrm{d}t} = Lu + f(u). \tag{7.120}$$

To prevent a possible violation of the (semi)discrete maximum principle, we will decompose the antidiffusive term $f(u)$ into numerical fluxes and limit the magnitude of these fluxes in regions where they threaten to create an undershoot or overshoot. The optimal values of the correction factors can be determined using the algebraic LED principle, as formulated in Definition 7.2, and a generalization of Zalesak's limiter that we used to control the amount of modulated dissipation in the Taylor–Galerkin FCT algorithm (Section 7.3.8).

In this section, we review the design philosophy behind algebraic flux correction [137] and present a family of FCT-like limiting techniques for continuous finite elements. In this introductory text, we focus on the numerical treatment of convective terms. For simplicity, we assume that the diffusive term is absent so that the matrix K represents the discrete convection operator. For a presentation of limiting techniques for Galerkin approximations to anisotropic diffusion operators, the interested reader is referred to [137, 143, 136].

7.5.1 ▪ Artificial Diffusion Operators

To extract the positivity-preserving "good part" (7.118) of the standard Galerkin approximation (7.117), we perform row-sum mass lumping on the left-hand side of the semidiscrete scheme and approximate the discrete transport operator K by its low-order counterpart

$$L = K + D, \tag{7.121}$$

where $D = \{d_{ij}\}$ is a symmetric matrix with zero row and column sums

$$\sum_j d_{ij} = \sum_i d_{ij} = 0, \qquad d_{ij} = d_{ji}. \tag{7.122}$$

Replacing M_C by M_L and K by L, we obtain a semidiscrete scheme of the form (7.118).

For each pair of nonzero off-diagonal entries k_{ij} and k_{ji}, the artificial diffusion coefficient $d_{ij} = d_{ji}$ should be chosen so that $l_{ij} = k_{ij} + d_{ij} \geq 0$ and $l_{ji} = k_{ji} + d_{ij} \geq 0$, as required by (7.103). Therefore, the lower bound for d_{ij} is given by [117, 131, 144, 214]

$$d_{ij} = \max\{-k_{ij}, 0, -k_{ji}\} = d_{ji} \qquad \forall j \neq i. \tag{7.123}$$

This default choice is just enough to eliminate all negative off-diagonal entries of the high-order operator K. Artificial diffusion coefficients that enforce positivity in this way were used to construct low-order schemes for FCT as early as the mid-1970s [17].

For the row sums of the matrix D to be zero, its diagonal entries are defined as

$$d_{ii} := -\sum_{j \neq i} d_{ij} \qquad \forall i. \tag{7.124}$$

By construction, $d_{ij} = d_{ji}$ for all i and j. Hence, the resulting matrix satisfies (7.122).

In a practical implementation, the elimination of negative off-diagonal entries is performed step by step without assembling the global matrix D. Instead, artificial diffusion can be built into the operator K in a loop over the edges of the sparsity graph (see Section 7.6.10). By definition, an edge $\vec{ij}$ is a pair of nodes $\{i, j\}$ that corresponds to a pair of nonzero off-diagonal coefficients k_{ij} and k_{ji}. The required solution update is as follows:

$$\begin{aligned} k_{ii} &:= k_{ii} - d_{ij}, \qquad & k_{ij} &:= k_{ij} + d_{ij}, \\ k_{ji} &:= k_{ji} + d_{ij}, \qquad & k_{jj} &:= k_{jj} - d_{ij}. \end{aligned} \tag{7.125}$$

Without loss of generality, the edges of the sparsity graph are oriented so that

$$k_{ij} \leq k_{ji}. \tag{7.126}$$

This orientation convention implies that node i is located "upwind" and corresponds to the row number of the negative off-diagonal entry to be eliminated (if any).

Equation (7.124) implies that the diagonal entries of $L=K+D$ are given by

$$l_{ii} := k_{ii} - \sum_{j\neq i} d_{ij}. \tag{7.127}$$

Furthermore, it can readily be seen that the row sums of K and $L=K+D$ are equal:

$$\sum_j l_{ij} = \sum_j (k_{ij}+d_{ij}) = \sum_j k_{ij}.$$

The ODE for the nodal value u_i can be written as

$$m_i \frac{\mathrm{d}u_i}{\mathrm{d}t} = \sum_{j\neq i} l_{ij}(u_j - u_i) + u_i \sum_j k_{ij}. \tag{7.128}$$

If K has zero row sums, then the low-order scheme is LED:

$$m_i \frac{\mathrm{d}u_i}{\mathrm{d}t} = \sum_{j\neq i} l_{ij}(u_j - u_i), \qquad l_{ij} \geq 0 \quad \forall j \neq i.$$

Otherwise, the second term on the right-hand side of (7.128) represents a nonvanishing discrete counterpart of the "reactive" term $-u\nabla\cdot\mathbf{v}$ which is responsible for compressibility effects. As in the continuous case, convective transport by a nonuniform velocity field may concentrate the mass in certain regions or create zones of low concentration. Then the low-order discretization (7.128) is no longer LED but it is still positivity-preserving.

Example 7.1. To clarify the implications of (7.125), consider the 1D convection equation

$$\frac{\partial u}{\partial t} + v\frac{\partial u}{\partial x} = 0 \qquad \text{in} \quad \Omega = (0,1), \tag{7.129}$$

where v is a positive constant. At the inlet, we prescribe the Dirichlet boundary condition

$$u(0) = 0.$$

On a uniform mesh of linear finite elements, the standard Galerkin method yields

$$K = \frac{1}{2}\begin{pmatrix} \cdots & & & & & \\ & v & 0 & -v & & \\ & & v & 0 & -v & \\ & & & v & 0 & -v \\ & & & & & \cdots \end{pmatrix}.$$

For any interior node, $m_i = h$, where h is the constant mesh size. Hence, the lumped-mass version of (7.102) is equivalent to the central difference scheme

$$\frac{\mathrm{d}u_i}{\mathrm{d}t} + v\,\frac{u_{i+1} - u_{i-1}}{2h} = 0.$$

Since $k_{ij} = -\frac{v}{2}$ for $j = i+1$, the artificial diffusion coefficient (7.123) is $d_{ij} = \frac{v}{2}$ and

$$L = \begin{pmatrix} \dots & & & & & \\ & v & -v & 0 & & \\ & & v & -v & 0 & \\ & & & v & -v & 0 \\ & & & & & \dots \end{pmatrix},$$

which corresponds to the first-order accurate upwind difference approximation

$$\frac{du_i}{dt} + v\,\frac{u_i - u_{i-1}}{h} = 0.$$

Thus, the elimination of negative off-diagonal entries from a skew-symmetric operator K can be interpreted as *discrete upwinding* [137]. For any pair of nodes i and $j = i+1$ numbered in accordance with (7.126), the grid point x_i lies upstream of x_j.

After the discretization in time by the standard θ–scheme, the upwind difference method proves positivity-preserving under the generalized CFL condition [144]

$$v\frac{\Delta t}{h} \leq \frac{1}{1-\theta}, \qquad 0 \leq \theta < 1. \tag{7.130}$$

According to this formula, there is no time step restriction for the backward Euler method ($\theta = 1$) which corresponds to first-order "upwinding" in time. ■

7.5.2 ▪ Conservative Flux Decomposition

The replacement of the high-order Galerkin scheme (7.117) by the perturbed system (7.118) guarantees positivity preservation but introduces a lot of numerical diffusion. As we have seen in Section 7.3.8), it is worthwhile to adjust the amount of artificial diffusion using a conservative flux decomposition. Hence, we will decompose (7.119) into a sum of numerical fluxes and constrain the magnitude of these fluxes in an adaptive manner.

The antidiffusive term (7.119) represents the difference between the residuals of systems (7.117) and (7.118). By definition of the matrices M_L and L, we have

$$f(u) = (M_L - M_C)\frac{du}{dt} - Du. \tag{7.131}$$

The matrices $M_C - M_L$ and D are symmetric with zero row sums. It follows that

$$(M_C u - M_L u)_i = \sum_j m_{ij} u_j - m_i u_i = \sum_{j\neq i} m_{ij}(u_j - u_i), \tag{7.132}$$

$$(Du)_i = \sum_j d_{ij} u_j = \sum_{j\neq i} d_{ij} u_j + d_{ii} u_i = \sum_{j\neq i} d_{ij}(u_j - u_i). \tag{7.133}$$

Note that the right-hand sides of (7.132) and (7.133) resemble that of an LED scheme.

By symmetry, the components of the sums over $j \neq i$ can be interpreted as numerical fluxes that describe the conservative mass exchange between a pair of nodes. The flux

$$f_{ij} = \left(m_{ij}\frac{d}{dt} + d_{ij}\right)(u_i - u_j) \qquad \forall j \neq i \tag{7.134}$$

represents the rate of antidiffusive mass transfer from node j into node i. In the fully discrete version, the time derivative is replaced with a finite difference.

The net antidiffusion received by node i admits the following decomposition:

$$f_i = \sum_{j\neq i} f_{ij}, \qquad f_{ji} = -f_{ij}. \tag{7.135}$$

Since $f_{ij} + f_{ij} = 0$ by definition, the antidiffusive term does not change the total mass of the discrete solution. The mass added to node i is subtracted from its neighbors.

7.5.3 ▪ Limited Antidiffusive Correction

Recall that the high-order Galerkin scheme (7.117) can be written in the equivalent form (7.120), where $f(u)$ compensates for the numerical diffusion built into (7.118). As shown above, each component of $f(u)$ admits a conservative decomposition into a sum of raw antidiffusive fluxes f_{ij}. Some of these fluxes are harmless, but others may create an undershoot or overshoot. Thus, the contribution of "bad" fluxes must be limited so as to keep the antidiffusive part LED for a given solution.

The flux-corrected counterpart of (7.120) is a semidiscrete problem of the form

$$M_L \frac{\mathrm{d}u}{\mathrm{d}t} = Lu + \bar{f}(u), \tag{7.136}$$

where the (nonlinear) term $\bar{f}(u)$ stands for the sum of limited antidiffusive fluxes

$$\bar{f}_i = \sum_{j\neq i} \bar{f}_{ij}, \qquad \bar{f}_{ji} = -\bar{f}_{ij}. \tag{7.137}$$

A well-designed flux limiter will produce $\bar{f}_{ij} = f_{ij}$ in smooth regions and $\bar{f}_{ij} = 0$ inside interior or boundary layers. The unconstrained Galerkin scheme (7.117) and its nonoscillatory low-order part (7.118) correspond to $\bar{f} = f$ and $\bar{f} = 0$, respectively.

In general, the best definition of $\bar{f}_{ij}$ satisfying the LED constraint is given by the solution of a constrained optimization problem [14]. A nonoptimal but cost-effective alternative is the multiplication by a solution-dependent correction factor

$$\bar{f}_{ij} := \alpha_{ij} f_{ij}, \qquad 0 \leq \alpha_{ij} \leq 1. \tag{7.138}$$

This sort of flux correction traces its origins to the FCT algorithm and forms the basis for the construction of our algebraic flux correction schemes.

The following criterion guarantees that the antidiffusive term (7.137) is LED for a given set of bounded nonnegative coefficients $q_{ij} \geq 0$:

$$\sum_{j\neq i} q_{ij} \min\{0, u_j - u_i\} \leq \sum_{j\neq i} \alpha_{ij} f_{ij} \leq \sum_{j\neq i} q_{ij} \max\{0, u_j - u_i\}. \tag{7.139}$$

The upper and lower bounds may consist of a single term associated with a local maximum or minimum,

$$u_i^{\max} := \max\{u_i, \max_{j\in S_i} u_j\}, \tag{7.140}$$

$$u_i^{\min} := \min\{u_i, \min_{j\in S_i} u_j\}. \tag{7.141}$$

Introducing $q_i := \sum_{j\neq i} q_{ij}$, we can replace (7.139) with the weakened LED constraint

$$q_i(u_i^{\min} - u_i) \leq \sum_{j\neq i} \alpha_{ij} f_{ij} \leq q_i(u_i^{\max} - u_i). \tag{7.142}$$

If u_i is a local maximum, then (7.139) and (7.142) imply the cancellation of all positive fluxes. Similarly, all negative fluxes are cancelled if u_i is a local maximum. Hence, the sum of $\bar{f}_{ij} := \alpha_{ij} f_{ij}$ cannot create an undershoot or overshoot at node i.

The above criteria provide a general framework for the design of algebraic flux correction schemes that differ in the definition of the LED bounds for the sum of limited antidiffusive fluxes. The best choice of q_{ij} and q_i is dictated by accuracy and efficiency considerations. Obviously, increasing the value of these parameters makes the bounds less restrictive. However, inordinately large values may cause divergence of iterative solvers for the resultant nonlinear system. For accuracy reasons, it is worthwhile to use a limiter which yields $\alpha_{ij} = 1$ whenever the solution varies linearly in a neighborhood of node i. This useful design principle is known as *linearity preservation* [14, 136, 137].

We will use a generalization of Zalesak's FCT algorithm [245] to calculate correction factors α_{ij} satisfying (7.142). A similar limiting strategy is used to enforce the LED bounds defined by (7.139) in algebraic flux correction schemes of TVD type [133, 135, 145]. In the following sections, we address the design of multidimensional flux limiters, as well as iterative solution of nonlinear systems produced by algebraic flux correction schemes.

7.6 ▪ Generalized FCT Algorithms

The classical FCT algorithms of Boris, Book, and Hain [18, 17, 19] belong to the class of *diffusion-antidiffusion* methods [47] based on the predictor-corrector approach:

1. Advance the solution in time with an explicit low-order scheme containing enough numerical diffusion to suppress undershoots and overshoots.

2. Correct the solution using antidiffusive fluxes limited in such a way that no new maxima or minima can form and existing extrema cannot grow.

The numerical diffusion of the low-order method makes it possible to maintain positivity and improves the phase accuracy of an explicit centered approximation to the convective term. The limited antidiffusive correction reduces the amplitude errors subject to an LED constraint of the form (7.142). In contrast to TVD methods [98, 99, 219], the upper and lower bounds for the FCT limiter are defined in terms of the low-order predictor and designed to accept as much antidiffusion as possible without violating the positivity constraint.

Zalesak's fully multidimensional FCT algorithm [245] is based on blending explicit high- and low-order approximations so as to constrain the maximum and minimum increments to each nodal value. The work of Zalesak has formed the basis for the development of all algebraic flux correction schemes to be presented below. The combination of FCT with finite elements and unstructured meshes dates back to the explicit algorithms of Parrott and Christie [184] and Löhner et al. [166, 167]. Some implicit FEM-FCT schemes were published by the first author and his coworkers [131, 134, 144, 142]. The rationale for the use of an implicit time discretization stems from the fact that the CFL stability condition becomes too restrictive in the case of nonuniform velocity fields and locally refined meshes.

In this section, we begin with a presentation of predictor-corrector FCT algorithms in which the antidiffusive fluxes are linearized about a provisional low-order solution. The linearized FCT scheme [134] is simple and efficient. It is recommended for semi-implicit treatment of evolutionary problems which call for the use of small time steps. We also present the nonlinear version of this scheme, which requires iterative flux correction. Finally, we summarize the pros and cons of the FCT approach to algebraic flux correction.

7.6.1 ▪ Linearized FCT Scheme

After the discretization in time by the two-level θ scheme, the constrained Galerkin discretization (7.136) produces a nonlinear algebraic system of the form

$$Au^{n+1} = Bu^n + \bar{f}, \tag{7.143}$$

where $\bar{f} = \bar{f}(u^{n+1}, u^n)$ denotes the limited antidiffusive term. The matrices

$$A = \frac{1}{\Delta t} M_L - \theta L \tag{7.144}$$

and

$$B = \frac{1}{\Delta t} M_L + (1-\theta) L \tag{7.145}$$

represent the nonoscillatory low-order part of the Galerkin scheme. If the velocity field is time-dependent, the coefficients of A and B may change as the solution evolves.

If the time step Δt is relatively small, it is worthwhile to linearize (7.143) using a predictor-corrector strategy. At the first step of the linearized FCT algorithm [134], we disregard the antidiffusive term $\bar{f}$ and solve the linear system

$$Au^L = Bu^n. \tag{7.146}$$

By construction, the off-diagonal entries of L are nonnegative. Hence, the off-diagonal coefficients of A and B have the right sign. The diagonal coefficients of these matrices are dominated by the positive contribution of the lumped mass matrix. By Theorem 7.10, our low-order scheme (7.146) is positivity-preserving under the CFL-like condition

$$\Delta t \le \frac{1}{\theta - 1} \frac{m_i}{l_{ii}} \qquad \forall i, \tag{7.147}$$

where m_i is a positive diagonal entry of M_L and l_{ii} is a negative diagonal entry of L. The discrete maximum principle holds if L has zero row sums.

The two-level θ-scheme can be replaced with any other time integration scheme, e.g., a strong stability-preserving (TVD) Runge–Kutta method [84, 85, 134]. Clearly, the time step restriction (if any) will depend on the time integration method.

The low-order predictor u^L is used to evaluate the raw antidiffusive fluxes

$$f_{ij} = m_{ij}(\dot{u}_i^L - \dot{u}_j^L) + d_{ij}(u_i^L - u_j^L), \qquad j \ne i, \tag{7.148}$$

where $\dot{u}^L$ is an approximation to the vector of time derivatives at the time level t^{n+1}.

For example, the semidiscrete low-order scheme (7.118) with $u := u^L$ yields

$$\dot{u}^L = M_L^{-1} L u^L. \tag{7.149}$$

This low-order approximation tends to be too diffusive. Using (7.117), we obtain

$$\dot{u}^L = M_C^{-1} K u^L, \tag{7.150}$$

which yields a more accurate approximation to f_{ij}. The well-conditioned mass matrix M_C can be "inverted" with a few cycles of the preconditioned Richardson iteration.

The raw antidiffusive fluxes f_{ij} are passed to the multidimensional FCT limiter (see Section 7.6.3), which returns a set of correction factors α_{ij}. This gives

$$\bar{f}_i = \sum_{j \neq i} \alpha_{ij} f_{ij}, \qquad 0 \leq \alpha_{ij} \leq 1. \tag{7.151}$$

After flux limiting, the final solution u^{n+1} is obtained with the explicit correction

$$M_L u^{n+1} = M_L u^L + \Delta t \bar{f}. \tag{7.152}$$

Due to the linearization about u^L, the unconstrained ($\alpha_{ij} := 1$) version of the above algorithm is no longer equivalent to the original Galerkin scheme [137].

7.6.2 ▪ Nonlinear FCT Scheme

Linearization errors are avoided if the nonlinear system (7.143) is solved in an iterative fashion. This approach leads to an FEM-FCT algorithm in which the antidiffusive fluxes f_{ij} and the corresponding correction factors α_{ij} are updated step by step until the residuals or relative changes become smaller than a prescribed tolerance.

Let $\{u^{(m)}\}$ be a sequence of successive approximations to the flux-corrected Galerkin solution u^{n+1}. A reasonable initial guess is $u^{(0)} = u^n$ or $u^{(0)} = 2u^n - u^{n-1}$. These settings correspond to the constant and linear extrapolation in time, respectively. Given the current iterate $u^{(m)}$ and the vector of approximate time derivatives

$$\dot{u}^{(m)} := \frac{u^{(m)} - u^n}{\Delta t}, \tag{7.153}$$

we recalculate the implicit part of the raw antidiffusive fluxes given by

$$\begin{aligned} f_{ij}^{(m)} &= m_{ij}(\dot{u}_i^{(m)} - \dot{u}_j^{(m)}) + \theta d_{ij}(u_i^{(m)} - u_j^{(m)}) \\ &\quad + (1-\theta) d_{ij}(u_i^n - u_j^n), \qquad j \neq i. \end{aligned} \tag{7.154}$$

Then we apply the FCT limiter (see Section 7.6.3) and solve the linear system

$$A u^{(m+1)} = B u^n + \bar{f}^{(m)}. \tag{7.155}$$

To prove positivity preservation, we split each solution update of the form (7.155) into three steps [131, 134]:

1. Compute an explicit low-order approximation $\tilde{u}^{(0)} \approx u^{n+1-\theta}$ by solving

$$M_L \tilde{u}^{(0)} = B u^n. \tag{7.156}$$

2. Apply limited antidiffusive fluxes to the intermediate solution $\tilde{u}^{(0)}$:

$$M_L \tilde{u}^{(m+1)} = M_L \tilde{u}^{(0)} + \Delta t \bar{f}^{(m)}. \tag{7.157}$$

3. Solve the linear system for the new approximation to u^{n+1}:

$$A u^{(m+1)} = M_L \tilde{u}^{(m+1)}. \tag{7.158}$$

Note that $\tilde{u}^{(0)}$ is independent of the iteration counter m, so it needs to be determined just once per time step (for $m = 0$). For this step to be positivity-preserving, the time step Δt must satisfy (7.147). The flux limiter presented in the next section 7.6.3 guarantees that

$$\tilde{u}^{(0)} \geq 0 \Rightarrow \tilde{u}^{(m+1)} \geq 0.$$

The last solution update is positivity-preserving by the M-matrix property of A. Thus

$$u^{(0)} \geq 0 \quad \Rightarrow \quad \tilde{u}^{(0)} \geq 0 \quad \Rightarrow \quad \tilde{u}^{(m+1)} \geq 0 \quad \Rightarrow \quad u^{(m+1)} \geq 0, \tag{7.159}$$

provided that the CFL-like condition (7.147) holds for the given Δt and $\theta \in (0,1]$ [137].

7.6.3 ▪ Zalesak's FCT Limiter

Zalesak's limiter introduced in Section 7.3.8 is used to calculate the correction factors α_{ij} in both linearized and nonlinear FCT schemes. Consider a solution update of the form

$$m_i u_i = m_i \tilde{u}_i + \Delta t \sum_{j \neq i} \alpha_{ij} f_{ij}, \tag{7.160}$$

where $\tilde{u}$ is a nonoscillatory intermediate solution. Let $u_i^{\max}$ and $u_i^{\min}$ denote the local extrema of $\tilde{u}$. The objective is to find the best value of α_{ij} such that

$$u_i^{\min} \leq u_i \leq u_i^{\max}. \tag{7.161}$$

This condition implies that (7.160) satisfies the local discrete maximum principle.

Prelimiting Step

The process of flux correction begins with the optional elimination of fluxes that have the same sign as $\tilde{u}_j - \tilde{u}_i$. Such fluxes flatten the solution profile instead of steepening it. As a consequence, spurious ripples may occur within the bounds allowed by the limiter [46]. In the original Boris–Book limiter [18], a wrong sign is reversed, and the magnitude of the antidiffusive flux is limited in the usual way. This fix works well for discontinuities but may distort a smooth profile. It is generally safer to cancel the "diffusive" fluxes by setting

$$f_{ij} := 0 \quad \text{if} \quad f_{ij}(\tilde{u}_j - \tilde{u}_i) > 0. \tag{7.162}$$

This optional adjustment is called *prelimiting* because it must be performed before the computation of the correction factors α_{ij} and flux limiting [46, 245].

Zalesak [245] argued that the effect of (7.162) is marginal and cosmetic in nature since the vast majority of antidiffusive fluxes have the right sign. This remark might have led many readers to disregard equations (14) and (14′) in [245]. Two decades later, the need for prelimiting of the form (7.162) was emphasized by DeVore [46], who explained its ramifications and demonstrated that it may lead to a marked improvement of accuracy. In our experience, prelimiting is particularly useful in the context of finite element approximations because the contribution of the consistent mass matrix may change the sign of the raw antidiffusive flux and render it diffusive. The cancellation of such outliers is essential for keeping the solution free of ripples that cannot be detected by the main limiter.

Limiting Strategy

In accordance with the LED criterion (7.142), the choice of the correction factors α_{ij} should ensure that positive antidiffusive fluxes cannot create an overshoot, while negative ones cannot create an undershoot. Assuming the worst-case scenario, we enforce condition (7.161) using the following implementation of Zalesak's FCT algorithm [245]:

1. Compute the sums of positive/negative antidiffusive fluxes into node i:

$$P_i^+ = \sum_{j \neq i} \max\{0, f_{ij}\}, \qquad P_i^- = \sum_{j \neq i} \min\{0, f_{ij}\}. \tag{7.163}$$

2. Determine the distance to a local maximum/minimum and the bounds

$$Q_i^+ = \frac{m_i}{\Delta t}(u_i^{\max} - \tilde{u}_i), \qquad Q_i^- = \frac{m_i}{\Delta t}(u_i^{\min} - \tilde{u}_i). \tag{7.164}$$

3. Evaluate the nodal correction factors for the net increment to node i:

$$R_i^+ = \min\left\{1, \frac{Q_i^+}{P_i^+}\right\}, \qquad R_i^- = \min\left\{1, \frac{Q_i^-}{P_i^-}\right\}. \tag{7.165}$$

4. Check the sign of the raw antidiffusive flux f_{ij} and multiply it by

$$\alpha_{ij} = \begin{cases} \min\{R_i^+, R_j^-\} & \text{if } f_{ij} > 0, \\ \min\{R_i^-, R_j^+\} & \text{if } f_{ij} < 0. \end{cases} \tag{7.166}$$

The above definition of α_{ij} guarantees that the sum of limited antidiffusive fluxes satisfies (7.142) with $q_i = \frac{m_i}{\Delta t}$. The LED property (7.161) follows from the estimate

$$u_i^{\min} = \tilde{u}_i + \frac{\Delta t}{m_i} Q_i^- \leq u_i \leq \tilde{u}_i + \frac{\Delta t}{m_i} Q_i^+ = u_i^{\max}.$$

The dependence of the bounds $Q_i^\pm$ on the time step Δt may have a positive or negative impact on the accuracy of the flux-corrected solution. On the one hand, the FCT constraints become less restrictive and, consequently, a larger portion of the raw antidiffusive flux f_{ij} is retained as the time step is refined. This makes FCT the method of choice for transient computations. On the other hand, the use of large time steps results in a loss of accuracy, and severe convergence problems may occur in steady-state computations [137].

7.6.4 ▪ Applicability of FCT

In our experience, FCT produces excellent results for strongly time-dependent problems. No other shock capturing technique performs better when it comes to solving an unsteady hyperbolic equation with linear finite elements [121]. The use of small time steps increases the amount of accepted antidiffusion and justifies the use of the predictor-corrector algorithm. The cost of an implicit FCT scheme depends on the choice of iterative methods, parameter settings, and stopping criteria. If the time step is very small, then the old solution provides a good initial guess. Since the left-hand-side matrix A is dominated by the contribution of M_L, the sparse linear systems can be solved using just a few iterations of the Jacobi or Gauss–Seidel method. Thus, the cost *per time step* approaches that of an explicit finite difference or finite volume scheme. As the time step increases, so does the number of iterations, and advanced linear algebra tools (smoothers, preconditioners, convergence acceleration techniques) may need to be employed. Moreover, the numerical diffusion of an FCT algorithm depends on the time step. In the next section, we introduce upwind-biased flux limiters of TVD type. The underlying limiting strategy is well suited for computations with large time steps and does not inhibit convergence to steady-state solutions [137].

7.6.5 ▪ Upwind-Biased Flux Correction

If the transport equation is stationary or weakly time-dependent, the use of mass lumping does not degrade the accuracy of the numerical solution. Neglecting the contribution of the consistent mass matrix to f_{ij}, we consider raw antidiffusive fluxes of the form

$$f_{ij} = d_{ij}(u_i - u_j), \qquad j \neq i, \tag{7.167}$$

where d_{ij} is the artificial diffusion coefficient defined by (7.123).

Zalesak's formula (7.166) for the correction factor α_{ij} is symmetric. In flux limiters of TVD type, a one-sided limiting strategy is employed [137, 145]. In accordance with our upwind-downwind edge orientation convention (7.126), we assume that $k_{ij} \le k_{ji}$. If

$$\bar{k}_{ji} := k_{ji} + (1-\alpha_{ij})d_{ij}$$

is nonnegative for all $\alpha_{ij} \in [0,1]$, it is sufficient to ensure that (7.139) holds for node i.

The process of upwind-biased flux correction begins with the prelimiting step

$$f_{ij} := (d_{ij} + \min\{0, k_{ji}\})(u_i - u_j), \qquad f_{ji} := -f_{ij}. \tag{7.168}$$

After this prelimiting, inequality constraints of the form (7.139) can be enforced as follows:

1. Compute the sums of positive and negative antidiffusive fluxes:
$$P_i^+ = \sum_{k_{ij} \le k_{ji}} \max\{0, f_{ij}\}, \qquad P_i^- = \sum_{k_{ij} \le k_{ji}} \min\{0, f_{ij}\}. \tag{7.169}$$

2. Compute the upper and lower bounds for the sums of limited fluxes:
$$Q_i^+ = \sum_{j \ne i} q_{ij} \max\{0, u_j - u_i\}, \qquad Q_i^- = \sum_{j \ne i} q_{ij} \min\{0, u_j - u_i\}. \tag{7.170}$$

3. Compute the nodal correction factors for positive/negative fluxes:
$$R_i^+ = \min\left\{1, \frac{Q_i^+}{P_i^+}\right\}, \qquad R_i^- = \min\left\{1, \frac{Q_i^-}{P_i^-}\right\}. \tag{7.171}$$

4. Multiply f_{ij} and f_{ji} by the nodal correction factor for the *upwind* node i:
$$k_{ij} \le k_{ji} \quad \Rightarrow \quad \alpha_{ij} = \begin{cases} R_i^+ & \text{if } f_{ij} \ge 0, \\ R_i^- & \text{if } f_{ij} < 0, \end{cases} \qquad \alpha_{ji} := \alpha_{ij}. \tag{7.172}$$

In contrast to FCT, the bounds $Q_i^\pm$ depend on the unknown solution. Hence, the correction factors α_{ij} must be calculated using an iterative defect correction scheme of the form (7.155). The accuracy and convergence behavior of this scheme depend on the definition of q_{ij} in (7.170). Choosing a value that is too large is as bad as choosing one that is too small. It is worthwhile to define the bounds so that a classical TVD scheme is recovered for linear convection in one dimension. As shown in [132], this is true for

$$q_{ij} := l_{ij} \ge 0 \qquad \forall j \ne i. \tag{7.173}$$

This definition guarantees that (i) limited antidiffusive fluxes have the same effect as the local extremum diminishing low-order part, and (ii) their contribution to the residual of the flux-corrected scheme is of the same or smaller magnitude.

In the nontrivial case, $d_{ij} > 0$ and $0 = l_{ij} < l_{ji}$ due to (7.126). Hence, the upper and lower bounds $Q_i^\pm$ consist of upstream edge contributions ($k_{ij} > k_{ji}$) only. The contribution of the edge $\overrightarrow{ij}$ can be inserted into the sums $P_i^\pm$ and $Q_j^\pm$ as follows:

$$\begin{aligned} P_i^+ &:= P_i^+ + \max\{0, f_{ij}\}, \qquad & Q_j^+ &:= Q_j^+ + \max\{0, l_{ji}(u_i - u_j)\}, \\ P_i^- &:= P_i^- + \min\{0, f_{ij}\}, \qquad & Q_j^- &:= Q_j^- + \min\{0, l_{ji}(u_i - u_j)\}. \end{aligned}$$

After the computation of the correction factors $\alpha_{ij} = R_i^\pm$ from (7.172), the limited antidiffusive fluxes are inserted into the global vector $\bar{f}$ that appears in (7.155):

$$\bar{f}_i := \bar{f}_i + \alpha_{ij} f_{ij}, \qquad \bar{f}_j := \bar{f}_j - \alpha_{ij} f_{ij}.$$

Alternatively, $Q_i^\pm$ may be defined in terms of the artificial diffusion coefficients [133]

$$q_{ij} := d_{ij} \geq 0 \qquad \forall j \neq i. \tag{7.174}$$

This version is designed to ensure that (i) limited antidiffusive fluxes have the same effect as artificial diffusion built into the low-order part, and (ii) their contribution to the residual of the flux-corrected scheme is of the same or smaller magnitude.

In a loop over edges, a raw antidiffusive flux of the form $f_{ij} = d_{ij}(u_i - u_j)$ is added to the upwind sum $P_i^\pm$ and to the upper/lower bounds for both nodes [133]:

$$Q_i^+ := Q_i^+ + \max\{0, -f_{ij}\}, \qquad Q_j^+ := Q_j^+ + \max\{0, f_{ij}\},$$
$$Q_i^- := Q_i^- + \min\{0, -f_{ij}\}, \qquad Q_j^- := Q_j^- + \min\{0, f_{ij}\}.$$

The above LED constraints may be weakened using bounds of the form

$$Q_i^+ = q_i(u_i^{\max} - u_i), \qquad Q_i^- = q_i(u_i^{\min} - u_i), \tag{7.175}$$

where

$$q_i = \sum_{j \neq i} q_{ij}. \tag{7.176}$$

In contrast to FCT, the resulting formula for $Q_i^\pm$ is independent of the time step Δt.

7.6.6 ▪ Gradient-Based Limiting

In a well-designed algebraic flux correction scheme, the correction factors α_{ij} should approach unity in regions where the solution is well resolved. In particular, no diffusion or antidiffusion should be applied, even on a nonuniform grid, if the solution varies linearly in the neighborhood of node i. This desirable property is known as linearity preservation.

One-dimensional TVD limiters are linearity-preserving due to the fact that $\Phi(1) = 1$ for all limiter functions presented in Chapter 6. Multidimensional generalizations based on (7.169)–(7.172) may fail to be linearity-preserving. In order to guarantee linearity preservation, one may construct linearity-preserving bounds for individual fluxes and define the global bounds $Q_i^\pm$ as the sum of the corresponding local bounds [136, 137]. The methodology to be presented in this section belongs to the family of edge-based stencil reconstruction methods that constrain the jumps of the gradient along the line connecting two nodes [117, 163, 170, 186].

A raw antidiffusive flux of the form $f_{ij} = d_{ij}(u_i - u_j)$ requires limiting if the difference between u_i and u_j is "too large." Introducing the limited slope

$$\bar{s}_{ij} := \alpha_{ij}(u_i - u_j), \qquad 0 \leq \alpha_{ij} \leq 1, \tag{7.177}$$

we define

$$\bar{f}_{ij} := d_{ij}\bar{s}_{ij} = \alpha_{ij} f_{ij}. \tag{7.178}$$

Thus, the multiplication of f_{ij} by α_{ij} is equivalent to replacing $u_i - u_j$ with $\bar{s}_{ij}$. An algorithm that produces $\bar{s}_{ij}$ rather than α_{ij} is called a *slope limiter.*

The definition of $\bar{s}_{ij}$ must guarantee that $\bar{f}_{ij}$ is LED. This will be the case if

$$s_{ij}^{\min} \le \bar{s}_{ij} \le s_{ij}^{\max}, \tag{7.179}$$

where the upper and lower bounds are given by

$$s_{ij}^{\max} = \gamma_{ij}(u_i^{\max} - u_i), \tag{7.180}$$

$$s_{ij}^{\min} = \gamma_{ij}(u_i^{\min} - u_i) \tag{7.181}$$

for some bounded $\gamma_{ij} \ge 0$. That is, each slope is constrained in the same manner as the sum of antidiffusive fluxes in (7.142). This approach is used in many edge-based extensions of 1D high-resolution schemes to unstructured meshes [117, 163, 170].

To construct the LED bounds $s_{ij}^{\max}$ and $s_{ij}^{\min}$, consider the linear approximation

$$u_i - u_j \approx s_{ij} := (\nabla u)_i \cdot (\mathbf{x}_i - \mathbf{x}_j). \tag{7.182}$$

The value of $(\nabla u)_i$ is obtained using numerical differentiation. A variety of gradient reconstruction techniques based on averaging or superconvergent patch recovery are available for this purpose. In our method, we use the lumped-mass L^2 projection

$$(\nabla u)_i = \frac{1}{m_i} \sum_k \mathbf{c}_{ik} u_k, \tag{7.183}$$

where m_i is a diagonal entry of the lumped mass matrix M_L, and $\mathbf{c}_{ik}$ is a vector-valued coefficient of the discrete gradient operator $\mathbf{C}$ given by (7.24). Since the gradients of Lagrange basis functions sum to zero, we have $\mathbf{c}_{ii} = -\sum_{k \ne i} \mathbf{c}_{ik}$. Thus

$$(\nabla u)_i = \frac{1}{m_i} \sum_{k \ne i} \mathbf{c}_{ik}(u_k - u_i). \tag{7.184}$$

We will use this representation to derive the LED bounds for the extrapolated slope s_{ij}, and then we will use these bounds to define γ_{ij} for (7.180) and (7.181).

Substituting (7.184) into the definition of s_{ij}, we obtain the estimates

$$s_{ij} \le \frac{1}{m_i} \sum_{k \ne i} |\mathbf{c}_{ik} \cdot (\mathbf{x}_i - \mathbf{x}_j)|(u_i^{\max} - u_i), \tag{7.185}$$

$$s_{ij} \ge \frac{1}{m_i} \sum_{k \ne i} |\mathbf{c}_{ik} \cdot (\mathbf{x}_i - \mathbf{x}_j)|(u_i^{\min} - u_i). \tag{7.186}$$

To make the bounds for s_{ij} less restrictive, we multiply them by 2 and define

$$\gamma_{ij} := \frac{2}{m_i} \sum_{k \ne i} |\mathbf{c}_{ik} \cdot (\mathbf{x}_i - \mathbf{x}_j)|. \tag{7.187}$$

This nonnegative coefficient is used to determine the bounds (7.180) and (7.181) for the slope limiter. The localized LED constraint (7.179) can be enforced by setting

$$\bar{s}_{ij} = \begin{cases} \min\{s_{ij}^{\max}, u_i - u_j\} & \text{if } u_i > u_j, \\ \max\{s_{ij}^{\min}, u_i - u_j\} & \text{if } u_i < u_j. \end{cases} \tag{7.188}$$

The one-sided limiting strategy is sufficient if the slope $\bar{s}_{ji} := -\bar{s}_{ij}$ cannot violate the LED principle for node j. In particular, this is the case if j is a node on the Dirichlet boundary or a downwind neighbor of node i (see above). In all other cases, the slope limiter must enforce not only (7.179) but also $s_{ji}^{\min} \leq \bar{s}_{ji} \leq s_{ji}^{\max}$. The proof of linearity preservation follows from the fact that (7.183) is exact for linear functions [137].

To maintain linearity preservation in a fully multidimensional upwind-biased flux limiter, we constrain the sum of antidiffusive fluxes using the sum of local bounds imposed on these fluxes. The corresponding global bounds $Q_i^{\pm}$ are of the form (7.175) with

$$q_i := \sum_{j \neq i} \gamma_{ij} d_{ij}. \tag{7.189}$$

This limiting strategy can also be used to constrain the contribution of the consistent mass matrix in applications to unsteady transport equations [136, 137].

Example 7.2. To illustrate the relationship of the linearity-preserving slope limiter to classical TVD schemes [98, 99, 219], consider a 1D mesh with uniform spacing h. In this case, the coefficients of (7.183) are given by $m_i = \Delta x$ and $c_{i\pm1/2} = \pm 1/2$.

The resulting formula for u_i' is equivalent to the second-order central difference

$$u_i' = \frac{1}{2}\left(\frac{u_i - u_{i-1}}{h} + \frac{u_{i+1} - u_i}{h}\right) = \frac{u_{i+1} - u_{i-1}}{2h}.$$

For any interior node, the local maxima and minima of the grid function are

$$u_i^{\max} = \max\{u_{i-1}, u_i, u_{i+1}\}, \qquad u_i^{\min} = \min\{u_{i-1}, u_i, u_{i+1}\}.$$

Furthermore, $\gamma_{ij} = 2$ for $j = i + 1$ since estimate (7.185)–(7.186) corresponds to

$$u_i^{\min} - u_i \leq \Delta x u_i' \leq u_i^{\max} - u_i.$$

The one-sided slope limiter (7.188) can be written as a single-line formula

$$\bar{s}_{ij} = minmod\{2(u_{i-1} - u_i), u_i - u_{i+1}\}.$$

The *minmod* limiter function returns the argument with the smallest magnitude if all arguments have the same sign and zero otherwise. That is,

$$minmod\{a, b, \dots\} = \begin{cases} \min\{a, b, \dots\} & \text{if } a > 0,\ b > 0, \dots, \\ \max\{a, b, \dots\} & \text{if } a < 0,\ b < 0, \dots, \\ 0 & \text{otherwise.} \end{cases}$$

It follows that the proposed slope limiter is activated only if two consecutive gradients have opposite signs or their magnitudes differ by a factor of 2 or more. ∎

7.6.7 ▪ Solution of Nonlinear Systems

All upwind-biased algebraic flux correction schemes lead to nonlinear systems of the form

$$A^{n+1} u^{n+1} = B^n u^n + \Delta t \bar{f}(u^{n+1}, u^n) \tag{7.190}$$

which must be solved iteratively. Consider a sequence of successive approximations,

$$u^{(m)} \approx u^{n+1}, \qquad m = 0, 1 \dots.$$

At the beginning of the first outer iteration, the value of $u^{(0)}$ is guessed making use of the previously computed data and initial/boundary conditions. A reasonable initial guess for an unsteady transport problem is $u^{(0)} = u^n$ or $u^{(0)} = 2u^n - u^{n-1}$. As already mentioned, these choices correspond to constant and linear extrapolation in time, respectively.

Given an approximation $u^{(m)}$, its successor $u^{(m+1)}$ can be calculated as follows:

$$A^{(m)}u^{(m+1)} = B^n u^n + \Delta t \bar{f}(u^{(m)}, u^n), \qquad m = 0,1,2,\ldots. \tag{7.191}$$

The sum of limited antidiffusive fluxes remains on the right-hand side and is evaluated using the current iterate $u^{(m)}$ in place of u^{n+1}. If the coefficients of the matrix $A^{(m)}$ depend on the solution, they also need to be recalculated. This approach is called *fixed-point iteration*, also known as the Picard iteration and the method of successive approximations.

A convenient way to enforce Dirichlet boundary conditions for a given node i is to replace the corresponding row of $A^{(m)}$ by that of the identity matrix (see Chapter 4)

$$a_{ij} := \delta_{ij} \qquad \forall j \tag{7.192}$$

and substitute the boundary value g_i for the ith element on the right-hand side [230].

In most cases, linear systems (7.191) are also solved iteratively. Due to the M-matrix property of the low-order operator $A^{(m)}$, a small number of inner iterations is usually sufficient. The number of outer iterations depends on the evolution of the residual

$$r^{(m)} = B^n u^n - A^{(m)}u^{(m)} + \Delta t \bar{f}(u^{(m)}, u^n). \tag{7.193}$$

All elements of $r^{(m)}$ associated with Dirichlet boundary nodes should be set to zero.

The Euclidean (or maximum) norm $||r^{(m)}||$ of the residual is a good indicator of how close the current approximation $u^{(m)}$ is to the final solution u^{n+1}. A typical set of stopping criteria for an iterative solution procedure like (7.191) is as follows:

$$||r^{(m+1)}|| \le \min\{\varepsilon_1, \varepsilon_2 ||r^{(0)}||\}, \qquad \frac{||u^{(m+1)} - u^{(m)}||}{||u^{(m+1)}||} < \varepsilon_3, \tag{7.194}$$

where ε_1, ε_2, and ε_3 are the tolerances prescribed for defects and relative changes.

Suppose that M outer iterations are required to meet the above stopping criteria and obtain $u^{n+1} := u^{(M)}$. Then the effective *rate of convergence* can be defined as

$$\rho_M = \left(\frac{||r^{(M)}||}{||r^{(0)}||} \right)^{\frac{1}{M}}.$$

In essence, this is the average factor by which the magnitude of the defect shrinks during a single outer iteration. Note that $\rho_M < 1$ is required for convergence.

The above methodology is also suitable for computing stationary solutions with

$$M_L = 0, \quad \theta = 1, \quad \Delta t = 1.$$

Alternatively, the solution can be marched to the steady state using the backward Euler method with large and, possibly, variable pseudo-time steps. The relative solution changes and/or convergence rates can be used to determine the optimal time step size adaptively. In steady-state computations, we initialize the solution by the low-order predictor $u^{(0)} := u^L$, which is inexpensive to compute and provides a good guess for further computations.

7.6.8 ▪ Defect Correction Schemes

Many iterative solution methods for system (7.190) can be formally written in the following generic form, which will be referred to as the *defect correction scheme*:

$$u^{(m+1)} = u^{(m)} + [\tilde{A}^{(m)}]^{-1} r^{(m)}, \qquad m = 0,1,2\ldots, \tag{7.195}$$

where $\bar{A}^{(m)}$ is a suitable "preconditioner" and $r^{(m)}$ is the residual given by (7.193). As before, the Dirichlet boundary conditions for node i are implemented by setting

$$\bar{a}_{ii} := 1, \quad \bar{a}_{ij} := 0 \quad \forall j \neq i,$$
$$u_i^{(m)} := g_i, \quad r_i^{(m)} := 0.$$

In practice, the "inversion" of $\bar{A}^{(m)}$ is performed by solving the linear subproblem

$$\bar{A}^{(m)} \Delta u^{(m+1)} = r^{(m)}, \qquad m = 0, 1, 2, \ldots. \tag{7.196}$$

After a few inner iterations, the increment $\Delta u^{(m+1)}$ is applied to the last iterate,

$$u^{(m+1)} = u^{(m)} + \Delta u^{(m+1)}. \tag{7.197}$$

The iteration process is terminated when the defect $r^{(m)}$ and/or the relative solution changes $\Delta u^{(m+1)}$ become sufficiently small in the sense of criteria (7.194).

The implementation of a defect correction cycle involves the following tasks:

- Assembly of $\bar{A}^{(m)}$ and $r^{(m)}$, imposition of Dirichlet boundary conditions.
- Monitoring the residual norms, relative changes, and convergence rates.
- Implicit computation of the solution increments $\Delta u^{(m+1)}$ from (7.196).
- Explicit computation of the new approximation $u^{(m+1)}$ from (7.197).

Ideally, the preconditioner $\bar{A}^{(m)}$ should be designed so that (i) matrix assembly is relatively fast, (ii) linear subproblems of the form (7.196) can be solved efficiently, and (iii) convergence is achieved with a small number of outer iterations. These requirements are often in conflict with one another, so some compromises need to be made.

If the time step Δt is very small, the approximate solution can be updated in a fully explicit fashion using (7.195) with a Jacobi-like diagonal preconditioner,

$$\bar{A} = \text{diag}\{m_i - \theta \Delta t\, l_{ii}\}. \tag{7.198}$$

The number of outer iterations for (7.195) preconditioned in this way can be as small as one if a good initial guess is available. Thus, the computational cost per time step might be comparable to that of a conditionally stable explicit algorithm.

At intermediate and large time steps, the default preconditioner for (7.195) is

$$\bar{A} = M_L - \theta \Delta t L. \tag{7.199}$$

Substitution of (7.193) and (7.199) into (7.195) reveals that the resulting defect correction scheme is equivalent to the standard fixed-point iteration given by (7.191).

The optimal choice of iterative solvers for sparse linear systems (7.196) with a nondiagonal and nonsymmetric matrix $\bar{A}^{(m)}$ also depends on the time step size and on the matrix properties. As long as Δt is relatively small, the low-order preconditioner (7.199) is dominated by the lumped mass matrix and a basic iteration of Jacobi, Gauss–Seidel, or successive overrelaxation (SOR) type may suffice. In this case, each inner iteration has approximately the same cost as one step of an explicit algorithm. As the time step and the condition number of the matrix increase, BiCGSTAB, GMRES, and multigrid methods are to be preferred. The Gauss–Seidel iteration and ILU factorization with Cuthill–McKee renumbering may serve as smoothers/preconditioners for intermediate and large time steps. Due to the M-matrix property of the low-order operator (7.199), its ILU factorization exists and is unique.

7.6.9 ▪ Anderson Acceleration

Since the cost of recalculating the correction factors for the flux limiter is rather high, slow convergence of an iterative method can make algebraic flux correction very expensive. The fixed-point defect correction scheme (7.195) generates a sequence of successive approximations, but only the last iterate $u^{(m-1)}$ is used when it comes to the computation of $u^{(m)}$. It turns out that inclusion of further previous iterates may dramatically improve the convergence behavior. A particularly useful convergence acceleration technique is known as *Anderson mixing* [2, 65, 66, 239]. As shown in [65], this approach is equivalent to the Broyden scheme for the inverse Jacobian but is easier to implement. On linear problems, the accelerated fixed-point iteration is related to the GMRES method [239].

Following Walker and Ni [239], we perform Anderson acceleration as follows:

For all $m \geq 1$ **do**

Compute $\tilde{u}^{(m)} := g(u^{(m-1)})$ using (7.195)

Store $\tilde{u}^{(m)}$ and $\Delta u^{(m)} := \tilde{u}^{(m)} - u^{(m-1)}$

Given $k \leq m$ iterates, determine the weights

$$\omega^{(m)} = (\omega_1^{(m)}, \ldots, \omega_k^{(m)})^T$$

by solving the constrained least-squares problem

$$\min_{\omega^{(m)}} \left\| \sum_{i=1}^{k} \omega_i^{(m)} \Delta u^{(m-k+i)} \right\|_2 \quad \text{s.t.} \sum_{i=1}^{k} \omega_i^{(m)} = 1$$

Set $u^{(m)} := \sum_{i=1}^{k} \omega_i^{(m)} \tilde{u}^{(m-k+i)}$

In practice, it is worthwhile to calculate the weights $\omega_i^{(m)}$ by solving an equivalent unconstrained least-squares problem [239]. Furthermore, Anderson acceleration may need to be restarted if the vectors $\Delta u^{(m)}$ become (almost) linearly dependent, or if the norm of $\Delta u^{(m)}$ is much greater than that of $\Delta u^{(m-1)}$. We refer the reader to [65, 66, 181, 239] for a discussion of various improvements and implementation details.

7.6.10 ▪ Edge-Based Data Structures

Assembly of sparse matrices and/or matrix-vector products represents one of the basic tasks involved in practical implementation of unstructured grid finite element methods. A lot of computer time is spent on retrieving information from large arrays and locating array elements which need to be updated. The search process depends on the links between individual elements, nodes, fluxes, and other data items. Hence, the overall performance of the code is strongly influenced by the choice of data structures and storage techniques.

In most finite element codes, global matrices are assembled by visiting all elements sequentially and collecting their contributions to the integrals that define the coefficients of discrete operators. The three basic operations are (i) *gathering* nodal data pertaining to each element, (ii) *processing* this data, and (iii) *scattering* the results back to nodes [163, p. 187]. To perform numerical integration, one needs to know the number and location of nodes that belong to a given element. Thus, a typical element-based data structure includes arrays in which the Cartesian coordinates of each node and the connectivity list of

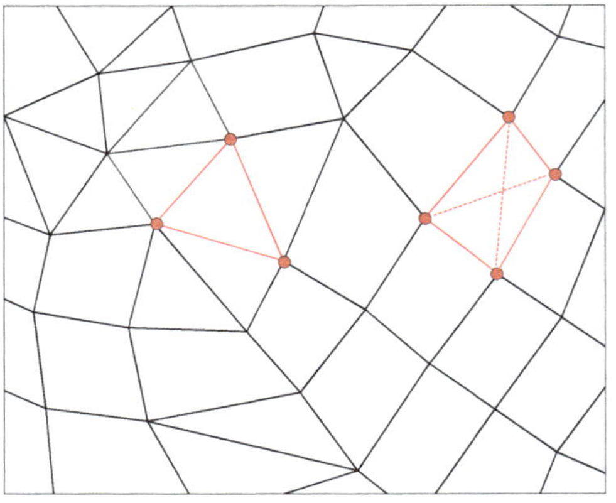

Figure 7.1. *Internodal links for linear and bilinear finite elements* [177].

each element are stored. After the matrix assembly, all information about the geometrical and topological properties of the mesh can be discarded, and the evaluation of right-hand sides and matrix-vector products can be performed using an edge-based data structure which operates with pairs of nodes, as in finite volume schemes. Such data structures offer algorithmic simplicity, low memory requirements, and a major reduction in indirect addressing [10, 164, 162, 186, 214]. Moreover, they are amenable to large-scale parallel computing [30, 163, 171].

The term "edge" refers to a pair of nodes associated with a pair of nonzero off-diagonal matrix entries, i.e., with an edge $\overrightarrow{ij}$ of the graph that represents the sparsity pattern (stencil) of the numerical scheme. In the finite element context, any pair of basis functions with overlapping supports defines such an edge. On a simplex mesh, the number of internodal links for linear finite elements equals the number of physical mesh edges. Bilinear approximations on rectangular or mixed meshes give rise to extra diagonal links; see Fig. 7.1.

To get started with edge-based data structures, consider a sparse matrix $A = \{a_{ij}\}$ with `neq` rows/columns and `nnz` entries that may assume nonzero values. If storage is allocated for a_{ij} and/or a_{ji} with $j > i$, then the node pair $\{i, j\}$ defines an edge in the above algebraic sense. The list of edges is stored in a 2D connectivity array $\texttt{kedge}(1:2, 1:\texttt{nedge})$ with two rows and `nedge` columns such that the two node numbers associated with the edge number $\texttt{iedge} = 1, \ldots, \texttt{nedge}$ are

$$
\begin{aligned}
i &= \texttt{kedge}(1, \texttt{iedge}), \\
j &= \texttt{kedge}(2, \texttt{iedge}).
\end{aligned}
$$

The columnwise storage of edge data is adopted since matrices are stored by columns in MATLAB, Fortran, and other programming environments.

Another array $\texttt{aedge}(1:2, 1:\texttt{nedge})$ is used to store the off-diagonal matrix entries

$$
\begin{aligned}
a_{ij} &= \texttt{aedge}(1, \texttt{iedge}), \\
a_{ji} &= \texttt{aedge}(2, \texttt{iedge}).
\end{aligned}
$$

The first row of `aedge` is associated with the upper triangular part ($j > i$), while the second row contains the nonzero entries of the lower triangular part ($j < i$). For symmetric matrices, a 1D array `aedge` is enough. A matrix with zero row/column sums is completely defined by the contents of `kedge` and `aedge`. Otherwise, an extra array `adiag(1 : neq)` is used to store the diagonal entries

$$a_{ii} = \texttt{adiag(i)}.$$

This 1D array is sufficient to represent a diagonal matrix such as M_L.

In summary, an elementary edge-based data structure includes the edge connectivity list and arrays that store the diagonal and/or off-diagonal part of a matrix with up to `nnz = neq + 2 * nedge` nonzero entries. In what follows, this representation of sparse matrices will be referred to as the compressed edge storage (CES) format.

The following examples illustrate some possibilities to evaluate the matrix-vector product Au and add the result to the residual or right-hand-side vector b. Here and below, pseudocode segments are written in the Fortran 90 terminology, which we believe is self-explanatory and readily portable to other programming languages.

Example 7.3. Given a nonsymmetric matrix `A = {adiag, aedge, kedge}` with nonvanishing row sums, its diagonal entries are processed in a loop over nodes,

```
do i = 1, neq
   b(i) = b(i) + adiag(i) * u(i)
end do
```

while the contribution of off-diagonal entries is included in a loop over edges,

```
do iedge = 1, nedge
   i = kedge(1, iedge)
   j = kedge(2, iedge)
   b(i) = b(i) + aedge(1, iedge) * u(j)
   b(j) = b(j) + aedge(2, iedge) * u(i)v
end do
```
■

Example 7.4. If the matrix A has zero row sums, then a loop over edges is enough:

```
do iedge = 1, nedge
   i = kedge(1, iedge)
   j = kedge(2, iedge)
   diff = u(j) - u(i)
   b(i) = b(i) + aedge(1, iedge) * diff
   b(j) = b(j) - aedge(2, iedge) * diff7
end do
```
■

Example 7.5. If the matrix A is symmetric with zero row sums, then it represents a discrete diffusion operator. Hence, a conservative flux decomposition exists.

```
do iedge = 1, nedge
   i = kedge(1, iedge)
   j = kedge(2, iedge)
   flux = aedge(iedge) * (u(j) - u(i))
   b(i) = b(i) + flux
   b(j) = b(j) - flux
end do
```

Other kinds of internodal fluxes can be inserted into the vector b in the same way. ■

In iterative solvers, preconditioners should also be applied edge by edge if the CES format is adopted. The design of such preconditioners is addressed in [31, 32]. A further reduction in indirect addressing can be achieved using the concepts of stars, superedges, and chains which represent extensions of the single-edge data structure [30, 162].

7.6.11 ▪ Compressed Row Storage

A complete transition to an edge-based data structure might be impractical in an existing finite element code that employs a different format for storage of sparse matrices in assembly routines and/or linear solvers. However, the edge-by-edge evaluation of internodal fluxes is nontrivial if the coefficients $\mathbf{c}_{ij}$ and d_{ij} are not available in the CES format. Moreover, the entries of global matrices may depend on edge data. Therefore, some parts of the code call for an edge-based implementation, and it is essential to have fast access to all matrix entries associated with a given edge. Alternatively, element matrices can be disassembled into edge contributions and combined to form edge matrices [30].

The compressed row storage (CRS) and compressed column storage (CCS) formats are often used in unstructured mesh codes. In either case, all nonzero entries of a matrix A are packed into a 1D array $\mathtt{aval(nnz)}$. This array is filled row by row or column by column in the CRS and CCS versions, respectively. In other words, the CRS form of a matrix A is equivalent to the CCS form of A^T. Hence, the decision as to which format is better suited for a given application is a matter of whether the matrix itself or its transposed is needed more frequently. In this text, we discuss the CRS format.

Given the array $\mathtt{aval(nnz)}$ of nonzero entries stored row by row, two auxiliary integer arrays, $\mathtt{kcol(1:nnz)}$ and $\mathtt{kptr(1:neq+1)}$, make it possible to find an entry with given row and column numbers. The value of $\mathtt{j} = \mathtt{kcol(ij)}$ is the column number of $a_{ij} = \mathtt{aval(ij)}$ for any index from the range $\mathtt{ij} = \mathtt{1},\dots,\mathtt{nnz}$. The beginning of the ith row is indicated by $\mathtt{ii} = \mathtt{kptr(i)}$. That is, if $\mathtt{i} = \mathtt{1},\dots,\mathtt{neq}$ is the row number of $a_{ij} = \mathtt{aval(ij)}$, then $\mathtt{kptr(i)} \le \mathtt{ij} \le \mathtt{kptr(i+1)} - \mathtt{1}$. The last element $\mathtt{kptr(neq+1)} = \mathtt{nnz} + \mathtt{1}$ of the row pointer array $\mathtt{kptr}$ corresponds to the position where the next row would begin. This entry is introduced to avoid treating the last row differently from the others.

For our purposes, it is worthwhile to store the diagonal entry of each row first so that $\mathtt{kcol(ii)} = \mathtt{i}$ if $\mathtt{ii} = \mathtt{kptr(i)}$. This convention provides fast access to the diagonal part of the matrix. The remaining column indices are stored in ascending order so that $\mathtt{kptr(i)} + \mathtt{1} \le \mathtt{ij} < \mathtt{ik} \le \mathtt{kptr(i+1)} - \mathtt{1}$ implies $\mathtt{kcol(ij)} < \mathtt{kcol(ik)}$. While this restriction is not a part of the standard CRS format, it plays a pivotal role in the development of the edge-based assembly algorithms to be presented below.

Example 7.6. Consider the following square matrix with $\mathtt{nnz} = 12$ nonzero entries:

$$A = \begin{bmatrix} \underline{1} & 2 & 0 & 7 \\ 2 & \underline{4} & 3 & 0 \\ 0 & 3 & \underline{6} & 5 \\ 7 & 0 & 5 & \underline{8} \end{bmatrix}.$$

In the sorted CSR format, this 4×4 matrix is represented by the three arrays

$$\begin{aligned} \mathtt{aval} &= (\underline{1},2,7,\underline{4},2,3,\underline{6},3,5,\underline{8},7,5), \\ \mathtt{kcol} &= (1,2,4,2,1,3,3,2,4,4,1,3), \\ \mathtt{kptr} &= (1,4,7,10,13). \end{aligned} \tag{7.200}$$

Note that the underlined diagonal entries of A are stored in positions indicated by $\mathtt{kptr}$ and followed by other elements of the same row traversed from left to right. ∎

Example 7.7. Given a matrix $\mathtt{A} = \{\mathtt{aval},\mathtt{kcol},\mathtt{kptr}\}$ in the CSR format, the following algorithm returns the sum of the matrix-vector product Au and a vector b:

```
do i = 1,neq
   do ij = kptr(i),kptr(i+1)-1
      j = kcol(ij)
      b(i) = b(i)+aval(ij)*u(j)
   end do
end do
```
(7.201) ∎

Example 7.8. If A has zero row sums, an alternative way to evaluate and add Au is

```
do i = 1,neq
   do ij = kptr(i)+1,kptr(i+1)-1
      j = kcol(ij)
      b(i) = b(i)+aval(ij)*(u(j)-u(i))
   end do
end do
```
(7.202) ∎

The edge-by-edge assembly/modification of the sparse matrix $\mathtt{A} = \{\mathtt{aval},\mathtt{kcol},\mathtt{kptr}\}$ or its conversion into the CES format $\mathtt{A} = \{\mathtt{adiag},\mathtt{aedge},\mathtt{kentry}\}$ involves visiting all node pairs $\{i,j\}$ with $j > i$ and searching for elements of `aval` that correspond to matrix entries with row/column indices i and j. The positions of the diagonal entries $a_{ii} =$ `aval(ii)` and $a_{jj} =$ `aval(jj)` are given by `ii = kptr(i)` and `jj = kptr(j)`, respectively. Let all node pairs $\{i,j\}$ be processed in the same order in which they are encountered in (7.201) and (7.202). Then the off-diagonal entry $a_{ij} =$ `aval(ij)` is readily available, but the location of $a_{ji} =$ `aval(ji)` is more difficult to find. In general, it can be determined by jumping to the beginning of row j and checking if `kcol(ji) = i` for `ji = kptr(j)+1,...,kptr(j+1)-1`. However, the fact that the elements of `kcol` are sorted makes it possible to retrieve the value of `ji` without additional searching.

Let `ksep(1:neq)` be an integer array that will serve as a pointer to the position of a_{ji} for the current node pair. The edge-by-edge processing of a matrix and its storage in the CES format can be performed by using the following pseudocode [177]:

```
iedge = 0
ksep(1:neq) = kptr(1:neq)
do i = 1,neq
   ii = kptr(i); adiag(i) = aval(ii)
   do ij = ksep(i)+1,kptr(i+1)-1
      j = kcol(ij); ksep(j) = ksep(j)+1
      ji = ksep(j); jj = kptr(j)
      ...
      iedge = iedge+1
      kedge(1,iedge) = i
      kedge(2,iedge) = j
      aedge(1,iedge) = aval(ij)
      aedge(2,iedge) = aval(ji)
      ...
   end do
end do
```
(7.203)

Initially, the separator array `ksep` points to the beginning of each row, where the diagonal entry is stored. For $i = 1$, the next row element (if any) belongs to the upper triangular matrix since there is no $j < 1$. All rows are visited sequentially in the outer loop. First, the value of $a_{ii} =$ `aval(ii)` is retrieved and packed into `adiag(i)`. Then the lists of edges and off-diagonal entries with row or column index i are extruded in the inner loop. Due to the above sorting convention, the column indices $j =$ `kcol(ij)` increase with `ij`. When a new entry $a_{ij} =$ `aval(ij)` with $j > i$ is encountered, the edge counter `iedge` and the entry `ksep(j)` are incremented by 1. The new value of `ji` $=$ `ksep(j)` marks the position of $a_{ji} =$ `aval(ji)`. When row j is reached in the outer loop, all rows with $i < j$ have already been processed, so that `ksep(j)` $+$ `1` is the position of the first row element from the upper triangular part. As an exercise that illustrates how `ksep` evolves, it is instructive to consider the sample matrix (7.200) and unroll the loops using paper and pencil.

The body of the inner loop may include further statements that involve manipulations with matrix entries indexed by i and j. In particular, the edge-by-edge assembly of `aval` for a discrete transport operator and/or the evaluation of numerical fluxes can be performed in this way. Since the values a_{ij} and a_{ji} are not stored contiguously in `aval`, the permanent "jumping" may incur a high overhead cost due to slow memory access. Thus, the CES format is well suited for storage and edge-by-edge processing of sparse matrices.

7.6.12 ▪ Element-Based Limiting

The first finite element FCT algorithms [166, 167] did not use edge-based data structures, and the limiter was designed to constrain *antidiffusive element contributions* rather than fluxes. The concepts of algebraic flux correction can be readily extended to such element-based schemes. In this section, we review the element-based implementation of FCT [144].

Let us consider the pure convection equation with weakly imposed inflow boundary conditions $u = u_{\text{in}}$. The corresponding variational formulation reads

$$\int_\Omega \left[w\frac{\partial u}{\partial t} - \nabla w \cdot (\mathbf{v}u) \right] \mathrm{d}\mathbf{x} + \int_{\Gamma_+} wu(\mathbf{v}\cdot\mathbf{n})\mathrm{d}s = -\int_{\Gamma_-} wu_{\text{in}}(\mathbf{v}\cdot\mathbf{n})\mathrm{d}s. \tag{7.204}$$

The global matrices M_C and K that appear in the standard Galerkin discretization (7.117) of this variational problem admit a natural decomposition into the element matrices

$$M_C^e = \{m_{ij}^e\}, \qquad K^e = \{k_{ij}^e\}.$$

In contrast to the edge-based FCT method, the construction of the underlying low-order scheme begins at the element matrix level. The lumped element mass matrix is given by

$$M_L^e = \left\{ m_i^e \delta_{ij} \right\}, \qquad m_i^e = \sum_j m_{ij}^e. \tag{7.205}$$

In a similar vein, the element matrix K^e is transformed into its low-order counterpart

$$L^e = K^e + D^e$$

by adding a local artificial diffusion operator $D^e = \{d_{ij}\}^e$ defined by

$$d_{ij}^e = \begin{cases} \max\{-k_{ij}^e, 0, -k_{ji}^e\} & \text{if } j \neq i, \\ -\sum_{j\neq i} d_{ij}^e & \text{if } j = i. \end{cases} \tag{7.206}$$

After the global matrix assembly, one obtains a low-order scheme of the form (7.118). The main difference compared to edge-based FCT is the fact that the antidiffusive part $f(u)$ of the high-order system (7.120) is assembled from antidiffusive element contributions,

$$f_e = (M_L^e - M_C^e)\frac{du_e}{dt} - D^e u_e, \tag{7.207}$$

where u_e denotes the vector of local degrees of freedom. Since the element matrices $M_L^e - M_C^e$ and D^e are symmetric with zero row and column sums, we have

$$\sum_i f_{e,i} = 0. \tag{7.208}$$

It follows that the multiplication of an antidiffusive element contribution f_e by a correction factor $\alpha_e \in [0,1]$ does not change the total mass of the numerical solution.

The element-based approach makes it possible to circumvent the CFL-like condition (7.147) by using the backward Euler time stepping in the low-order scheme

$$\left[\frac{1}{\Delta t}M_L - L^{n+1}\right]u^L = \frac{1}{\Delta t}M_L u^n \tag{7.209}$$

and the Crank–Nicolson time stepping in the high-order scheme

$$\left[\frac{1}{\Delta t}M_C - \frac{1}{2}K^{n+1}\right]u^H = \left[\frac{1}{\Delta t}M_C + \frac{1}{2}K^n\right]u^n. \tag{7.210}$$

The antidiffusive element contributions associated with this combination are given by

$$f_e = \frac{1}{\Delta t}(M_L^e - M_C^e)(u_e^H - u_e^n) + \frac{1}{2}\left[K^{e,n+1}u_e^H + K^{e,n}u_e^n\right] - L^{e,n+1}u_e^L. \tag{7.211}$$

Due to integration by parts in the variational formulation (7.204), the element contributions $K^e u_e$ and $L^e u_e$ to the global vectors Ku and Lu are also massless unless the element borders the outflow boundary Γ_+ of the domain Ω. Hence, the multiplication of f_e by a correction factor α^e is conservative away from outflow boundaries. Let

$$\bar{f}_e = \alpha_e f_e, \qquad 0 \le \alpha_e \le 1,$$

denote the limited counterpart of f_e. The constrained Galerkin solution is given by

$$u_i^{n+1} = u_i^L + \frac{\Delta t}{m_i}\sum_e \bar{f}_{e,i}, \tag{7.212}$$

where $m_i = \sum_j m_{ij}$ is a diagonal entry of the global lumped mass matrix M_L.

The element-based version [144, 166, 167] of Zalesak's multidimensional FCT limiter [245] is designed to enforce the local maximum principle

$$u_i^{\min} \le u_i^{n+1} \le u_i^{\max},$$

where $u_i^{\max}$ and $u_i^{\min}$ are the local extrema of the low-order predictor:

$$u_i^{\max} = \max\{u_j^L : m_{ij} \neq 0\},$$
$$u_i^{\min} = \min\{u_j^L : m_{ij} \neq 0\}.$$

Assuming the worst-case scenario, the correction factors α_e are calculated as follows:

$$P_i^+ = \sum_e \max\{0, f_{e,i}\}, \qquad P_i^- = \sum_e \min\{0, f_{e,i}\},$$

$$Q_i^+ = \frac{m_i}{\Delta t}(u_i^{\max} - u_i^L), \qquad Q_i^- = \frac{m_i}{\Delta t}(u_i^{\min} - u_i^L),$$

$$R_i^+ = \min\left\{1, \frac{Q_i^+}{P_i^+}\right\}, \qquad R_i^- = \min\left\{1, \frac{Q_i^-}{P_i^-}\right\},$$

$$\alpha_e = \min_i \alpha_{e,i}, \qquad \alpha_{e,i} = \begin{cases} R_i^+ & \text{if } f_{e,i} > 0, \\ 1 & \text{if } f_{e,i} = 0, \\ R_i^- & \text{if } f_{e,i} < 0. \end{cases}$$

In the 1D case, this algorithm is equivalent to Zalesak's limiter for edge-based FCT.

For a more detailed presentation of the element-based FCT algorithm, we refer the reader to [144]. The element-based counterparts of other algebraic flux correction schemes have a similar algorithmic structure. We remark that the design of upwind-biased limiters for element contributions requires a decomposition of f_e into upwind and downwind components.

7.6.13 ▪ Numerical Examples

In this section, we apply the linearized FCT algorithm [134] and the linearity-preserving upwind-biased flux limiter [136] to two test problems with known analytical solutions. The third example illustrates the use of FCT in the context of constrained initialization.

Solid Body Rotation

The solid body rotation benchmark [158, 245] is often used to evaluate numerical advection schemes. The problem to be solved is the linear continuity equation

$$\frac{\partial u}{\partial t} + \nabla \cdot (\mathbf{v}u) = 0 \qquad \text{in } \Omega = (0,1) \times (0,1). \tag{7.213}$$

The prescribed velocity $\mathbf{v}$ describes a counterclockwise rotation about the center of Ω:

$$\mathbf{v}(x,y) = (0.5 - y, x - 0.5). \tag{7.214}$$

After each full revolution, the exact solution u coincides with the initial data u_0. Hence, the challenge of this test is to preserve the shape of u_0.

Following LeVeque [158], we simulate solid body rotation of the profile displayed in Fig. 7.2. The geometry of each body is described by a given function $G(x,y)$ defined on a circle of radius $r_0 = 0.15$ centered at some point $(x_0, y_0) \in \Omega$. Let

$$r(x,y) = \frac{1}{r_0}\sqrt{(x-x_0)^2 + (y-y_0)^2}$$

be the normalized distance from the point (x_0, y_0). Then $r(x,y) \le 1$ inside the circle.

The slotted cylinder is centered at the point $(x_0, y_0) = (0.5, 0.75)$ and

$$G(x,y) = \begin{cases} 1 & \text{if } |x - x_0| \ge 0.025 \text{ or } y \ge 0.85, \\ 0 & \text{otherwise.} \end{cases}$$

The sharp cone is centered at $(x_0, y_0) = (0.5, 0.25)$, and its shape is given by

$$G(x,y) = 1 - r(x,y).$$

The smooth hump is centered at $(x_0, y_0) = (0.25, 0.5)$, and the shape function is

$$G(x,y) = \frac{1 + \cos(\pi r(x,y))}{4}.$$

In the rest of the domain, the solution to (7.213) is initialized by zero, and homogeneous Dirichlet boundary conditions are prescribed at the inlets.

The snapshots presented in Figs. 7.2–7.5 show the shape of the solution at the final time $T = 2\pi$, which corresponds to one full rotation. All computations were performed on a uniform mesh of 128×128 bilinear elements using the Crank–Nicolson time-stepping with the time step $\Delta t = 10^{-3}$. The results obtained with $\alpha_{ij} := 1$ and $\alpha_{ij} := 0$ are displayed in Figs. 7.3 and 7.4, respectively. As expected, the unconstrained Galerkin solution exhibits spurious oscillations, while its low-order counterpart is too diffusive. The

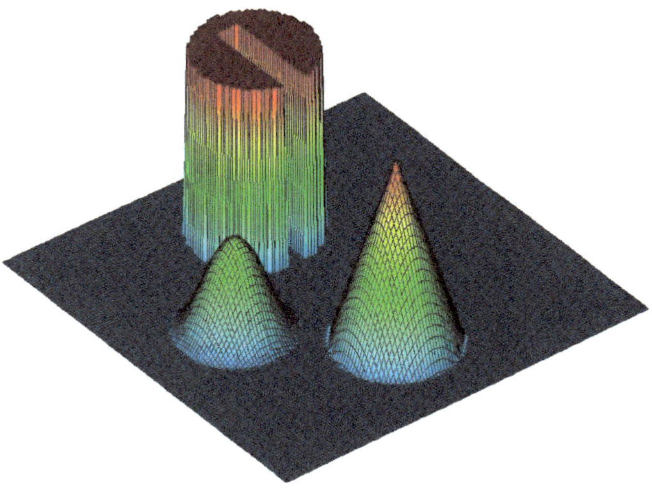

Figure 7.2. *Solid body rotation: initial data / exact solution at $t = 2\pi$ [137].*

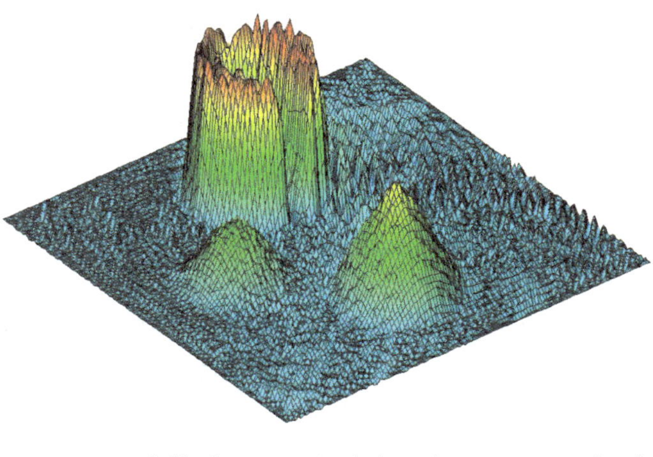

Figure 7.3. *Solid body rotation: Galerkin solution at $t = 2\pi$ [137].*

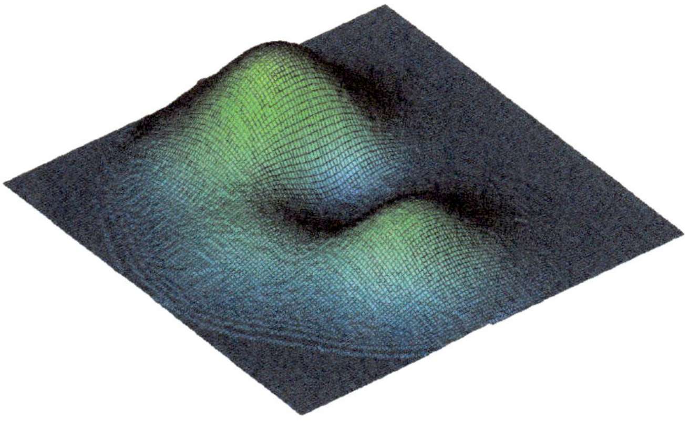

Figure 7.4. *Solid body rotation: low-order solution at* $t = 2\pi$ [137].

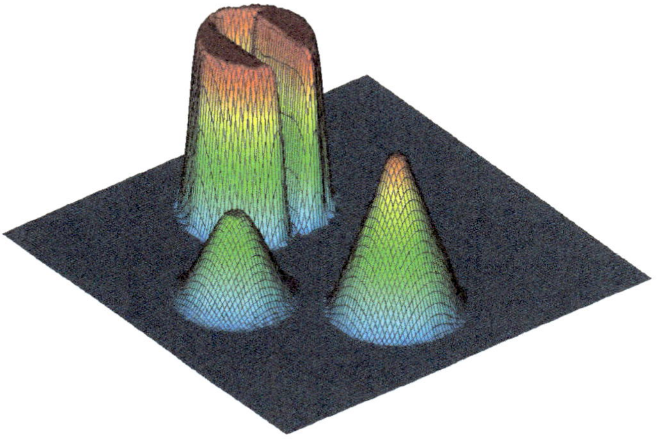

Figure 7.5. *Solid body rotation: FEM-FCT solution at* $t = 2\pi$ [137].

solution shown in Fig. 7.5 was computed using linearized FCT (see Section 7.6.1) with $\dot{u}^L$ given by (7.150).

John and Schmeyer [121] used the solid body rotation test for a detailed comparative study of FCT and conventional discontinuity capturing techniques for finite elements. FCT was the only method that did not produce undershoots or overshoots in this test.

Circular Convection

The second test problem is taken from [106]. Consider the hyperbolic PDE

$$\nabla \cdot (\mathbf{v}u) = 0 \qquad \text{in } \Omega = (-1,1) \times (0,1), \tag{7.215}$$

which describes steady circular convection if the velocity field is defined as

$$\mathbf{v}(x,y) = (y, -x).$$

The exact solution and inflow boundary conditions for this test are given by

$$u(x,y) = \begin{cases} G(r) & \text{if } 0.35 \le r = \sqrt{x^2+y^2} \le 0.65, \\ 0 & \text{otherwise,} \end{cases}$$

where $G(r)$ is a given function that defines the shape of the solution profile.

To evaluate the performance of the upwind-biased linearity-preserving flux limiter for smooth data and discontinuous solutions, we consider the following shape functions:

$$G_1(r) = \cos^2\left(5\pi \frac{2r+1}{3}\right), \qquad G_2(r) \equiv 1.$$

As before, computations are performed on a uniform mesh of bilinear finite elements which is successively refined to perform a grid convergence study.

The exact solution to the circular convection problem is constant along the streamlines of the stationary velocity field. Fig. 7.6 displays the results for $G = G_1$ and $G = G_2$ computed using the linearity-preserving upwind-biased flux limiter with $h = 1/64$.

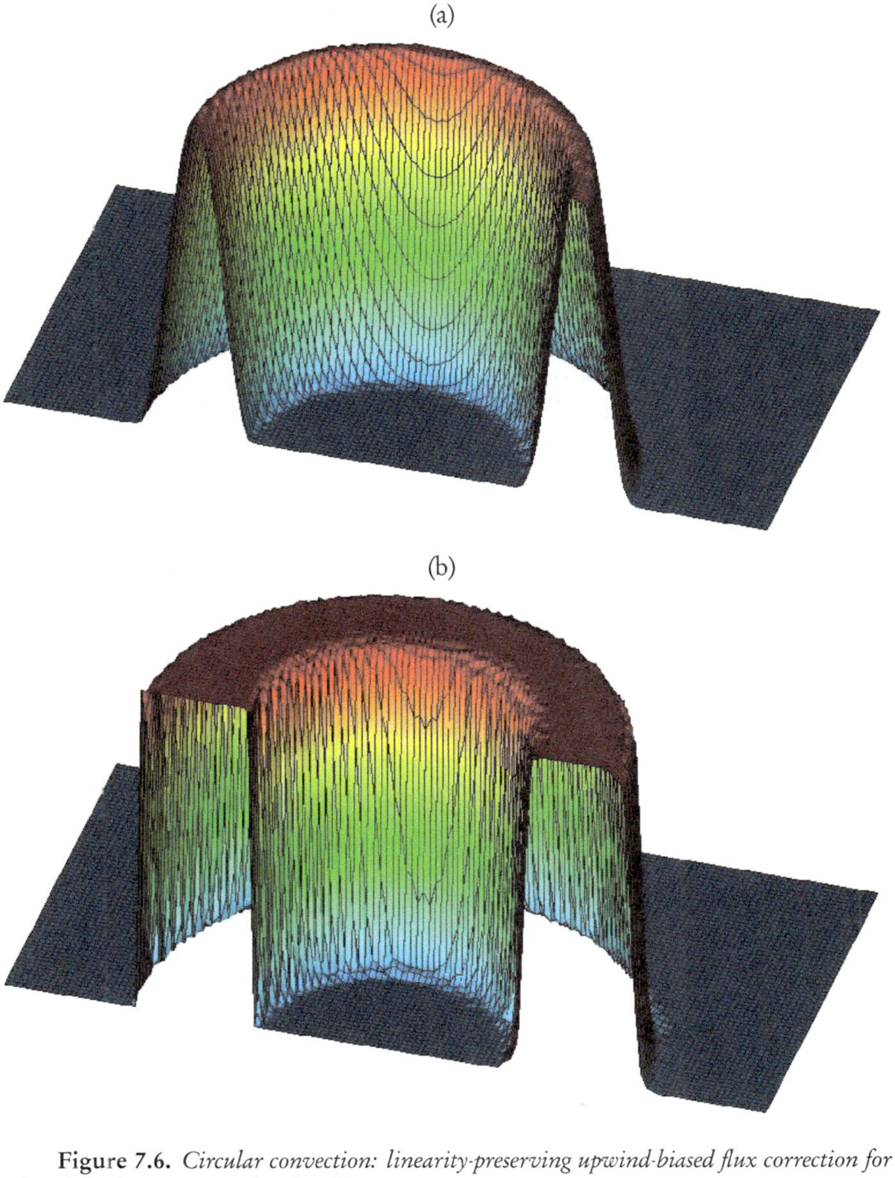

Figure 7.6. *Circular convection: linearity-preserving upwind-biased flux correction for* (a) *smooth and* (b) *discontinuous data* [137].

Constrained Interpolation

Conservative initialization plays an important role in CFD. The Galerkin L^2 projection (7.183) with $u = u_0$ is conservative but not monotone. On the other hand, the lumped-mass version is less accurate for smooth data. This is the rationale for the use of the FCT limiter presented in Section 7.3.8. The data to be projected may also represent a numerical solution calculated on a different mesh or using another set of basis functions. The transfer of information from one space to another is an integral part of adaptive finite element methods. The need for conservative and bounded interpolation arises whenever mesh coarsening, remeshing, or rezoning are performed.

To assess the accuracy of the FCT-based conservative projection scheme in the context of constrained interpolation, we triangulate the square domain $\Omega = (0,1)^2$ as shown in Fig. 7.7(a) and prescribe the following solution values at the vertices:

$$u_i = \begin{cases} 1.0 & \text{if } 0.3 \leq r_i, \\ 0.01 & \text{otherwise,} \end{cases}$$

where $r_i = \sqrt{(0.5-x_i)^2+(0.5-y_i)^2}$ is the distance from the center $(x_0, y_0) = (0.5, 0.5)$. The background mesh is constructed in such a way that a number of grid points lie on the circle of radius $r = 0.3$ where u is discontinuous; see Fig. 7.8(a).

The consistent, lumped, and constrained L^2 projection schemes are employed to transfer the above solution onto the generalized tensor product mesh shown in Fig. 7.7(b). This mesh is constructed by regular refinement of a quadrilateral 3×3 grid whose inner points are displaced in a nonsymmetric fashion. The interpolated solution profiles are depicted in Fig. 7.8(b)–(d). The consistent L^2 projection produces overshoots and undershoots of about 8.5%. In contrast, the nodal values obtained with the lumped-mass L^2 projection and FCT-constrained interpolation vary between 0.01 and 1.0. In the latter method, the diffusive mass lumping error is compensated for by a limited antidiffusive correction, which makes it possible to resolve discontinuities in a crisp and nonoscillatory manner [137].

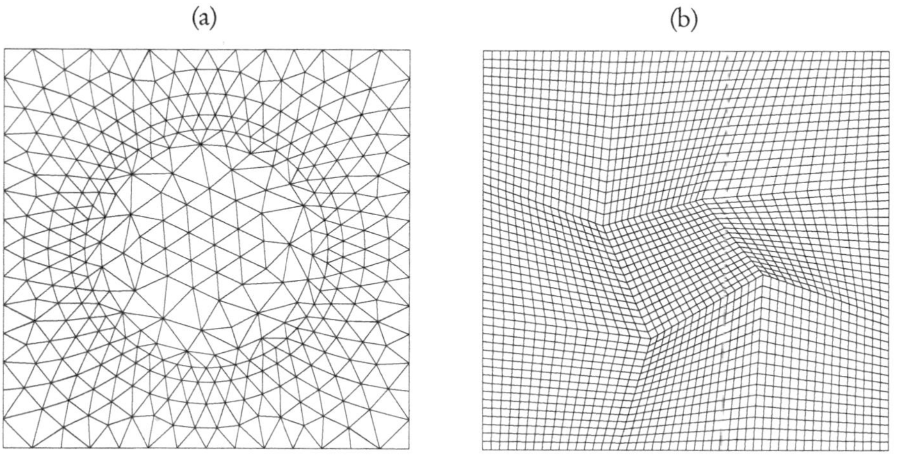

Figure 7.7. *Constrained interpolation: original* (a) *and target* (b) *meshes* [141].

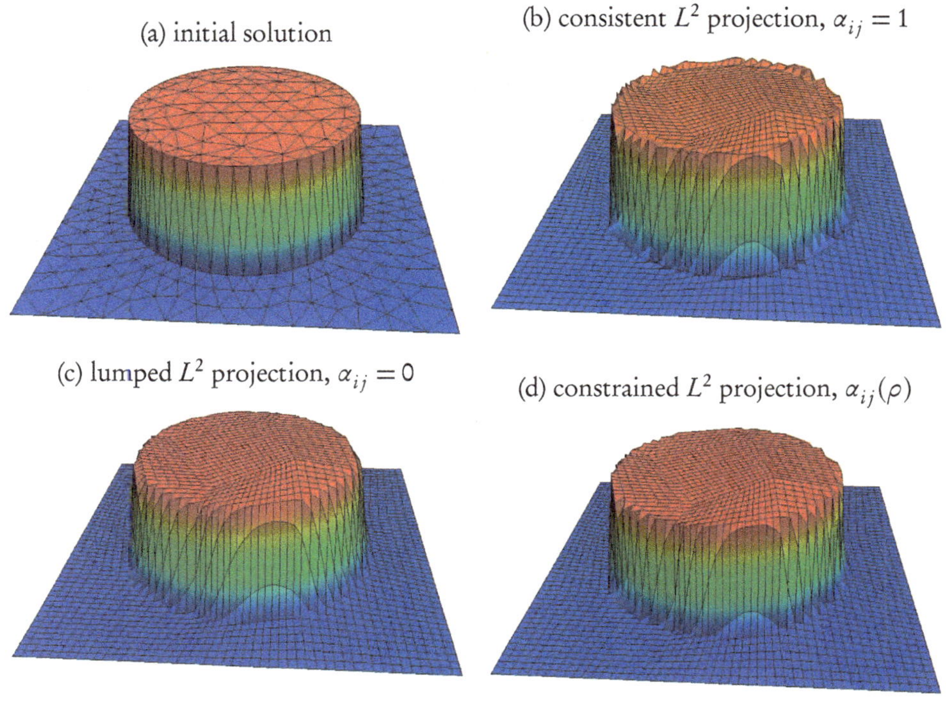

Figure 7.8. *Constrained interpolation: initial and interpolated solutions* [141].

7.7 ▪ Summary

In this chapter, we discussed the design of multidimensional finite element methods for convection-dominated transport problems. A survey of representative Petrov–Galerkin and Taylor–Galerkin schemes was included to introduce the key ideas. The reader may want to consult the excellent texts by Donea and Huerta [50] and Löhner [163] for further details and numerical examples. The presented material does not cover the whole range of available stabilization and discontinuity capturing techniques. In particular, semi-Lagrangian and characteristic methods, including those based on space-time finite element discretizations, have not even been mentioned. Likewise, no reference was made to stabilization by bubble functions, variational multiscale, or least-squares methods. For a detailed presentation of these approaches we refer the reader to the literature [37, 51, 70, 75, 122, 180, 249].

The algebraic flux correction paradigm presented in this chapter provides a set of tools that make it possible to enforce positivity constraints and discrete maximum principles in an adaptive way. The presented algorithms are based on a generalization of classical high-resolution schemes. All of them exhibit a common structure, but the actual flux/slope limiting procedure can be tailored to specific properties of a given transport problem.

In the context of finite element methods, an upwind-biased limiting strategy is the method of choice if the Péclet number is large and the mass lumping error is negligible/acceptable. The inclusion of the consistent mass matrix and the use of symmetric FCT limiters are to be recommended for transient computations with small time steps. The best choice of iterative solvers for linear and nonlinear systems is also problem-dependent.

The presentation of algebraic flux correction schemes in this chapter was aimed primarily at finite element practitioners who feel that positivity preservation is more important than the Galerkin orthogonality when it comes to numerical solution of problems with steep fronts. Instead of manipulating the variational formulation and fitting a free parameter, we have shown that artificial diffusion operators can be constructed at the discrete level so as to control the contribution of matrix entries associated with antidiffusive fluxes.

The unavoidable errors due to low-order artificial diffusion that cannot be removed in the neighborhood of steep fronts can be compensated for by means of local mesh refinement. The aspects of goal-oriented a posteriori error estimation and mesh adaptation for flux-corrected finite element approximations to transport equations are addressed in [140].

Chapter 8

Advanced Methods for Incompressible Flow Problems

Flow models based on the incompressible Navier–Stokes equations are widely used in CFD. In this chapter, we focus on the design of stable finite element approximations, fractional-step algorithms, and efficient iterative methods for discrete saddle-point problems. We also address the coupling with the transport equations of the k–ε turbulence model and the implementation of wall laws in finite element codes.

8.1 ▪ The Navier–Stokes Equations

The laminar flow of an incompressible fluid is governed by the Navier–Stokes equations

$$\frac{\partial \mathbf{u}}{\partial t} + \mathbf{u}\cdot\nabla\mathbf{u} = -\nabla p + \nu\Delta\mathbf{u}, \tag{8.1}$$

$$\nabla\cdot\mathbf{u} = 0, \tag{8.2}$$

where $\mathbf{u}$ is the velocity, p is the (modified) pressure, and $\nu = \mu/\rho$ is the kinematic viscosity.

In contrast to compressible flow models, the pressure p is not a thermodynamic variable satisfying an equation of state but a Lagrange multiplier that forces the velocity $\mathbf{u}$ to satisfy the incompressibility constraint. Furthermore, the Navier–Stokes system (8.1)–(8.2) does not contain an explicit equation for calculating the pressure. The momentum equation (8.1) and continuity equation (8.2) must be solved simultaneously to determine $\mathbf{u}$ and p.

We will solve (8.1)–(8.2) in a bounded domain Ω subject to the initial condition

$$\mathbf{u}(\mathbf{x},t) = \mathbf{u}_0(\mathbf{x}) \qquad \forall\mathbf{x}\in\Omega. \tag{8.3}$$

On the boundary $\Gamma = \Gamma_D \cup \Gamma_N$, we impose the velocity-stress boundary conditions

$$\mathbf{u} = \mathbf{g} \quad \text{on } \Gamma_D, \tag{8.4}$$

$$-p\mathbf{n} + \nu(\mathbf{n}\cdot\nabla\mathbf{u}) = \mathbf{h} \quad \text{on } \Gamma_N, \tag{8.5}$$

where $\mathbf{n}$ is the unit outward normal to Γ and $\mathbf{h}$ is the prescribed traction.

The initial data and Dirichlet boundary values must satisfy the compatibility condition

$$\mathbf{n}\cdot\mathbf{u}_0 = \mathbf{n}\cdot\mathbf{g} \quad \text{on } \Gamma_D. \tag{8.6}$$

In the case of a pure Dirichlet problem ($\Gamma_D = \Gamma$, $\Gamma_N = \emptyset$), the pressure p is defined up to an arbitrary constant because it does not appear in the boundary condition (8.4) and only its derivatives appear in the momentum equation (8.1). To obtain a well-posed problem, it is necessary to fix the value of p at a single node or impose the zero-mean constraint

$$\int_\Omega p \,\mathrm{d}\mathbf{x} = 0. \tag{8.7}$$

Integrating (8.2) over Ω and using the divergence theorem, one obtains

$$\int_\Omega \nabla \cdot \mathbf{u} \,\mathrm{d}\mathbf{x} = \int_\Gamma \mathbf{n} \cdot \mathbf{u} \,\mathrm{d}s = 0. \tag{8.8}$$

By (8.4), the normal components of $\mathbf{g}$ must satisfy the solvability condition

$$\int_\Gamma \mathbf{n} \cdot \mathbf{g} \,\mathrm{d}s = 0, \tag{8.9}$$

which implies that the convective flux across the inflow boundary ($\mathbf{n} \cdot \mathbf{g} < 0$) must be balanced by the flux across the outflow boundary ($\mathbf{n} \cdot \mathbf{g} > 0$).

In the presence of Neumann boundaries ($\Gamma_N \neq \emptyset$), the pressure is uniquely defined and condition (8.8) holds regardless of the Dirichlet boundary values prescribed on Γ_D.

It may appear odd that the pressure p does not need to be initialized and no pressure boundary condition is imposed explicitly. However, it turns out that p satisfies an elliptic boundary value problem known as the *pressure Poisson equation* (PPE) [86, 87, 207]. Assuming sufficient regularity, the application of the divergence operator to (8.1) yields

$$-\Delta p = \frac{\partial(\nabla \cdot \mathbf{u})}{\partial t} + \nabla \cdot (\mathbf{u} \cdot \nabla \mathbf{u} - \nu \Delta \mathbf{u}) \qquad \text{in } \Omega. \tag{8.10}$$

By virtue of the incompressibility constraint $\nabla \cdot \mathbf{u} = 0$, this Poisson equation reduces to

$$-\Delta p = \nabla \cdot (\mathbf{u} \cdot \nabla \mathbf{u} - \nu \Delta \mathbf{u}) \qquad \text{in } \Omega. \tag{8.11}$$

Taking the normal component of (8.1), one obtains the Neumann boundary condition

$$\mathbf{n} \cdot \nabla p = \mathbf{n} \cdot \left(\nu \Delta \mathbf{u} - \mathbf{u} \cdot \nabla \mathbf{u} - \frac{\partial \mathbf{u}}{\partial t} \right) \qquad \text{on } \Gamma. \tag{8.12}$$

The system of equations (8.1) and (8.10) is equivalent to (8.1)–(8.2) under the above regularity assumptions. Note that the PPE-based formulation still represents a coupled problem for $\mathbf{u}$ and p since the right-hand side of the momentum equation depends on the pressure gradient, while the right-hand side of the PPE depends on the derivatives of the velocity field. At the discrete level, the velocity and pressure can be decoupled using the operator splitting techniques presented in Section 8.2.2. This approach leads to efficient fractional-step algorithms in which p is calculated by solving a simplified PPE.

8.1.1 ▪ Finite Element Discretization

The numerical solution of the Navier–Stokes system (8.1)–(8.2) requires a careful choice of discretization techniques. We will begin with the choice of the time-stepping scheme and use the FEM when it comes to the discretization of space derivatives.

The forward Euler time discretization of the momentum equation (8.1) reads

$$\frac{\mathbf{u}^{n+1} - \mathbf{u}^n}{\Delta t} + \mathbf{u}^n \cdot \nabla \mathbf{u}^n = -\nabla p^n + \nu \Delta \mathbf{u}^n. \tag{8.13}$$

The resulting velocity field $\mathbf{u}^{n+1}$ may fail to satisfy the incompressibility constraint

$$\nabla \cdot \mathbf{u}^{n+1} = 0, \tag{8.14}$$

and the pressure p^n can only be determined by solving the PPE.

Due to the above, the pressure must be treated implicitly. The use of the backward Euler time-stepping scheme leads to the following coupled problem:

$$\frac{\mathbf{u}^{n+1} - \mathbf{u}^n}{\Delta t} + \mathbf{u}^{n+1} \cdot \nabla \mathbf{u}^{n+1} = -\nabla p^{n+1} + \nu \Delta \mathbf{u}^{n+1}, \tag{8.15}$$

$$\nabla \cdot \mathbf{u}^{n+1} = 0. \tag{8.16}$$

Let $\mathbf{W}$ and Q denote the functional spaces in which velocity and pressure reside. The variational form of the semidiscrete problem is derived using vector-valued test functions $\mathbf{w} \in \mathbf{W}$ and scalar-valued test functions $q \in Q$ for (8.15) and (8.16), respectively. Integrating over Ω and invoking the boundary conditions (8.4)–(8.5), we obtain

$$\begin{aligned} \int_\Omega \mathbf{w} \cdot \left(\frac{\mathbf{u}^{n+1} - \mathbf{u}^n}{\Delta t} \right) d\mathbf{x} + \int_\Omega \mathbf{w} \cdot (\mathbf{u}^{n+1} \cdot \nabla \mathbf{u}^{n+1}) \, d\mathbf{x} &= \int_\Omega (\nabla \cdot \mathbf{w}) p^{n+1} \, d\mathbf{x} \\ - \nu \int_\Omega \nabla \mathbf{w} \cdot \nabla \mathbf{u}^{n+1} \, d\mathbf{x} &+ \int_{\Gamma_N} \mathbf{w} \cdot \mathbf{h}^{n+1} \, ds \qquad \forall \mathbf{w} \in \mathbf{W}, \end{aligned} \tag{8.17}$$

$$\int_\Omega q \nabla \cdot \mathbf{u}^{n+1} \, d\mathbf{x} = 0 \qquad \forall q \in Q. \tag{8.18}$$

Introducing a pair of finite-dimensional approximation spaces $\mathbf{W}_h = \operatorname{span}\{\varphi_1, \ldots, \varphi_N\}$ and $\mathbf{Q}_h = \operatorname{span}\{\psi_1, \ldots, \psi_M\}$, the numerical solution to (8.17)–(8.18) is defined by

$$\mathbf{u}_h = \sum_{j=1}^{N} u_j \varphi_j, \qquad \mathbf{p}_h = \sum_{k=1}^{M} p_k \psi_k. \tag{8.19}$$

The discretized form of the variational problem (8.17)–(8.18) can be written as

$$a(\mathbf{w}_h, \mathbf{u}_h^{n+1}, \mathbf{u}_h^{n+1}) + b(p_h^{n+1}, \mathbf{w}_h) = f(\mathbf{w}_h) \qquad \forall \mathbf{w}_h \in \mathbf{W}_h, \tag{8.20}$$

$$b(q_h, \mathbf{u}_h^{n+1}) = 0 \qquad \forall q_h \in Q_h, \tag{8.21}$$

where

$$a(\mathbf{w}, \mathbf{v}, \mathbf{u}) = \frac{1}{\Delta t} \int_\Omega \mathbf{w} \cdot \mathbf{u} \, d\mathbf{x} + \int_\Omega \mathbf{w} \cdot (\mathbf{v} \cdot \nabla \mathbf{u}) \, d\mathbf{x} + \nu \int_\Omega \nabla \mathbf{w} \cdot \nabla \mathbf{u} \, d\mathbf{x},$$

$$b(q, \mathbf{w}) = -\int_\Omega q \nabla \cdot \mathbf{u} \, d\mathbf{x}, \qquad f(\mathbf{w}) = \frac{1}{\Delta t} \int_\Omega \mathbf{w} \cdot \mathbf{u}^n \, d\mathbf{x} + \int_{\Gamma_N} \mathbf{w} \cdot \mathbf{h}^{n+1} \, ds.$$

The fully discrete from of the coupled nonlinear problem has the structure

$$\begin{bmatrix} A & B \\ B^T & 0 \end{bmatrix} \begin{bmatrix} u \\ p \end{bmatrix} = \begin{bmatrix} f \\ 0 \end{bmatrix}, \tag{8.22}$$

where $u = [u_1, \ldots, u_N]^T$ and $p = [p_1, \ldots, p_M]^T$ are the vectors of unknown degrees of freedom, A is an $N \times N$ matrix that depends on $\mathbf{u}^{n+1}$, B is an $N \times M$ matrix with fixed coefficients, and f is a load vector that depends on $\mathbf{u}^n$ and $\mathbf{h}^{n+1}$. We have

$$a_{ij} = a(\varphi_i, \mathbf{u}^{n+1}, \varphi_j), \qquad b_{ik} = b(\varphi_i, \psi_k), \qquad f_i = f(\varphi_i). \tag{8.23}$$

The approximation spaces $\mathbf{W}_h$ and Q_h must be chosen so as to guarantee the existence of a unique solution to the nonlinear system (8.22). In particular, their dimensions must satisfy the necessary condition $M < N$; i.e., the number of velocity unknowns must be greater than the number of pressure unknowns. Otherwise, the discrete velocities become overconstrained and can be determined by solving $B^T u = 0$ without taking the momentum equation $Au + Bp = f$ into account. In the case $M > N$, the system matrix

$$G = \begin{bmatrix} A & B \\ B^T & 0 \end{bmatrix} \tag{8.24}$$

is singular due to the fact that $\operatorname{rank}(B) \leq N < M$. Even if a solution to (8.22) exists, it must be nonunique. For the matrix G to be invertible, the rank of the discrete gradient operator B must be equal to M; i.e., all columns of B must be linearly independent.

8.1.2 ▪ Stability Considerations

Numerical solutions to the incompressible Navier–Stokes equations may exhibit spurious oscillations in the velocity and/or pressure fields. The velocity oscillations are caused by the numerical instability of the convective terms at high Reynolds numbers. This side effect can be cured using the stabilization techniques presented in Chapters 6 and 7 using the scalar transport equation as a model problem. In our experience, linear stabilization (e.g., by streamline diffusion) is sufficient. The use of sophisticated discontinuity capturing or flux limiters is not required because no shocks can exist in viscous incompressible flows and small velocity fluctuations can be tolerated. In the turbulent flow regime, the discontinuity capturing effect is provided by the eddy viscosity of the turbulence model.

The arising of spurious checkerboard modes in the pressure field is caused by incompatibility of the finite-dimensional approximation spaces $\mathbf{W}_h$ and Q_h. Instabilities of this kind may occur not only at high Reynolds numbers but even in the absence of the convective term. The mixed finite element discretization of the linear Stokes problem

$$-\nu\Delta\mathbf{u} + \nabla p = 0, \tag{8.25}$$

$$\nabla \cdot \mathbf{u} = 0 \tag{8.26}$$

leads to a saddle-point problem of the form (8.22), where A is the stiffness matrix associated with $-\nu\Delta$. The global system matrix G is symmetric but indefinite and possibly singular. A unique solution to system (8.22) exists if and only if the block B has full rank [50].

Since A is invertible, the solution to the discretized momentum equation is given by

$$u = A^{-1}[f - Bp]. \tag{8.27}$$

Substituting this formula into $B^T u = 0$, one obtains the Schur complement equation

$$B^T A^{-1} B p = B^T A^{-1} f, \tag{8.28}$$

which represents a discrete counterpart of the PPE. The matrix

$$S := B^T A^{-1} B$$

is symmetric. It is positive definite if $\operatorname{Ker} B = \{0\}$. Note that A^{-1} is a full matrix, and so is S. In the case $\operatorname{Ker} B \neq \{0\}$, the velocity is uniquely defined by (8.27), but spurious pressure modes may exist due to the nonuniqueness of the solution to (8.28).

A finite element discretization of the Stokes problem is stable if the approximation spaces $\mathbf{W}_h$ and Q_h satisfy the *Ladyzhenskaya-Babuška-Brezzi* (LBB) condition [79]

$$\min_{q_h \in Q_h} \max_{\mathbf{w}_h \in \mathbf{W}_h} \frac{(q_h, \nabla \cdot \mathbf{w}_h)}{||q_h|| ||\nabla \mathbf{w}_h||} \geq \gamma > 0, \tag{8.29}$$

where $(\cdot,\cdot)$ is the L^2 scalar product, $||\cdot||$ is the L^2 norm, and γ is a mesh-independent constant γ. The lowest-order finite element approximations satisfying the inf-sup condition are the nonconforming Crouzeix–Raviart ($\tilde{P}_1/P_0$) and Rannacher–Turek ($\tilde{Q}_1/Q_0$) elements [42, 199]. Stable approximations with a continuous pressure field include the Taylor-Hood (P_2/P_1) element and the Bercovier–Pironneau (P_1-iso-P_2/P_1) element. For a comprehensive review of stable finite element pairs, we refer the reader to Quarteroni and Valli [195].

If the use of equal-order interpolations is desired, additional stabilization terms must be included to ensure uniqueness of the solution to the discrete problem. In the method proposed by Brezzi and Pitkäranta [22], the incompressibility constraint is replaced by

$$\nabla \cdot \mathbf{u} = h^2 \Delta p, \tag{8.30}$$

where h is the local mesh size. This modification is weakly consistent since the perturbation error is $\mathcal{O}(h^2)$. The Galerkin least-squares (GLS) method developed by Hughes, Franca, and Balestra [109] leads to a strongly consistent Petrov–Galerkin approximation in which the stabilization terms are defined as weighted residuals [74]. We used this kind of stabilization in Section 5.2.6.

In the pressure gradient projection (PGP) method [13, 38], the variational form (8.18) of the incompressibility constraint is augmented by the stabilization term $(\delta \nabla q_h, \mathbf{g}_h - \nabla p_h)$, where δ is a free parameter and $\mathbf{g}_h$ is the L^2 projection of ∇p_h into $\mathbf{W}_h$.

Burman and Hansbo [27] stabilized an equal-order P_1P_1 approximation by adding interior penalty terms. Many other stabilization techniques have been developed in the literature for circumventing the LBB condition [50]. In a generic *pseudocompressibility method*, the zero block of (8.24) is replaced by a regular matrix. The discrete momentum equation may also be modified to suppress convective and/or pressure instabilities.

A common drawback of stabilized methods is the presence of free parameters and the lack of a fail-safe strategy for choosing the values of these parameters.

8.1.3 ▪ Stabilized FEM

In this section, we outline the GLS and PGP methods for the linearized time-dependent Navier-Stokes equations

$$\frac{\partial \mathbf{u}}{\partial t} + \mathbf{v} \cdot \nabla \mathbf{u} = -\nabla p + \nu \Delta \mathbf{u}, \tag{8.31}$$

$$\nabla \cdot \mathbf{u} = 0, \tag{8.32}$$

where $\mathbf{v} \approx \mathbf{u}$ is a "frozen" velocity field from the previous outer iteration or time step.

Galerkin Least-Squares Method

The GLS method is derived by using modified test functions of the form [74]

$$\tilde{\mathbf{w}} = \mathbf{w} + \tau[\mathbf{v} \cdot \nabla \mathbf{w} - \nu \Delta \mathbf{w} + \nabla q] \tag{8.33}$$

and

$$\tilde{q} = q + \delta \nabla \cdot \mathbf{w}. \tag{8.34}$$

The sign of $\nu\Delta\mathbf{w}$ in the definition of $\tilde{\mathbf{w}}$ may be reversed [56, 74]. Clearly, the "+" and "−" versions are equivalent in the case of a piecewise-linear velocity approximation.

The stabilized version of the fully discrete problem (8.20)–(8.21) can be written as

$$a(\mathbf{w}_h,\mathbf{v},\mathbf{u}_h^{n+1})+(\mathbf{w}_h,\nabla p_h^{n+1})+s_\tau(\mathbf{w}_h,\mathbf{u}_h^{n+1},p_h^{n+1})=\tilde{f}(\mathbf{w}_h) \qquad \forall \mathbf{w}_h\in\mathbf{W}_h,$$
$$b(q_h,\mathbf{u}_h^{n+1})+s_\delta(q_h,\mathbf{u}_h^{n+1},p_h^{n+1})=f(\tau\nabla q_h) \qquad \forall q_h\in Q_h,$$

where $a(\cdot,\cdot,\cdot)$ and $b(\cdot,\cdot)$ correspond to the standard Galerkin discretization,

$$\tilde{f}(\mathbf{w})=f(\mathbf{w}+\tau[\mathbf{v}\cdot\nabla\mathbf{w}\pm\nu\Delta\mathbf{w}]),$$
$$f(\mathbf{w})=\frac{1}{\Delta t}(\mathbf{w},\mathbf{u}_h^n)+\int_{\Gamma_N}\mathbf{w}\cdot\mathbf{h}_\nu^{n+1}\,ds.$$

In contrast to (8.20)–(8.21), the pressure term is not integrated by parts. Hence, the traction vector $\mathbf{h}_\nu=\nu\mathbf{n}\cdot\nabla\mathbf{u}$ is determined by the viscous flux of momentum across Γ_N.

The GLS stabilization terms for the momentum and continuity equations are defined by

$$\begin{aligned} s_\tau(\mathbf{w}_h,\mathbf{u}_h^{n+1},p_h^{n+1})=&\,a(\tau[\mathbf{v}\cdot\nabla\mathbf{w}_h\pm\nu\Delta\mathbf{w}_h],\mathbf{v},\mathbf{u}_h^{n+1})\\ &+(\tau[\mathbf{v}\cdot\nabla\mathbf{w}_h\pm\nu\Delta\mathbf{w}_h],\nabla p_h^{n+1})\\ &+(\delta\nabla\cdot\mathbf{w}_h,\nabla\cdot\mathbf{u}_h^{n+1}) \end{aligned} \tag{8.35}$$

and

$$s_\delta(q_h,\mathbf{u}_h^{n+1},p_h^{n+1})=a(\tau\nabla q_h,\mathbf{v},\mathbf{u}_h^{n+1})+(\tau\nabla q_h,\nabla p_h^{n+1}). \tag{8.36}$$

The matrix form of the linearized algebraic system exhibits the following structure:

$$\left(\begin{bmatrix} A & B\\ B^T & 0\end{bmatrix}+\begin{bmatrix} S_{uu} & S_{up}\\ S_{pu} & S_{pp}\end{bmatrix}\right)\begin{bmatrix} u\\ p\end{bmatrix}=\begin{bmatrix}\tilde{f}\\ 0\end{bmatrix}. \tag{8.37}$$

The inf-sup stability is provided by the diagonal block S_{pp}, which is associated with the weak form $(\tau\nabla q,\nabla p)$ of the Brezzi–Pitkäranta stabilization term $-\tau\Delta p$.

Pressure Gradient Projection

In the PGP method [13, 38], the stabilized form of the continuity equation is given by

$$b(q_h,\mathbf{u}_h^{n+1})+s_\delta(q_h,p_h^{n+1},\mathbf{g}_h^{n+1})=0 \qquad \forall q_h\in Q_h, \tag{8.38}$$

where

$$s_\delta(q_h,p_h^{n+1},\mathbf{g}_h^{n+1})=(\delta\nabla q_h,\mathbf{g}_h^{n+1}-\nabla p_h^{n+1}) \tag{8.39}$$

and

$$(\mathbf{w}_h,\mathbf{g}_h^{n+1})=(\mathbf{w}_h,\nabla p_h^{n+1}) \qquad \forall\mathbf{w}_h\in\mathbf{W}_h. \tag{8.40}$$

The convective term may be stabilized using any of the methods presented in Chapter 7. This makes the PGP approach an attractive alternative to GLS, but the size of the linear systems is larger since the projected gradient $\mathbf{g}_h^{n+1}$ is coupled to $\mathbf{u}^{n+1}$ and p^{n+1}. For a presentation of efficient iterative solution techniques for the PGP system, we refer the reader to [13, 39].

8.2 ▪ Fractional-Step Methods

The numerical solution of the nonlinear saddle-point problem (8.22) is a rather challenging task. As an alternative to strongly coupled iterative solvers, efficient fractional-step methods have been developed for the time-dependent Navier–Stokes equations. The design of such methods is based on the divide-and-conquer strategy: the unwieldy coupled problem is decomposed into a sequence of smaller and simpler subproblems. Such decompositions can be constructed at the continuous or discrete level. The decoupling of the basic phenomena (nonlinear convection, viscous diffusion, and incompressibility) leads to simple and efficient algorithms in which the velocity and pressure are updated in a segregated manner.

8.2.1 ▪ Operator Splitting Methods

Many fractional-step methods for the Navier–Stokes equations are constructed using some sort of operator splitting. In Section 6.2.7, we used operator splitting techniques to decompose a 1D convection-diffusion equation into pure convection and pure diffusion problems.

Let us now consider a multidimensional initial-value problem of the form

$$\begin{cases} \dfrac{\partial u}{\partial t} + (\mathscr{L}_1 + \mathscr{L}_2)u = f, & t \in (t^n, t^{n+1}), \\ u(t^n) := u^n, \quad u^{n+1} := u(t^{n+1}), \end{cases} \tag{8.41}$$

where $\mathscr{L}_1$ and $\mathscr{L}_2$ are differential operators associated with different physical phenomena (e.g., convection and diffusion) or different directions (e.g., derivatives w.r.t. x and y).

The first-order accurate Marchuk–Yanenko method splits (8.41) into two subproblems:

$$\begin{cases} \dfrac{\partial u}{\partial t} + \mathscr{L}_1 u = 0, & t \in (t^n, t^{n+1}), \\ u(t^n) := u^n, \quad u^{n+1/2} := u(t^{n+1}). \end{cases} \tag{8.42}$$

$$\begin{cases} \dfrac{\partial u}{\partial t} + \mathscr{L}_2 u = f, & t \in (t^n, t^{n+1}), \\ u(t^n) := u^{n+1/2}, \quad u^{n+1} := u(t^{n+1}). \end{cases} \tag{8.43}$$

Importantly, the imposition of boundary conditions must ensure the well-posedness of each subproblem. For example, only inflow boundary conditions may be prescribed for hyperbolic equations, whereas in elliptic and parabolic problems a boundary condition must be prescribed at each boundary point. Thus, the choice of boundary conditions is an important issue if the splitting is performed at the differential (rather than algebraic) level.

Marchuk–Yanenko splitting makes it possible to choose the discretization techniques for each subproblem individually, making use of existing methods and software. For instance, it is possible to treat convective transport explicitly and diffusive transport implicitly. Moreover, local time stepping (alias *substepping*) may be used to advance the solution of a given subproblem from the time level t^n to the time level t^{n+1}. The modularity of the code simplifies programming and debugging, but the use of operator splitting may have an adverse effect on the accuracy and robustness of the numerical algorithm.

The price to be paid for the simplicity of the Marchuk–Yanenko method is an $\mathcal{O}(\Delta t)$ splitting error. Actually, the real rates of convergence may be higher if subproblems are discretized in time using higher-order methods. The three-step Strang splitting,

$$\begin{cases} \dfrac{\partial u}{\partial t} + \mathscr{L}_1 u = 0, & t \in (t^n, t^{n+1/2}), \\ u(t^n) := u^n, & \tilde{u}^{n+1/2} := u(t^{n+1/2}), \end{cases} \tag{8.44}$$

$$\begin{cases} \dfrac{\partial u}{\partial t} + \mathscr{L}_2 u = f, & t \in (t^n, t^{n+1}), \\ u(t^n) := \tilde{u}^{n+1/2}, & u^{n+1/2} := u(t^{n+1}), \end{cases} \tag{8.45}$$

$$\begin{cases} \dfrac{\partial u}{\partial t} + \mathscr{L}_1 u = 0, & t \in (t^{n+1/2}, t^{n+1}), \\ u(t^{n+1/2}) := u^{n+1/2}, & u^{n+1} := u(t^{n+1}), \end{cases} \tag{8.46}$$

is second-order accurate and just as easy to implement as the Marchuk–Yanenko method.

A popular fractional-step time-stepping method is Glowinski's θ-scheme [80, 81] in which one of the two operators is treated explicitly and the other implicitly at each step:

$$\frac{u^{n+\theta} - u^n}{\theta \Delta t} + \mathscr{L}_1 u^{n+\theta} = f^n - \mathscr{L}_2 u^n, \tag{8.47}$$

$$\frac{u^{n+1-\theta} - u^{n+\theta}}{(1-2\theta)\Delta t} + \mathscr{L}_2 u^{n+1-\theta} = f^{n+\theta} - \mathscr{L}_1 u^{n+\theta}, \tag{8.48}$$

$$\frac{u^{n+1} - u^{n+1-\theta}}{\theta \Delta t} + \mathscr{L}_1 u^{n+1} = f^{n+1-\theta} - \mathscr{L}_2 u^{n+1-\theta}, \tag{8.49}$$

where $0 < \theta < \frac{1}{2}$ is a splitting parameter. This fractional-step method is second-order accurate in time for $\theta = 1 - \frac{\sqrt{2}}{2}$ and first-order accurate otherwise.

In algebraic splitting methods, the differential operators $\mathscr{L}_1$ and $\mathscr{L}_2$ are replaced by their discrete counterparts, i.e., by matrices produced by a finite difference, finite volume, or finite element discretization. In contrast to differential splitting methods, the boundary conditions are already built into these matrices, but the splitting must be constructed in a way which guarantees that the matrix associated with each subproblem is invertible.

For a detailed introduction to operator splitting methods, we refer the reader to the texts by Glowinski [80], Quarteroni and Valli [195], and Hundsdorfer and Verwer [114].

Operator splitting methods developed for scalar equations of the form (8.41) can be extended to the case of more than two operators and to PDE systems. For example, Glowinski's fractional-step θ-scheme makes it possible to split the incompressible Navier–Stokes equations (8.1)–(8.2) into two linear Stokes problems and one nonlinear convection-diffusion problem (*viscous Burgers* equation). The three subproblems are as follows:

1. Stokes problem:

$$\begin{cases} \frac{\mathbf{u}^{n+\theta} - \mathbf{u}^n}{\theta \Delta t} - \eta \nu \Delta \mathbf{u}^{n+\theta} + \nabla p^{n+\theta} = (1-\eta)\nu \Delta \mathbf{u}^n - \mathbf{u}^n \cdot \nabla \mathbf{u}^n, \\ \nabla \cdot \mathbf{u}^{n+\theta} = 0. \end{cases}$$

2. Burgers equation:

$$\frac{\mathbf{u}^{n+1-\theta} - \mathbf{u}^{n+\theta}}{(1-2\theta)\Delta t} - (1-\eta)\nu \Delta \mathbf{u}^{n+1-\theta} + \mathbf{u}^{n+1-\theta} \cdot \nabla \mathbf{u}^{n+1-\theta} = \eta \nu \Delta \mathbf{u}^{n+\theta} - \nabla p^{n+\theta}.$$

3. Stokes problem:

$$\begin{cases} \frac{\mathbf{u}^{n+1}-\mathbf{u}^{n+1-\theta}}{\theta\Delta t} - \eta\nu\Delta\mathbf{u}^{n+1} + \nabla p^{n+1} = (1-\eta)\nu\Delta\mathbf{u}^{n+1-\theta} - \mathbf{u}^{n+1-\theta}\cdot\nabla\mathbf{u}^{n+1-\theta}, \\ \nabla\cdot\mathbf{u}^{n+1} = 0, \end{cases}$$

where $0 < \eta < \frac{1}{2}$. In this context, the Stokes operator ($\mathscr{L}_1$) is associated with a linear saddle-point problem which can be solved efficiently, e.g., using a variational conjugate gradient method [80]. The Burgers operator ($\mathscr{L}_2$) is associated with a nonlinear scalar conservation law which can be solved, e.g., using the fixed-point iteration method

$$\frac{\mathbf{u}^{(m+1)} - \mathbf{u}^{n+\theta}}{(1-2\theta)\Delta t} - (1-\eta)\nu\Delta\mathbf{u}^{(m+1)} + \mathbf{u}^{(m)}\cdot\nabla\mathbf{u}^{(m+1)} = \eta\nu\Delta\mathbf{u}^{n+\theta} - \nabla p^{n+\theta},$$

where $\mathbf{u}^{(m)}$ denotes an approximation to $\mathbf{u}^{n+1-\theta}$ initialized by $\mathbf{u}^{(0)} := \mathbf{u}^{n+\theta}$. Faster convergence of nonlinear iterations can be achieved using the Newton method or an adaptive defect correction scheme. Both approaches are described in detail by Turek [230].

We remark that the converged solution $\mathbf{u}^{n+1-\theta}$ is generally not divergence-free but the incompressibility constraint $\nabla\cdot\mathbf{u}^{n+1} = 0$ is enforced at the second Stokes step.

8.2.2 ▪ Continuous Projection Schemes

In the above fractional-step θ-scheme, the splitting is designed to separate the nonlinearity and incompressibility. The intimate coupling of the velocity and pressure still exists within the linear Stokes problems and requires the use of advanced linear algebra. In this section, we present a family of *projection schemes* which lead to fractional-step algorithms with segregated treatment of the velocity and pressure. The simplest representative of such Navier–Stokes solvers is the Chorin–Temam projection method [34, 35, 221, 222], which can be interpreted as a differential Marchuk–Yanenko splitting. In the case of pure Dirichlet boundary conditions ($\mathbf{u} = \mathbf{g}$ on Γ), the velocity and pressure are updated as follows:

1. Solve the viscous Burgers equation

$$\begin{cases} \frac{\mathbf{v}^{n+1}-\mathbf{u}^{n}}{\Delta t} + \mathbf{v}^{n+1}\cdot\nabla\mathbf{v}^{n+1} - \nu\Delta\mathbf{v}^{n+1} = 0 \quad \text{in } \Omega, \\ \mathbf{v}^{n+1} = \mathbf{g}^{n+1} \quad \text{on } \Gamma. \end{cases} \tag{8.50}$$

2. Solve the inviscid flow problem

$$\begin{cases} \frac{\mathbf{u}^{n+1}-\mathbf{v}^{n+1}}{\Delta t} + \nabla p^{n+1} = 0 \quad \text{in } \Omega, \\ \nabla\cdot\mathbf{u}^{n+1} = 0 \quad \text{in } \Omega, \\ \mathbf{n}\cdot\mathbf{u}^{n+1} = \mathbf{n}\cdot\mathbf{g}^{n+1} \quad \text{on } \Gamma. \end{cases} \tag{8.51}$$

No boundary condition is imposed on the tangential component of $\mathbf{u}^{n+1}$ since this would lead to an ill-posed problem. It follows that tangential slip may occur on solid walls.

At the first step of the Chorin–Temam method, the pressure gradient and the incompressibility constraint are neglected. As a consequence, the intermediate velocity $\mathbf{v}^{n+1}$ may fail to satisfy the incompressibility constraint. At the second step, the pressure gradient projects $\mathbf{v}^{n+1}$ into the space of divergence-free functions. By the Helmholtz theorem of vector calculus, a differentiable vector field $\mathbf{v}$ can be decomposed into a divergence-free component $\mathbf{u}$ and a curl-free component $\nabla\varphi$. In the above projection scheme, the final solution $\mathbf{u} := \mathbf{u}^{n+1}$ represents the divergence-free component of $\mathbf{v} := \mathbf{v}^{n+1}$, while the curl-free component is the gradient of $\varphi := \Delta t\, p$; see the book by Quartapelle [192] for a concise presentation of the mathematical theory behind such projection schemes.

In practice, the pressure p^{n+1} is calculated by solving a Poisson equation which is derived in the same way as (8.11). Taking the divergence and the normal component of the first equation in (8.51), one obtains the following representation of the projection step:

$$\begin{cases} -\Delta p^{n+1} = -\frac{1}{\Delta t}\nabla\cdot\mathbf{v}^{n+1} & \text{in } \Omega, \\ \mathbf{n}\cdot\nabla p^{n+1} = 0 & \text{on } \Gamma, \end{cases} \tag{8.52}$$

$$\mathbf{u}^{n+1} = \mathbf{v}^{n+1} - \Delta t \nabla p^{n+1}. \tag{8.53}$$

Since the solution to a pure Neumann problem for the Poisson equation is nonunique, the additive constant must be defined, e.g. using the zero-mean condition (8.7). The homogeneous Neumann boundary condition is nonphysical and inconsistent with (8.12), so the pressure approximation is wrong in a boundary layer of width $\mathcal{O}(\sqrt{\nu\Delta t})$. Since projection methods are designed to simulate high Reynolds number flows with small time steps, the pressure boundary layer is quite thin, and the overall accuracy is acceptable.

As shown by Rannacher [198], the Chorin–Temam method is equivalent to a pressure stabilization scheme written in terms of the nonsolenoidal intermediate velocities. Indeed, substituting $\mathbf{u}^n = \mathbf{v}^n - \Delta t \nabla p^n$ into the viscous Burgers problem (8.50), one obtains

$$\begin{cases} \dfrac{\mathbf{v}^{n+1}-\mathbf{v}^n}{\Delta t} + \mathbf{v}^{n+1}\cdot\nabla\mathbf{v}^{n+1} - \nu\Delta\mathbf{v}^{n+1} = -\nabla p^n & \text{in } \Omega, \\ \nabla\cdot\mathbf{v}^{n+1} - \Delta t\Delta p^{n+1} = 0 & \text{in } \Omega, \\ \mathbf{v}^{n+1} = \mathbf{g}^{n+1} & \text{on } \Gamma. \end{cases} \tag{8.54}$$

Remarkably, the continuity equation has the same structure as (8.30). It can be shown that equal-order finite element interpolations are stable for time steps satisfying

$$\Delta t \geq C h^{k+1},$$

where C is a positive constant and k stands for the polynomial degree [92, 190].

The Chorin–Temam projection method is only first-order accurate in time, and steady-state solutions depend on the value of the (pseudo-)time step Δt. If $\mathbf{v}^{n+1} = \mathbf{v}^n$ for sufficiently large n, then it follows that $p^{n+1} = p^n$ and $\mathbf{u}^{n+1} = \mathbf{v}^{n+1} - \Delta t\nabla p^{n+1}$.

A second-order accurate projection scheme can be constructed by adding the gradient of p^n to the viscous Burgers equation and solving a Poisson equation for the pressure increment. The following algorithm is known as the van Kan scheme [190, 237]:

1. Solve the viscous Burgers equation

$$\begin{cases} \dfrac{\mathbf{v}^{n+1}-\mathbf{u}^n}{\Delta t} + \dfrac{1}{2}(\mathbf{v}^{n+1}\cdot\nabla\mathbf{v}^{n+1} - \nu\Delta\mathbf{v}^{n+1}) \\ \qquad + \frac{1}{2}(\mathbf{u}^n\cdot\nabla\mathbf{u}^n - \nu\Delta\mathbf{u}^n) = -\nabla p^n & \text{in } \Omega, \\ \mathbf{v}^{n+1} = \mathbf{g}^{n+1} & \text{on } \Gamma. \end{cases} \tag{8.55}$$

2. Solve the inviscid flow problem

$$\begin{cases} \dfrac{\mathbf{u}^{n+1}-\mathbf{v}^{n+1}}{\Delta t} + \dfrac{1}{2}\nabla(p^{n+1}-p^n) = 0 & \text{in } \Omega, \\ \nabla\cdot\mathbf{u}^{n+1} = 0 \quad \text{in } \Omega, \\ \mathbf{n}\cdot\mathbf{u}^{n+1} = \mathbf{n}\cdot\mathbf{g}^{n+1} \quad \text{on } \Gamma. \end{cases} \tag{8.56}$$

This algorithm is based on the Helmholtz decomposition $\mathbf{v}^{n+1} = \mathbf{u}^{n+1} + \Delta t \nabla q^{n+1}$, where

$$q^{n+1} := \frac{1}{2}(p^{n+1}-p^n)$$

is determined by solving a Poisson equation. The projection step (8.56) is equivalent to

$$\begin{cases} -\Delta q^{n+1} = -\frac{1}{\Delta t}\nabla\cdot\mathbf{v}^{n+1}, \\ \mathbf{n}\cdot\nabla q^{n+1} = 0 \quad \text{on } \Gamma, \end{cases} \tag{8.57}$$

$$p^{n+1} = p^n + 2q^{n+1}, \tag{8.58}$$
$$\mathbf{u}^{n+1} = \mathbf{v}^{n+1} - \Delta t \nabla q^{n+1}. \tag{8.59}$$

By definition of q^{n+1}, the pressure field satisfies the Neumann boundary condition

$$\mathbf{n}\cdot\nabla p^{n+1} = \mathbf{n}\cdot\nabla p^n \qquad \text{on } \Gamma,$$

which is better than $\mathbf{n}\cdot\nabla p^{n+1} = 0$. The width of the pressure boundary layer is $\mathcal{O}(\sqrt{\nu\Delta t})$.

The continuity equation written in terms of the intermediate velocities becomes

$$\nabla\cdot\mathbf{v}^{n+1} = \frac{(\Delta t)^2}{2}\Delta\left(\frac{p^{n+1}-p^n}{\Delta t}\right) \approx \frac{(\Delta t)^2}{2}\Delta\left(\frac{\partial p}{\partial t}\right).$$

In contrast to the first-order projection scheme, the amount of intrinsic stabilization is generally insufficient to warrant the use of equal-order interpolations. On the positive side, steady-state solutions are independent of Δt since $p^{n+1} = p^n$ implies $\mathbf{u}^{n+1} = \mathbf{v}^{n+1}$.

The following algorithm unites the Chorin–Temam and van Kan projection schemes:

1. Solve the viscous Burgers equation

$$\begin{cases} \dfrac{\mathbf{v}^{n+1}-\mathbf{u}^n}{\Delta t} + \theta(\mathbf{v}^{n+1}\cdot\nabla\mathbf{v}^{n+1} - \nu\Delta\mathbf{v}^{n+1}) \\ \qquad + (1-\theta)(\mathbf{u}^n\cdot\nabla\mathbf{u}^n - \nu\Delta\mathbf{u}^n) = -\nabla\tilde{p}^n \quad \text{in } \Omega, \\ \mathbf{v}^{n+1} = \mathbf{g}^{n+1} \quad \text{on } \Gamma. \end{cases} \tag{8.60}$$

2. Solve the pressure correction equation

$$\begin{cases} -\Delta q^{n+1} = -\frac{1}{\Delta t}\nabla\cdot\mathbf{v}^{n+1} \quad \text{in } \Omega, \\ \mathbf{n}\cdot\nabla q^{n+1} = 0 \quad \text{on } \Gamma. \end{cases} \tag{8.61}$$

3. Correct the pressure

$$p^{n+1} = p^n + \alpha q^{n+1}. \tag{8.62}$$

4. Project the velocity
$$\mathbf{u}^{n+1} = \mathbf{v}^{n+1} - \Delta t \nabla q^{n+1}. \tag{8.63}$$

In Chorin's method, we have $\tilde{p}^n = 0$, $\theta = 1$, $\alpha = 0$. The van Kan scheme corresponds to $\tilde{p}^n = p^n$, $\theta = \frac{1}{2}$, $\alpha = 2$. The Burgers equation can be split into a convection step and a viscous diffusion step [80, 154]. The Marchuk–Yanenko splitting is defined by

$$\begin{cases} \dfrac{\partial \mathbf{u}}{\partial t} + \mathbf{u}\cdot\nabla\mathbf{u} = 0 \quad \text{in } \Omega, \quad t \in (t^n, t^{n+1}), \\ \mathbf{u} = \mathbf{g} \quad \text{on } \Gamma_- := \{\mathbf{x} \in \Gamma : \mathbf{n}\cdot\mathbf{u} < 0\}, \\ \mathbf{u}(t^n) := \mathbf{u}^n, \quad \mathbf{v}^{n+1/2} := \mathbf{u}(t^{n+1}), \end{cases} \tag{8.64}$$

$$\begin{cases} \dfrac{\partial \mathbf{u}}{\partial t} = -\nabla \tilde{p}^n + \nu\Delta\mathbf{u} \quad \text{in } \Omega, \quad t \in (t^n, t^{n+1}), \\ \mathbf{u} = \mathbf{g} \quad \text{on } \Gamma, \\ \mathbf{u}(t^n) := \mathbf{v}^{n+1/2}, \quad \mathbf{v}^{n+1} := \mathbf{u}(t^{n+1}). \end{cases} \tag{8.65}$$

A conditionally stable explicit approximation to the hyperbolic Burgers equation (8.64) can be constructed, e.g., using the Lax–Wendroff method that we used to discretize linear convection problems in Chapters 6 and 7. The second-order Taylor expansion yields

$$\mathbf{v}^{n+1/2} = \mathbf{u}^n + \Delta t\ \mathbf{u}_t^n + \frac{1}{2}(\Delta t)^2 \mathbf{u}_{tt}^n, \tag{8.66}$$

where $\mathbf{u}_t = -\mathbf{u}\cdot\nabla\mathbf{u}$ by (8.64) and $\mathbf{u}_{tt}$ is obtained by differentiating $\mathbf{u}_t$ thus:

$$\mathbf{u}_{tt} = -\mathbf{u}_t \cdot \nabla\mathbf{u} - \mathbf{u}\cdot\nabla\mathbf{u}_t = (\mathbf{u}\cdot\nabla\mathbf{u})\cdot\nabla\mathbf{u} + \mathbf{u}\cdot\nabla(\mathbf{u}\cdot\nabla\mathbf{u}). \tag{8.67}$$

The explicit treatment of the convective term eliminates the need for using an iterative method to solve the nonlinear problem. If the global time step Δt violates the stability condition, local time stepping may be used to calculate $\mathbf{v}^{n+1/2}$.

The parabolic viscous diffusion equation (8.64) must be discretized using an implicit scheme since a fully explicit time discretization would require the use of impractically small time steps (see Chapter 6). The generic discretization of (8.64) is given by

$$\frac{\mathbf{v}^{n+1} - \mathbf{v}^{n+1/2}}{\Delta t} + \theta\Delta\mathbf{v}^{n+1} + (1-\theta)\Delta\mathbf{v}^{n+1/2} = -\nabla\tilde{p}^n. \tag{8.68}$$

This scheme is second-order accurate in time for $\theta = \frac{1}{2}$ but the overall accuracy of the fractional-step method is $O(\Delta t)$ due to the first-order operator splitting error.

The space derivatives that appear in the continuous projection methods may be discretized using finite difference, finite volume, or finite element methods. The Galerkin finite element discretization of the viscous Burgers equation (8.60) leads to

$$M_C \frac{v^{n+1} - u^n}{\Delta t} - \theta[K + D]v^{n+1} = (1-\theta)[K+D]u^n - B\tilde{p}^n, \tag{8.69}$$

where M_C is the (velocity) mass matrix, $K = K(v^{n+1})$ is the discrete convection operator, D is the discrete Laplacian multiplied by ν, and B is the discrete gradient.

The discrete problems associated with (8.66) and (8.68) have the following structure:

$$M_C \frac{v^{n+1/2} - u^n}{\Delta t} = \left[K + \frac{\Delta t}{2} S\right] u^n, \tag{8.70}$$

$$M_C \frac{v^{n+1} - v^{n+1/2}}{\Delta t} - \theta D v^{n+1} - (1-\theta) D v^{n+1/2} = -B\tilde{p}^n, \tag{8.71}$$

where S is the matrix associated with the finite element discretization of (8.67) [154].

The discretization of the Poisson equation (8.66) leads to the linear system

$$Lq^{n+1} = \frac{1}{\Delta t} B^T v^{n+1}, \tag{8.72}$$

where L is the discrete Laplacian w.r.t. the finite element space Q_h. Since the entries of this matrix are scalar products of the form $(\nabla\psi_i, \nabla\psi_j) = \int_\Omega \nabla\psi_i \cdot \nabla\psi_j \, d\mathbf{x}$, the pressure must be approximated by piecewise-polynomial basis functions of degree $k \geq 1$.

The variational form of equation (8.63) is defined by the L^2 projection

$$\int_\Omega \mathbf{w}_h \cdot \mathbf{u}_h^{n+1} \, d\mathbf{x} = \int_\Omega \mathbf{w}_h \cdot (\mathbf{v}_h^{n+1} - \Delta t \nabla q^{n+1}) \, d\mathbf{x} \qquad \forall \mathbf{w}_h \in \mathbf{W}_h. \tag{8.73}$$

Thus, the velocity correction step gives rise to the well-conditioned linear system

$$M_C u^{n+1} = M_C v^{n+1} - \Delta t B q^{n+1}. \tag{8.74}$$

If the velocity space $\mathbf{W}_h$ corresponds to a piecewise-linear or multilinear finite element approximation, the use of row-sum mass lumping reduces (8.74) to

$$M_L u^{n+1} = M_L v^{n+1} - \Delta t B q^{n+1}. \tag{8.75}$$

Since $M_L = \{\delta_{ij} \sum_j m_{ij}\}$ is a diagonal matrix, this velocity update is fully explicit.

Due to the weak imposition of the Neumann boundary condition $\mathbf{n} \cdot \nabla q = 0$, the projected velocity $\mathbf{u}_h^{n+1}$ may violate the boundary condition $\mathbf{n} \cdot \mathbf{u}_h^{n+1} = \mathbf{n} \cdot \mathbf{v}_h^{n+1}$ on Γ. In particular, a nonphysical flow through a solid wall may occur if $\mathbf{n} \cdot \nabla q_h \neq 0$. This alarming side effect can be avoided by adding a symmetric penalty term of the form

$$\alpha \int_\Gamma (\mathbf{n} \cdot \nabla w)(\mathbf{n} \cdot \nabla q) \, ds$$

to the weak form of the Poisson equation [130]. The modified discrete problem is given by

$$(-L + P) q^{n+1} = -\frac{1}{\Delta t} B^T v^{n+1}, \tag{8.76}$$

where P is a symmetric matrix associated with the contribution of the penalty term.

The above discrete problems have unique solutions for any choice of the finite element pair $(\mathbf{W}_h, Q_h)$. The solution to problem (8.72) satisfies the linear system

$$B^T v^{n+1} = \Delta t L q^{n+1},$$

which represents a stabilized form of $B^T v^{n+1} = 0$. The weak compressibility of intermediate velocities makes equal-order interpolations stable for sufficiently large time steps. However, spurious checkerboard pressure modes may occur if Δt is chosen too small.

8.2.3 ▪ Discrete Projection Schemes

Instead of discretizing the Poisson equation, one may discretize the underlying inviscid flow problem and derive a Schur complement equation for the pressure. The projection of intermediate velocities into the space of discretely divergence-free functions is defined by

$$\begin{cases} M_C \frac{u^{n+1} - v^{n+1}}{\Delta t} + B q^{n+1} = 0, \\ B^T u^{n+1} = 0. \end{cases} \tag{8.77}$$

Since M_C is regular, the first step corresponds to the discrete Helmholtz decomposition

$$v^{n+1} = u^{n+1} + \Delta t M_C^{-1} B q^{n+1}.$$

It follows that $B^T u^{n+1} = B^T[v^{n+1} - \Delta t M_C^{-1} B q^{n+1}] = 0$, whence

$$B^T M_C^{-1} B q^{n+1} = \frac{1}{\Delta t} B^T v^{n+1}. \tag{8.78}$$

This Schur complement equation represents a mixed finite element discretization of (8.61).

The matrix $B^T M_C^{-1} B$ is full (because M_C^{-1} is). If the lumped mass matrix M_L is regular, the Schur complement operator may be approximated by the sparse matrix

$$C = B^T M_L^{-1} B.$$

The choice of the finite element pair $(\mathbf{W}_h, Q_h)$ must guarantee that the Schur complement matrix is invertible. In contrast to continuous projection methods, the use of a piecewise-constant pressure approximation is admissible, but C is singular if the Q_1Q_1 element is employed. A possible remedy is the use of the pressure gradient projection (PGP) stabilization method [13], which adds a symmetric positive definite matrix to C and leads to an inf-sup stable approximation. The Q_1Q_0 element is "slightly unstable but highly usable" according to Gresho and Sani [87]. The nonconforming Rannacher–Turek ($\tilde{Q}_1Q_0$) element is stable and well suited for the design of efficient multigrid solvers [229, 230, 233].

Discrete projection schemes can be interpreted as algebraic splitting methods for the coupled problem (8.22). The main algorithmic steps are as follows [87, 229]:

1. Solve the viscous Burgers equation
$$A v^{n+1} = f^n - B\tilde{p}^n. \tag{8.79}$$

2. Solve the Schur complement equation
$$B^T M_L^{-1} B q^{n+1} = \frac{1}{\Delta t} B^T v^{n+1}. \tag{8.80}$$

3. Correct the pressure
$$p^{n+1} = p^n + \alpha q^{n+1}. \tag{8.81}$$

4. Project the velocity
$$u^{n+1} = v^{n+1} - \Delta t M_L^{-1} B q^{n+1}. \tag{8.82}$$

As in the case of continuous projection methods, the computation of the intermediate velocity v^{n+1} may be split into a pure convection step and a viscous diffusion step.

8.2.4 ▪ Schur Complement Methods

The family of global pressure Schur complement methods introduced by Turek [229, 230] makes it possible to obtain a coupled solution to the discrete saddle-point problem

$$Au + Bp = f, \qquad B^T u = 0 \tag{8.83}$$

using a sequence of discrete projection cycles, which corresponds to an iterative method for solving the consistent pressure Schur complement equation

$$B^T A^{-1} B p = B^T A^{-1} f. \tag{8.84}$$

If the convective term is linearized, then the matrix A is independent of u, the pressure p can be determined by solving (8.84) and used to calculate the velocity u from $Au = f - Bp$. In the fully nonlinear version, the pressure Schur complement operator $S = B^T A^{-1} B$ depends on the velocity, so a number of outer iterations must be performed.

The practical implementation of the two-step algorithm also requires a number of inner iterations. Since the matrix A^{-1} is full, the assembly and storage of S would be prohibitively expensive in terms of CPU time and memory requirements. Thus, it is imperative to solve the pressure Schur complement (PSC) equation in an iterative way. For instance, consider a preconditioned Richardson's method based on the basic iteration

$$p^{(l)} = p^{(l-1)} + \alpha C^{-1} \left[\frac{1}{\Delta t} B^T A^{-1} f - S p^{(l-1)} \right], \tag{8.85}$$

where $l = 1, \ldots, L$ is the iteration counter, α is a relaxation parameter, C is an approximation to S, and the expression in the brackets is the residual of the PSC equation.

By definition of S, an equivalent form of the pressure correction equation (8.86) is

$$p^{(l)} = p^{(l-1)} + \alpha C^{-1} \frac{1}{\Delta t} B^T A^{-1} \left[f - \Delta t B p^{(l-1)} \right]. \tag{8.86}$$

If the time step Δt is small, then A is dominated by the contribution of the velocity mass matrix $M_C \approx M_L$. Therefore, the mixed Laplacian operator $C = B^T M_L^{-1} B$ constitutes a usable preconditioner for the Richardson iteration (8.86).

Of course, the matrices A and C are "inverted" by solving a linear system. Thus, the implementation of (8.86) can be split into the following basic tasks:

1. Given the pressure $p^{(l-1)}$, solve the viscous Burgers equation

$$Au^{(l)} = f - \Delta t B p^{(l-1)}. \tag{8.87}$$

2. Given the velocity $u^{(l)}$, solve the discrete Poisson equation

$$B^T M_L^{-1} B q^{(l)} = \frac{1}{\Delta t} B^T u^{(l)}. \tag{8.88}$$

3. Add the pressure increment $q^{(l)}$ to the current approximation:

$$p^{(l)} = p^{(l-1)} + \alpha q^{(l)}. \tag{8.89}$$

To enforce $B^T u = 0$, perform the divergence-free L^2 projection

$$u = u^{(l)} - \Delta t M_L^{-1} B q^{(l)}. \tag{8.90}$$

The projection step is included because the intermediate velocity $u^{(l)}$ is calculated using an approximate pressure $p^{(l-1)}$ and is generally not (discretely) divergence-free. Multiplying the velocity correction equation (8.90) by B^T and using (8.88), we obtain

$$B^T u = B^T u^{(l)} - \Delta t B^T M_L^{-1} B q^{(l)} = 0. \tag{8.91}$$

If just one basic iteration is performed, algorithm (8.87)–(8.90) has the structure of a classical projection scheme for the time-dependent incompressible Navier–Stokes equations. In particular, a discrete counterpart of Chorin's method is obtained with the trivial initial guess $p^{(0)} = 0$. The choice $p^{(0)} = p(t^n)$ leads to a discrete version of the van Kan scheme.

The advantages of iterative PSC solvers include [233]

- applicability to discontinuous pressure approximations,
- consistent treatment of boundary conditions (no splitting),
- alleviation of spurious boundary layers for the pressure,
- convergence to the fully coupled solution as l increases,
- possibility of using other global PSC preconditioners.

The PSC schemes implemented as discrete projection methods are more efficient than coupled solvers in the high Reynolds number regime. If the discrete evolution operator A is dominated by the mass matrix, it is sufficient to perform just one PSC iteration per time step. The number of inner iterations for the viscous Burgers equation (8.87) can also be as small as 1 since u^n is a good initial guess. If an optimized multigrid method is used to solve the pressure Poisson problem (8.87), the total cost per time step is just a small fraction of that for a coupled solver. However, the sparse matrix $B^T M_L^{-1} B$ may become a poor approximation to $B^T A^{-1} B$ at large time steps. The design of more sophisticated additive preconditioners for (8.86) is discussed in the books by Turek [230] and Elman, Silvester, and Wathen [61]. The reader is encouraged to consult these texts.

8.2.5 ▪ Inexact LU Factorizations

Another approach to the design of segregated algorithms for solving the Navier–Stokes system (8.83) is based on approximate block LU factorizations of the system matrix [187]

$$G = \begin{bmatrix} A & B \\ B^T & 0 \end{bmatrix} = \begin{bmatrix} A & 0 \\ B^T & -B^T A^{-1} B \end{bmatrix} \begin{bmatrix} I & A^{-1}B \\ 0 & I \end{bmatrix} = LU. \tag{8.92}$$

This representation of G suggests the use of a two-step block elimination algorithm:

- Backward elimination (U-step):

$$\begin{bmatrix} v \\ q \end{bmatrix} = \begin{bmatrix} I & A^{-1}B \\ 0 & I \end{bmatrix} \begin{bmatrix} u \\ p \end{bmatrix} \quad \Rightarrow \quad \begin{cases} u = v - A^{-1}Bp, \\ p = q. \end{cases}$$

- Forward elimination (L-step):

$$\begin{bmatrix} A & 0 \\ B^T & -B^T A^{-1} B \end{bmatrix} \begin{bmatrix} v \\ q \end{bmatrix} = \begin{bmatrix} f \\ 0 \end{bmatrix} \quad \Rightarrow \quad \begin{cases} Av = f, \\ B^T A^{-1} B q = B^T v. \end{cases}$$

This solution strategy leads to the following fractional-step method:

1. Solve the viscous Burgers equation
$$Av = f.$$
2. Solve the Schur complement equation
$$B^T A^{-1} B p = B^T v.$$
3. Correct the velocity field
$$Au = Av - Bp.$$

The assembly of the full matrix $S = B^T A^{-1} B$ can be avoided using a GMRES-like iterative solver that operates with matrix-vector products. The computation of $B^T A^{-1} B y$

would involve an iterative solution of the linear system $Ax = By$ followed by the matrix-vector multiplication $B^T x$. This procedure must be repeated as many times as necessary to reach the prescribed tolerance for the residual of the PSC equation. Hence, the cost per time step is likely to be very high even if multigrid acceleration is employed.

In many cases, the matrix-free "inversion" of S is impractical. The cost per time step is always the same, although a good initial guess is available when Δt is small. More efficient algorithms can be designed using inexact block LU factorizations of the form [194]

$$\tilde{G} = \begin{bmatrix} A & 0 \\ B^T & -B^T H_1 B \end{bmatrix} \begin{bmatrix} I & H_2 B \\ 0 & I \end{bmatrix} = \begin{bmatrix} A & \tilde{B} \\ B^T & Q \end{bmatrix} \approx G, \tag{8.93}$$

where

$$H_1 \approx A^{-1} \approx H_2, \qquad \tilde{B} = AH_2B.$$

A discretely divergence-free velocity approximation ($B^T u = 0$) is obtained with

$$H_1 = H = H_2, \qquad \tilde{B} = AHB.$$

This method results in a perturbation of the momentum equation since $\tilde{B} \neq B$.

If conservation of momentum is desired, the exact gradient must be employed. Let

$$H_1 \neq H_2 = A^{-1}, \qquad \tilde{B} = B.$$

This version corresponds to an artificial compressibility method with $B^T u + Qp = 0$, where the matrix $Q = B^T(A^{-1} - H_1)B$ acts as a pressure stabilization operator.

In the case $A^{-1} \neq H_1 \neq H_2 \neq A^{-1}$ neither mass nor momentum is conserved since $\tilde{B} \neq B$ and $Q \neq 0$. Such nonconservative approximations may provide enhanced stability.

For a given choice of H_1 and H_2, the fractional-step algorithm becomes the following:

1. Solve the viscous Burgers equation
$$Av = f.$$
2. Solve the Schur complement equation
$$B^T H_1 B p = B^T v.$$
3. Correct the velocity field
$$u = v - H_2 B p.$$

The choice $H_1 = \Delta t M_L^{-1} = H_2$ leads to the discrete version of Chorin's projection method.

The *Yoshida scheme* [194] employs an inexact block LU factorization with

$$H_1 = \Delta t M_L^{-1}, \qquad H_2 = A^{-1}, \qquad Q = B^T(\Delta t M_L^{-1} - A^{-1})B$$

and leads to the following projection-like fractional-step method:

$$Av = f, \qquad B^T M_L^{-1} B p = \frac{1}{\Delta t} B^T v, \qquad Au = Av - Bp. \tag{8.94}$$

Of course, the matrix $B^T M_L^{-1} B$ must be nonsingular for this algorithm to work. Equal-order interpolations can be stabilized by using inexact factorizations of the form [194]

$$\tilde{G} = \begin{bmatrix} A & 0 \\ B^T & -\Delta t(B^T M_L^{-1} B + \epsilon \tilde{M}) \end{bmatrix} \begin{bmatrix} I & A^{-1}B \\ 0 & I \end{bmatrix},$$

where $\tilde{M}$ denotes the pressure mass matrix and $\epsilon > 0$ is a regularization parameter.

8.3 • Implementation of the k–ε Model

Turbulence plays an important role in many incompressible flow problems. Since direct numerical simulation (DNS) of turbulent flows is unaffordable for Reynolds numbers of practical interest, eddy viscosity models based on the Reynolds-averaged Navier–Stokes (RANS) equations are commonly employed in CFD codes.

This section describes a numerical implementation of the k–ε model (see Section 3.4) which has been in use since the 1970s. To model the effect of unresolved velocity fluctuations, the viscous part of the Navier–Stokes equations is replaced with

$$\nabla\cdot(\nu+\nu_T)[\nabla\mathbf{u}+(\nabla\mathbf{u})^T],$$

where ν_T is the turbulent eddy viscosity. In the standard k–ε model [152, 176], ν_T depends on the turbulent kinetic energy k and its dissipation rate ε as follows:

$$\nu_T=C_\mu\frac{k^2}{\varepsilon},\qquad C_\mu=0.09.$$

The evolution of k and ε is governed by the convection-diffusion-reaction equations

$$\frac{\partial k}{\partial t}+\nabla\cdot\left(\mathbf{u}k-\left(\nu+\frac{\nu_T}{\sigma_k}\right)\nabla k\right)=P_k-\varepsilon, \tag{8.95}$$

$$\frac{\partial \varepsilon}{\partial t}+\nabla\cdot\left(\mathbf{u}\varepsilon-\left(\nu+\frac{\nu_T}{\sigma_\varepsilon}\right)\nabla \varepsilon\right)=\frac{\varepsilon}{k}(C_1P_k-C_2\varepsilon), \tag{8.96}$$

where $P_k=\frac{\nu_T}{2}|\nabla\mathbf{u}+\nabla\mathbf{u}^T|^2$ is responsible for the production of k. The involved empirical constants are given by $C_1=1.44,\ C_2=1.92,\ \sigma_k=1.0,\ \sigma_\varepsilon=1.3$.

The above equations are nonlinear and strongly coupled, which makes them very sensitive to the choice of numerical algorithms. In particular, the discretization procedure must be positivity-preserving because negative values of the eddy viscosity would produce numerical instabilities and eventually result in a crash of the code [139].

8.3.1 • Positivity-Preserving Linearization

In our implementation of the k–ε model [139], the incompressible Navier–Stokes equations are discretized using the nonconforming $\tilde{Q}_1/Q_0$ element pair. Standard Q_1 elements are employed for k and ε. The discretization of (8.95)–(8.96) yields [139, 146]

$$A_k(\mathbf{u},\nu_T)k=f_k, \tag{8.97}$$

$$A_\epsilon(\mathbf{u},\nu_T)\epsilon=f_\epsilon. \tag{8.98}$$

The use of algebraic flux correction for the convective terms is not sufficient for positivity preservation. Indeed, nonphysical negative values can also be produced by the right-hand sides f_k and f_ϵ. As shown by Patankar [185], a *negative slope linearization* of sink terms makes it possible to maintain positivity at the discrete level.

To write the equations of the k–ε model in the desired form, we introduce

$$\gamma=\frac{\varepsilon}{k}.$$

The negative slope linearization of (8.95)–(8.96) is based on the representation [159]

$$\frac{\partial k}{\partial t}+\nabla\cdot\left(\mathbf{u}k-\left(\nu+\frac{\nu_T}{\sigma_k}\right)\nabla k\right)+\gamma k=P_k, \tag{8.99}$$

$$\frac{\partial \varepsilon}{\partial t}+\nabla\cdot\left(\mathbf{u}\varepsilon-\left(\nu+\frac{\nu_T}{\sigma_\varepsilon}\right)\nabla \varepsilon\right)+C_2\gamma\varepsilon=\gamma C_1 P_k, \tag{8.100}$$

where ν_T and γ are evaluated using the solution from the previous outer iteration [146].

After solving the linearized equations (8.99) and (8.100), the new values of k and ε are used to calculate the linearization parameter γ for the next outer iteration, if any. The associated eddy viscosity ν_T is bounded below by a certain fraction of the laminar viscosity $0<\nu_{\min}\le\nu$ and above by $\nu_{\max}=l_{\max}\sqrt{k}$, where $l_{\max}$ is the maximum admissible mixing length (the size of the largest eddies, e.g., the width of the domain).

In the algorithm proposed in [139], the turbulent eddy viscosity

$$\nu_T=\max\{\nu_{\min}, l_*\sqrt{k}\} \tag{8.101}$$

is defined in terms of the limited mixing length

$$l_*=\begin{cases} C_\mu\frac{k^{3/2}}{\varepsilon} & \text{if } C_\mu k^{3/2}<\varepsilon l_{\max},\\ l_{\max} & \text{otherwise.}\end{cases} \tag{8.102}$$

The corresponding linearization parameter γ is given by

$$\gamma=C_\mu\frac{k}{\nu_T}. \tag{8.103}$$

The above representation makes it possible to avoid division by zero and obtain bounded nonnegative coefficients without manipulating the values of k and ε directly.

8.3.2 ▪ Initial Conditions

It is not always easy to find reasonable initial values for the k–ε model. If the velocity is initialized by zero, it takes the flow some time to become turbulent. Therefore, it is worthwhile to use a constant eddy viscosity ν_0 during a startup phase that ends at a certain time $t_*>0$. The values to be assigned to k and ε at $t=t_*$ depend on the choice of ν_0 and on the mixing length $l_0\in[l_{\min}, l_{\max}]$, where the threshold parameter $l_{\min}$ is related to the size of the smallest admissible eddies. Given the values of ν_0 and l_0, we define

$$k_0=\left(\frac{\nu_0}{l_0}\right)^2, \qquad \varepsilon_0=C_\mu\frac{k_0^{3/2}}{l_0}. \tag{8.104}$$

Alternatively, the initial values of k and ε can be determined a zero-equation turbulence model or an extension of the boundary conditions into the interior of the domain.

8.3.3 ▪ Boundary Conditions

The k–ε turbulence model is very sensitive to the choice and numerical implementation of boundary conditions. In particular, an improper near-wall treatment may lead to unrealistic solutions. The right choice of inflow values is also important in some applications. For this reason, we discuss the imposition of boundary conditions in some detail.

On the inflow boundary Γ_{in}, the values of all variables are commonly prescribed:

$$\mathbf{u} = \mathbf{g}, \qquad k = c_\infty |\mathbf{u}|^2, \quad \varepsilon = C_\mu \frac{k^{3/2}}{l_0} \quad \text{on} \quad \Gamma_{\text{in}}, \tag{8.105}$$

where $c_\infty \in [0.03, 0.1]$ is a constant and $|\mathbf{u}|$ denotes the magnitude of the velocity vector.

At the outlet Γ_{out}, the normal derivatives of all variables are set equal to zero:

$$\mathbf{n} \cdot [\nabla \mathbf{u} + \nabla \mathbf{u}^T] = 0, \qquad \mathbf{n} \cdot \nabla k = 0, \quad \mathbf{n} \cdot \nabla \varepsilon = 0 \qquad \text{on} \quad \Gamma_{\text{out}}. \tag{8.106}$$

In the context of finite element methods, the normal derivatives (diffusive fluxes) appear in the surface integrals that result from integration by parts in the variational form of the governing equations. These integrals do not need to be assembled if homogeneous Neumann ("do-nothing") boundary conditions of the form (8.106) are prescribed.

On a fixed solid wall Γ_w, the velocity must satisfy the no-penetration condition

$$\mathbf{n} \cdot \mathbf{u} = 0 \qquad \text{on} \quad \Gamma_w. \tag{8.107}$$

In laminar flow models, the tangential velocity of the fluid is also set equal to zero, so that the *no-slip condition* $\mathbf{u} = 0$ holds on the wall Γ_w. To avoid the need for resolving the viscous boundary layer in turbulent flow simulations, the boundary condition for the tangential direction is frequently given in terms of the wall shear stress

$$\mathbf{t}_w = \mathbf{n} \cdot \bar{\bar{\sigma}} - (\mathbf{n} \cdot \bar{\bar{\sigma}} \cdot \mathbf{n})\mathbf{n}, \qquad \bar{\bar{\sigma}} = \nu[\nabla \mathbf{u} + \nabla \mathbf{u}^T]. \tag{8.108}$$

If $\mathbf{t}_w$ is prescribed on Γ_w, then equation (8.107) is called the *free-slip condition* because the tangential velocity is defined implicitly and its value is generally unknown.

The practical implementation of the free-slip condition is nontrivial, unless the boundary of the domain is aligned with the axes of the Cartesian coordinate system. In contrast to the no-slip condition, (8.107) constrains a linear combination of several velocity components whose boundary values are unknown. Therefore, standard implementation techniques do not work. The free-slip condition can be implemented using element-by-element transformations to a local coordinate system aligned with the wall [64]. However, this strategy requires substantial modifications of the code. In our algorithm [139], we constrain the normal velocities to zero in an iterative way using projections of the form [139]

$$\mathbf{u} := \mathbf{u} - (\mathbf{n} \cdot \mathbf{u})\mathbf{n}.$$

8.3.4 ▪ Wall Functions

To complete the problem statement, we still have to prescribe the wall shear stress $\mathbf{t}_w$, as well as the boundary conditions for k and ε on the wall Γ_w. The equations of the standard k–ε model are invalid in the near-wall region, where the Reynolds number is low and viscous effects are dominant. To bridge the gap between the no-slip boundaries and the region of turbulent flow, analytical solutions to the boundary layer equations are frequently used to determine the values of $\mathbf{t}_w$, k, and ε near the wall. The logarithmic wall laws (see Section 3.4.4) that hold at a small distance y from the wall Γ_w are as follows:

$$\mathbf{t}_w = -u_\tau^2 \frac{\mathbf{u}}{|\mathbf{u}|}, \qquad k = \frac{u_\tau^2}{\sqrt{C_\mu}}, \qquad \varepsilon = \frac{u_\tau^3}{\varkappa y}, \tag{8.109}$$

where $\varkappa = 0.41$ is the von Karman constant. The friction velocity u_τ is given by

$$\frac{|\mathbf{u}|}{u_\tau} = \frac{1}{\varkappa} \log y^+ + \beta, \qquad y^+ = \frac{u_\tau y}{\nu}. \tag{8.110}$$

The value of the parameter β depends on the wall roughness ($\beta = 5.2$ for smooth walls). The above logarithmic relationship is valid for $11.06 \leq y^+ \leq 300$. The use of wall functions implies that a thin boundary layer of width y is removed, and the equations of the k–ε model should be solved in the reduced domain. Since the local Reynolds number y^+ is proportional to y, the wall distance should be chosen carefully. It is common to apply the wall laws (8.109) at the first internal node or integration point. However, the so-defined y depends on the mesh size and may fall into the viscous sublayer where (8.110) is invalid.

Another possibility is to adapt the mesh so that the location of boundary nodes always corresponds to a fixed value of y^+ which should be as small as possible for accuracy reasons. Taking the smallest value for which the logarithmic law still holds, one can neglect the width of the removed boundary layer and avoid mesh adaptation [90, 159]. In this case, the nodes located on the wall Γ_w should be treated as if they were shifted by the distance $y = \frac{y^+ \nu}{u_\tau}$ in the normal direction. As explained in [90], the smallest wall distance for the definition of y^+ corresponds to the point where the logarithmic layer meets the viscous sublayer. At this point, the linear relation $y^+ = \frac{|\mathbf{u}|}{u_\tau}$ and the logarithmic law (8.110) must hold, whence

$$y^+ = \frac{1}{\varkappa} \log y^+ + \beta. \tag{8.111}$$

This nonlinear equation can be solved iteratively. The resulting value of the parameter y^+ (for the default settings $\varkappa = 0.41$, $\beta = 5.2$) is given by $y_*^+ \approx 11.06$.

The relationship between y_*^+ and the friction velocity u_τ becomes very simple:

$$u_\tau = \frac{|\mathbf{u}|}{y_*^+}. \tag{8.112}$$

On the other hand, the wall boundary condition for k implies that

$$u_\tau = C_\mu^{0.25} \sqrt{k}. \tag{8.113}$$

Following Grotjans and Menter [90], we use a combination of the above to define

$$\mathbf{t}_w = -\frac{u_\tau}{y_*^+} \mathbf{u}, \qquad u_\tau = \max\left\{ C_\mu^{0.25} \sqrt{k}, \frac{|\mathbf{u}|}{y_*^+} \right\}. \tag{8.114}$$

This definition of $\mathbf{t}_w$ is consistent with (8.109) and prevents the momentum flux from going to zero at separation/stagnation points [90]. The natural boundary condition for the wall shear stress is used to evaluate the surface integral

$$\int_{\Gamma_w} \mathbf{t}_w \cdot \mathbf{w} \, ds = -\int_{\Gamma_w} \frac{u_\tau}{y_*^+} \mathbf{u} \cdot \mathbf{w} \, ds, \tag{8.115}$$

where $\mathbf{w}$ is the test function for the Galerkin weak form of the momentum equation.

By (8.109), the wall function for the turbulent eddy viscosity ν_T is given by

$$\nu_T = C_\mu \frac{k^2}{\varepsilon} = \varkappa u_\tau y = \varkappa y_*^+ \nu. \tag{8.116}$$

Note that this relation is satisfied automatically if the wall functions for k and ε are implemented in the strong sense. However, the use of Dirichlet boundary conditions implies that the values of k and ε depend on the flow field $\mathbf{u}$ via the friction velocity $u_\tau = \frac{|\mathbf{u}|}{y_*^+}$ but there is no feedback. This results in an unrealistic one-way coupling.

For the boundary values of k and ϵ to be "free" and influence the tangential velocity via relations (8.114)–(8.115), the wall functions must be implemented in a weak sense. Differentiating (8.109), one obtains the Neumann boundary conditions [90]

$$\mathbf{n}\cdot\nabla k = -\frac{\partial k}{\partial y} = 0, \qquad \mathbf{n}\cdot\nabla\varepsilon = -\frac{\partial \varepsilon}{\partial y} = \frac{u_\tau^3}{\varkappa y^2} = \frac{\varepsilon}{y}. \tag{8.117}$$

The unknown wall distance y can be expressed in terms of the turbulent eddy viscosity $\nu_T = \varkappa u_\tau y$, which yields a natural boundary condition of Robin type:

$$\mathbf{n}\cdot\nabla\varepsilon = \frac{\varkappa u_\tau}{\nu_T}\varepsilon, \qquad u_\tau = C_\mu^{0.25}\sqrt{k}. \tag{8.118}$$

The surface integrals associated with the Neumann boundary condition are given by

$$\int_{\Gamma_w} \frac{\nu_T}{\sigma_k}(\mathbf{n}\cdot\nabla k) w \, ds = 0, \tag{8.119}$$

$$\int_{\Gamma_w} \frac{\nu_T}{\sigma_\varepsilon}(\mathbf{n}\cdot\nabla\varepsilon) w \, ds = \int_{\Gamma_w} \frac{\varkappa u_\tau}{\sigma_\varepsilon}\varepsilon w \, ds. \tag{8.120}$$

Alternatively, the strong form of the wall law $\varepsilon = \frac{u_\tau^3}{\varkappa y} = \frac{u_\tau^4}{\varkappa y_*^+ \nu}$ may be used to prescribe a Dirichlet boundary condition for ε or evaluate the right-hand side of (8.120).

If the wall functions for ε and/or k are prescribed in a weak sense, it is essential to calculate ν_T and P_k using the strong form of the wall law. That is, the correct value of the turbulent eddy viscosity is given by (8.116), while the production term

$$P_k = \frac{u_\tau^3}{\varkappa y} = \frac{u_\tau^4}{\varkappa y_*^+ \nu} \tag{8.121}$$

is in equilibrium with the dissipation rate. The friction velocity u_τ is defined by (8.114).

8.3.5 ▪ Chien's Low Reynolds Number k–ε Model

Logarithmic laws provide a reasonably accurate description of the near-wall region, avoiding the need for costly integration to the wall. The derivation is valid only for flat-plate boundary layers and developed flow conditions, but wall functions of the form (8.109) are frequently used in more general settings with considerable success. The assumption that the viscous sublayer is very thin is justified in many cases. However, it is no longer safe to apply the wall functions on Γ_w if the wall distance associated with y_*^+ becomes too large.

A robust, albeit costly, alternative to wall laws is the use of *damping functions* that provide a smooth transition from laminar to turbulent flow. In Chien's low Reynolds number k–ε model [33], the turbulent eddy viscosity is redefined thus:

$$\nu_T = C_\mu f_\mu \frac{k^2}{\tilde{\varepsilon}}, \qquad f_\mu = 1 - \exp(-0.0115 y^+), \tag{8.122}$$

$$\tilde{\varepsilon} = \varepsilon - 2\nu\frac{k}{y^2}. \tag{8.123}$$

This popular model is supported by the DNS results, which indicate that the ratio $f_\mu = \frac{\nu_T \tilde{\varepsilon}}{C_\mu k^2}$ is not a constant but a function approaching zero at the wall.

The following modification of (8.99)–(8.100) is used in Chien's model [33]:

$$\frac{\partial k}{\partial t} + \nabla \cdot \left(\mathbf{u}k - \left(\nu + \frac{\nu_T}{\sigma_k} \right) \nabla k \right) + \alpha k = P_k, \tag{8.124}$$

$$\frac{\partial \tilde{\varepsilon}}{\partial t} + \nabla \cdot \left(\mathbf{u}\tilde{\varepsilon} - \left(\nu + \frac{\nu_T}{\sigma_\varepsilon} \right) \nabla \tilde{\varepsilon} \right) + \beta \tilde{\varepsilon} = \gamma C_1 f_1 P_k, \tag{8.125}$$

where the coefficients and damping functions are given by

$$\begin{aligned} \alpha &= \gamma + \frac{2\nu}{y^2}, \quad \beta = C_2 f_2 \gamma + \frac{2\nu}{y^2} \exp(-0.5 y^+), \\ \gamma &= \frac{\tilde{\varepsilon}}{k}, \quad f_1 = 1, \quad f_2 = 1 - 0.22 \exp\left(\frac{k^2}{6\nu\tilde{\varepsilon}} \right)^2. \end{aligned} \tag{8.126}$$

In contrast to wall functions, the boundary conditions on Γ_w are very simple:

$$\mathbf{u} = 0, \quad k = 0, \quad \tilde{\varepsilon} = 0 \quad \text{on} \quad \Gamma_w. \tag{8.127}$$

Note that the sink terms in (8.124) and (8.125) have positive coefficients, as required by Patankar's rule [185]. The value of y^+ is a function of the friction velocity:

$$y^+ = \frac{u_\tau y}{\nu}, \qquad u_\tau = \max\left\{ C_\mu^{0.25} \sqrt{k}, \sqrt{|\mathbf{t}_w|} \right\}. \tag{8.128}$$

The wall shear stress $\mathbf{t}_w$ is calculated using (8.108). In [139], the brute-force approach is used to compute the wall distance y. An efficient alternative is the use of PDE-based methods developed for calculating distance functions in level set methods [233].

8.3.6 ▪ Numerical Examples

To verify the above implementation of the k–ε turbulence model, a numerical study for two standard test problems was performed in [139, 233]. The first case study is used to validate the code for Chien's low Reynolds number $k-\varepsilon$ (LRKE) model. In the second example, we use the LRKE solution to evaluate the results obtained with logarithmic wall functions implemented as Dirichlet (DIRBC) and Neumann (NEUBC) boundary conditions.

Channel Flow Problem

In the first numerical example, we simulate the turbulent channel flow at $\mathrm{Re}_\tau = 395$ based on the friction velocity u_τ, half of the channel width d, and kinematic viscosity ν. The reference data for this well-known benchmark problem are provided by the DNS results of Kim, Moin, and Moser [127]. In order to obtain the developed flow conditions required for validation, the inflow and outflow boundary conditions for the reduced domain were swapped repeatedly so as to emulate periodic boundary conditions.

The equations of the LRKE model are solved with the 3D code on a hexahedral mesh of 50,000 elements. Due to the need for high resolution, local mesh refinement is performed in the near-wall region, as shown in Fig. 8.1. The distance from the wall boundary

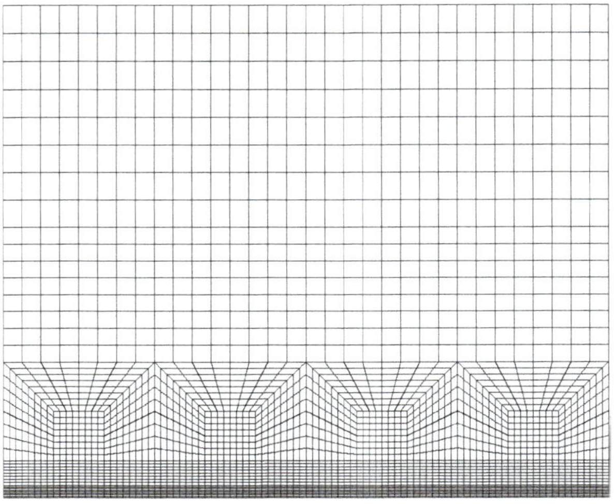

Figure 8.1. *Channel flow: local mesh refinement in the boundary layer* [139].

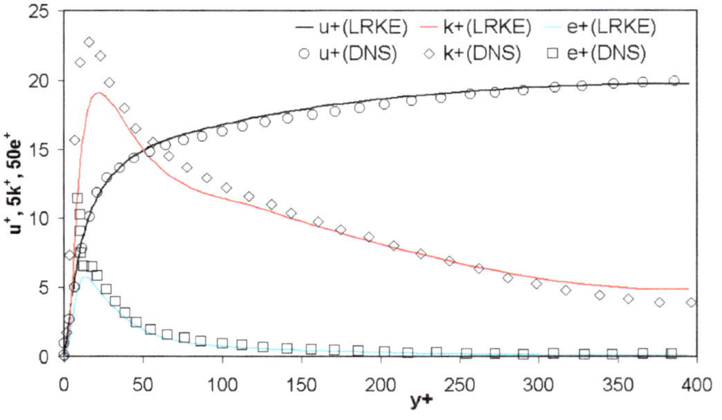

Figure 8.2. *Channel flow: LRKE solutions vs. Kim's DNS results for* $\mathrm{Re}_\tau = 395$ [139].

to the nearest interior point corresponds to $y^+ \approx 2$. The numerical results for this test are presented in Fig. 8.2. The profiles of the nondimensional quantities

$$u^+ = \frac{u_x}{u_\tau}, \quad k^+ = \frac{k}{u_\tau^2}, \quad \varepsilon^+ = \frac{\varepsilon\nu}{u_\tau^4}$$

are in good agreement with the DNS results [127] for this benchmark. The calculated profiles of u^+ and ε^+ are particularly close to the reference data.

Backward Facing Step

In the second example, we simulate the turbulent flow past a backward facing step in three dimensions. The definition of the Reynolds number $\mathrm{Re} = 47{,}625$ is based on the step height H, mean inflow velocity u_{mean}, and kinematic viscosity ν. The objective is to evaluate the performance of the k–ε model with three different kinds of near-wall treatment: LRKE vs. DIRBC and NEUBC implementation of wall functions.

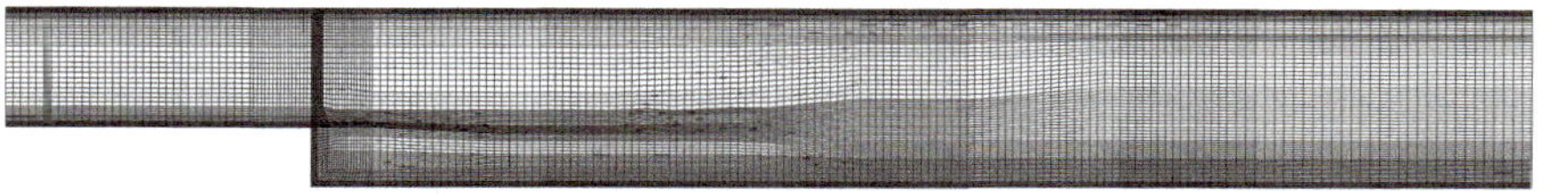

Figure 8.3. *Backward facing step: a 2D view of the computational mesh in the xy-plane* [139].

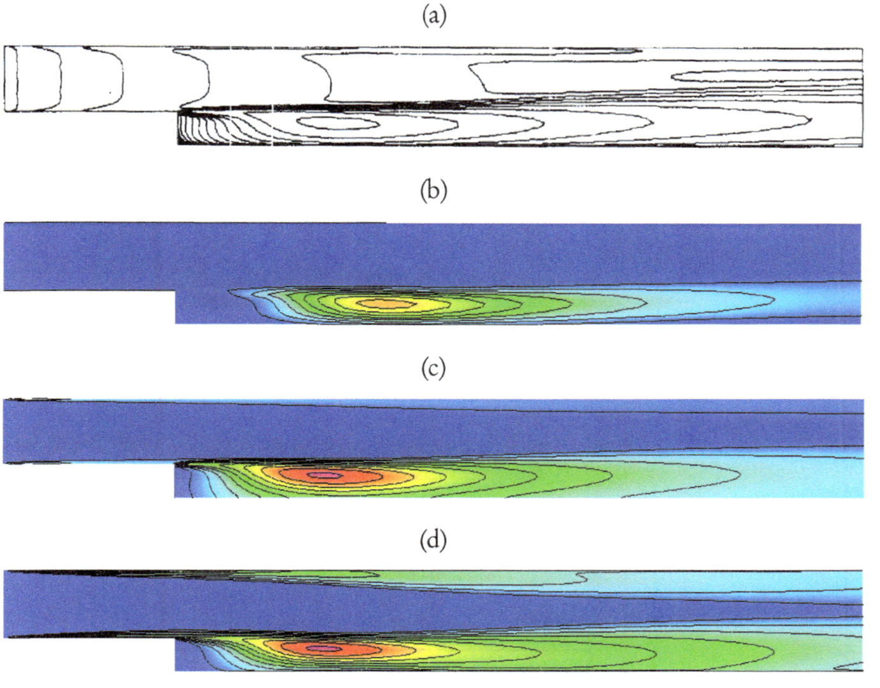

Figure 8.4. *Backward facing step: steady-state distribution of k for* Re = 47,625. (a) *Reference solution* [115], (b) *DIRBC solution,* (c) *NEUBC solution,* (d) *LRKE solution* [139].

All simulations are performed on the same mesh that consists of approximately 260,000 hexahedral elements. Local mesh refinement is performed in the near-wall region and behind the step (see Fig. 8.3). A comparison of the steady-state solutions for the turbulent kinetic energy k and eddy viscosity ν_T with the reference solution from [115] is presented in Figs. 8.4 and 8.5. Significant differences between the solutions computed using the strong and weak forms of logarithmic wall functions are observed even in the "eyeball norm." The DIRBC version was found to produce disappointing results in this test, whereas the accuracy of the NEUBC and LRKE solutions is comparable.

An important evaluation criterion for this popular test problem is the recirculation length defined as $L_R = x_r/H$. For the DIRBC implementation, this integral quantity can be readily inferred from the distribution of the skin friction coefficient

$$c_f = \frac{u_\tau^2}{u_{\text{mean}}^2}\frac{u_x}{|u_x|}$$

on the bottom wall (see Fig. 8.6). The recirculation length predicted by LRKE and NEUBC is underestimated ($L_R \approx 5.4$). The computational results published in the literature exhibit the same trend ($5.0 < L_R < 6.5$; see [90, 115, 224]). On the other hand, DIRBC yields

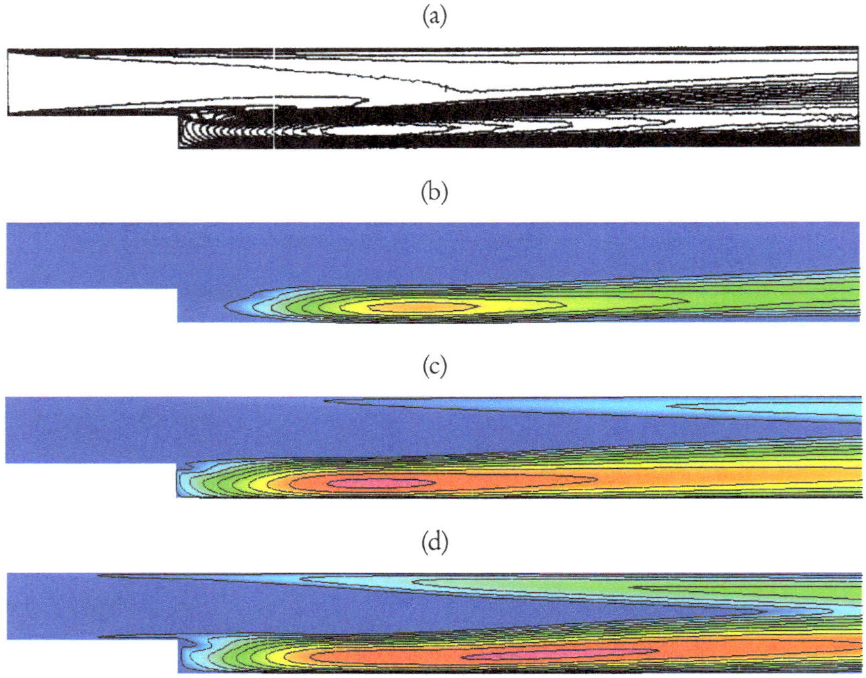

Figure 8.5. *Backward facing step: steady-state distribution of ν_T for* Re = 47,625. (a) *Reference solution* [115], (b) *DIRBC solution,* (c) *NEUBC solution,* (d) *LRKE solution* [139].

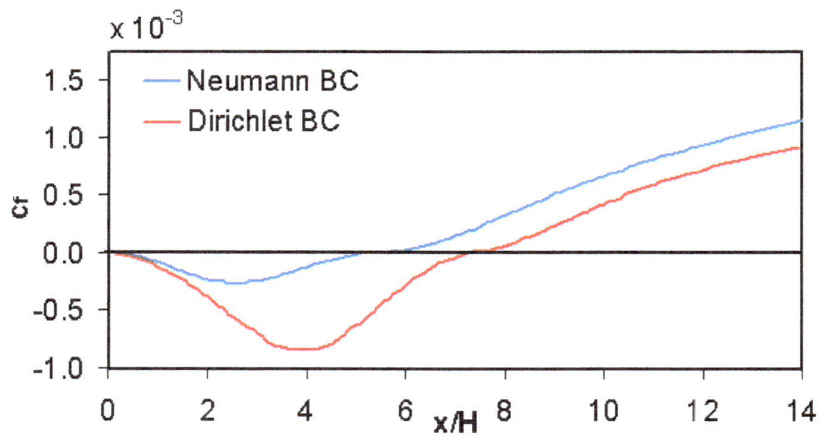

Figure 8.6. *Backward facing step: distribution of c_f along the lower wall,* Re = 47,625 [139].

$L_R \approx 7.1$, which matches the experimentally measured recirculation length ($L_R \approx 7.1$; see [126]). Unfortunately, this perfect agreement turns out to be pure coincidence.

In Fig. 8.7, the calculated velocity profiles for 6 different distances from the step are compared to one another and to the experimental data from Kim's thesis [126]. The corresponding profiles of k and ε are displayed in Fig. 8.8 and Fig. 8.9, respectively. This comparative study indicates that NEUBC yields essentially the same results as Chien's low Reynolds number model, whereas the use of DIRBC leads to a significant discrepancy,

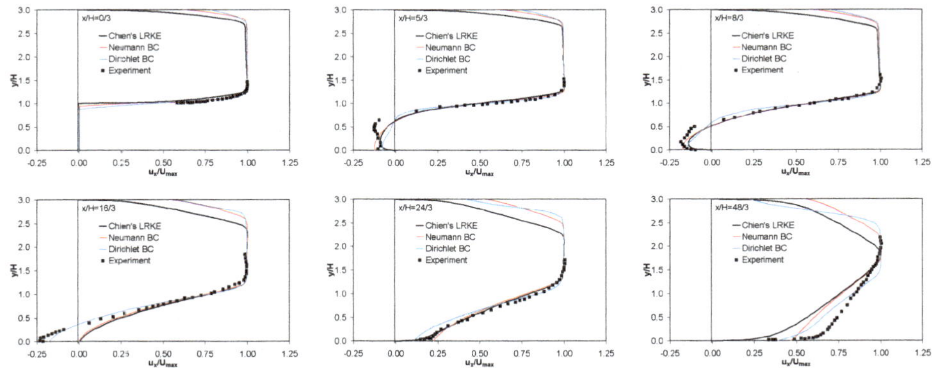

Figure 8.7. *Backward-facing step: profiles of* u_x *for* 6 *different distances* x/H *from the step* [139].

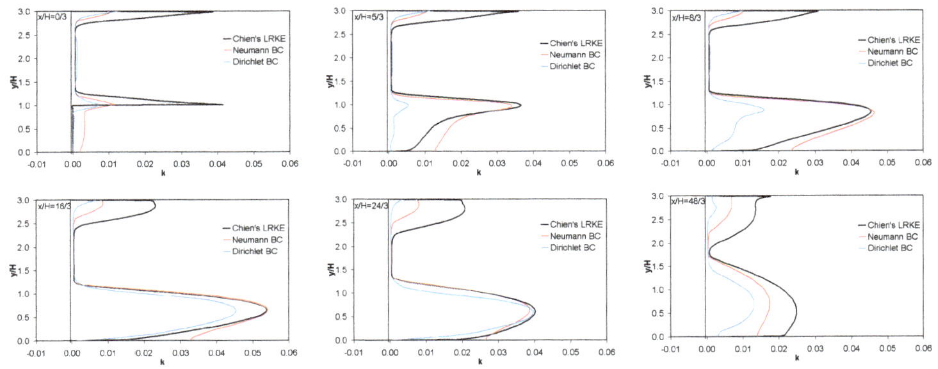

Figure 8.8. *Backward-facing step: profiles of k for* 6 *different distances* x/H *from the step* [139].

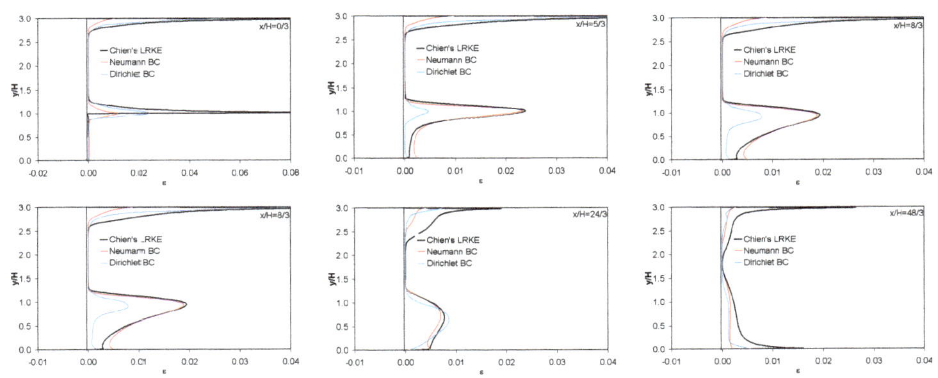

Figure 8.9. *Backward-facing step: profiles of* ε *for* 6 *different distances* x/H *from the step* [139].

especially at small distances from the step. The presented profiles of ε do not suffer from spurious undershoots, which are frequently observed in other computations. This can be attributed to the positivity-preserving treatment of the convective terms and sinks [233].

8.4 • Summary

In this chapter, we presented some useful numerical tools for handling advanced flow models based on the incompressible Navier–Stokes equations. Our brief introduction covers just a small fraction of the algorithms currently in use. For further reading on advanced incompressible flow solvers, we refer the reader to Blasco, Codina, and Huerta [13], Cuvelier, Segal, and van Steenhoven [44], Donea and Huerta [50], Elman, Silvester, and Wathen [61], Glowinski [80], Gresho and Sani [87], Quarteroni and Valli [195], and Turek [230]. The mathematical theory of finite element methods for the Navier–Stokes equations is covered in the books by Boffi, Brezzi, and Fortin [15], Brezzi and Fortin [21], Girault and Raviart [79], and Layton [155]. The mathematical foundations of projection methods are presented in the monographs by Prohl [190] and Quartapelle [192]. The analysis of the k–ε turbulence model can be found in the book by Mohammadi and Pironneau [176]. We encourage the reader to consult the above texts for a focused presentation of selected methods.

Chapter 9

Further Reading

In this guidebook, we introduced the equations of fluid mechanics, as well as a distillation of finite element methods for solving scalar convection-diffusion equations and incompressible flow problems. Although we have covered a large number of topics in a fairly self-contained manner, some questions that the reader might have will certainly remain unanswered. Many answers can be found in the references that we have cited throughout the book. Therefore, we conclude this text with a short list of recommended readings on assorted topics:

- Equations of fluid dynamics [3, 9, 105, 148]
- CFD introductions [3, 71, 104, 182, 185, 69, 241]
- Numerical methods for PDEs [151, 195]
- Numerical linear algebra [83, 93, 124, 201, 242]
- Mesh generation [29, 78, 76, 225, 163]
- FEM introductions [44, 91, 122, 129, 217, 249]
- Advanced FEMs [1, 8, 20, 45, 101, 218]
- FEMs for CFD [50, 160, 163, 188, 195, 250]
- Convection-diffusion equations [60, 70, 114, 180, 200, 205]
- Hyperbolic conservation laws [138, 156, 157, 226, 228]
- Operator splitting methods [80, 114, 195]
- Incompressible flows [44, 58, 61, 80, 87, 155, 192, 195, 230]
- Compressible flows [4, 68, 149, 157, 163, 138, 226]
- Turbulent flows [103, 153, 155, 176, 243]
- Large eddy simulation [69, 88, 119, 204]
- Multiphase flows [43, 57, 102, 128, 197]
- Free and moving boundaries [89, 183, 215]

- Fluid-structure interaction [11, 50, 163]
- Domain decomposition [174, 196, 227, 244]

This list is certainly incomplete. Computational fluid dynamics is a rapidly evolving research field which produces a vast amount of new literature every year. We invite the reader to become a member of the growing CFD community and contribute to the development of advanced finite element methods for numerical simulation of fluid flows.

Bibliography

[1] M. Ainsworth and J.T. Oden, *A Posteriori Error Estimation in Finite Element Analysis.* John Wiley & Sons, 2000. (Cited on p. 293)

[2] D.G. Anderson, Iterative procedures for nonlinear integral equations. *J. Assoc. Comput. Machinery* **12** (1965) 547–560. (Cited on p. 251)

[3] J.D. Anderson, Jr., *Computational Fluid Dynamics. The Basics with Applications.* McGraw-Hill, 1995. (Cited on pp. 18, 20, 30, 43, 127, 138, 195, 196, 293)

[4] J.D. Anderson, Jr., *Modern Compressible Flow: With Historical Perspective.* McGraw-Hill, 3rd edition, 2003. (Cited on p. 293)

[5] F. Angrand and A. Dervieux, Some explicit triangular finite element schemes for the Euler equations. *Int. J. Numer. Methods Fluids* **4** (1984) 749–764. (Cited on p. 224)

[6] F. Angrand, A. Dervieux, L. Loth, and G. Vijayasundaram, Simulation of Euler transonic flows by means of explicit finite element type schemes. *INRIA Research Report* **250**, 1983. (Cited on p. 224)

[7] P. Arminjon and A. Dervieux, Construction of TVD-like artificial viscosities on 2-dimensional arbitrary FEM grids. *INRIA Research Report* **1111**, 1989. (Cited on p. 217)

[8] W. Bangerth and R. Rannacher, *Adaptive Finite Element Methods for Differential Equations.* Birkhäuser, 2003. (Cited on p. 293)

[9] G.K. Batchelor, *An Introduction to Fluid Dynamics.* Cambridge University Press, 2000. (Cited on pp. 54, 293)

[10] J.D. Baum and R. Löhner, Numerical simulation of pilot/seat ejection from an F-16. *AIAA Paper* **93-0783**, 1993. (Cited on p. 252)

[11] Yu. Bazilevs, K. Takizawa, and T.E. Tezduyar, *Computational Fluid-Structure Interaction: Methods and Applications.* John Wiley & Sons, 2013. (Cited on p. 294)

[12] H. Blank, M. Rudgyard, and A. Wathen, Stabilized finite element methods for steady incompressible flow. *Comput. Methods Appl. Mech. Engrg.* **174** (1999) 91–105. (Cited on pp. 222, 223)

[13] J. Blasco, R. Codina, and A. Huerta, *Analysis of Fractional Step Finite Element Methods for the Incompressible Navier-Stokes Equations.* Monograph CIMNE **38**, 1997. (Cited on pp. 269, 270, 278, 292)

[14] P. Bochev, D. Ridzal, G. Scovazzi, and M. Shashkov, Constrained-optimization-based data transfer: A new perspective on flux correction. In: D. Kuzmin, R. Löhner, S. Turek (eds.), *Flux-Corrected Transport: Principles, Algorithms, and Applications.* Springer, 2nd edition, 2012, pp. 345–398. (Cited on pp. 239, 240)

[15] D. Boffi, F. Brezzi, and M. Fortin, *Mixed Finite Element Methods and Applications.* Springer, 2013. (Cited on p. 292)

[16] D.L. Book, The conception, gestation, birth, and infancy of FCT. In: D. Kuzmin, R. Löhner, S. Turek (eds.), *Flux-Corrected Transport: Principles, Algorithms, and Applications.* Springer, Berlin, 2005, pp. 5–28. (Cited on p. 205)

[17] D.L. Book, J.P. Boris, and K. Hain, Flux-Corrected Transport: II. Generalizations of the method. *J. Comput. Phys.* **18** (1975) 248–283. (Cited on pp. 203, 236, 240)

[18] J.P. Boris and D.L. Book, Flux-Corrected Transport: I. SHASTA, a fluid transport algorithm that works. *J. Comput. Phys.* **11** (1973) 38–69. (Cited on pp. 203, 240, 243)

[19] J.P. Boris and D.L. Book, Flux-Corrected Transport: III. Minimal-error FCT algorithms. *J. Comput. Phys.* **20** (1976) 397–431. (Cited on pp. 203, 240)

[20] S. Brenner and R. Scott, *The Mathematical Theory of Finite Element Methods.* Springer, 3rd edition, 2008. (Cited on p. 293)

[21] F. Brezzi and M. Fortin, *Mixed and Hybrid Finite Element Methods.* Springer, 1991. (Cited on p. 292)

[22] F. Brezzi and J. Pitkäranta, On the stabilization of finite element approximations of the Stokes equations. In: *Efficient Solutions of Elliptic Systems*, Notes Numer. Fluid Mech. **10**, Vieweg & Sohn, Braunschweig, 1984, pp. 11–19. (Cited on p. 269)

[23] A.N. Brooks and T.J.R. Hughes, Streamline upwind Petrov-Galerkin formulations for convection dominated flows with particular emphasis on the incompressible Navier-Stokes equations. *Comput. Methods Appl. Mech. Engrg.* **32** (1982) 199–259. (Cited on pp. 121, 158, 220, 221)

[24] E. Burman, A unified analysis for conforming and nonconforming stabilized finite element methods using interior penalty. *SIAM J. Numer. Anal.* **43** (2005) 2012–2033. (Cited on p. 227)

[25] E. Burman and A. Ern, Stabilized Galerkin approximation of convection-diffusion-reaction problems: Discrete maximum principle and convergence. *Math. Comput.* **74** (2005) 1637–1652. (Cited on p. 227)

[26] E. Burman and P. Hansbo, Edge stabilization for Galerkin approximations of convection-diffusion-reaction problems. *Comput. Methods Appl. Mech. Engrg.* **193** (2004) 1437–1453. (Cited on p. 227)

[27] E. Burman and P. Hansbo, Edge stabilization for the generalized Stokes problem: A continuous interior penalty method. *Comput. Methods Appl. Mech. Engrg.* **195** (2006) 2393–2410. (Cited on p. 269)

[28] D.S. Burnett, *Finite Element Analysis from Concepts to Applications.* AT&T Bell Laboratories, Whippany, New Jersey, Addison-Wesley, 1987. (Cited on pp. 79, 80)

[29] G.F. Carey, *Computational Grids: Generation, Adaptation, and Solution Strategies.* Taylor & Francis, 1997. (Cited on pp. 16, 18, 293)

[30] L. Catabriga and A.L.G.A. Coutinho, Implicit SUPG solution of Euler equations using edge-based data structures. *Comput. Methods Appl. Mech. Engrg.* **191** (2002) 3477–3490. (Cited on pp. 252, 254)

[31] L. Catabriga, M.A.D. Martins, A.L.G.A. Coutinho, and J.L.D. Alves, An edge-based preconditioner for non-symmetric finite element equations. In: *Proceedings of the 18th Ibero-Latin American Congress on Computational Methods in Engineering.* Brazil, 1997, Vol. 3, pp. 1249–1255. (Cited on p. 253)

[32] L. Catabriga, M.A.D. Martins, A.L.G.A. Coutinho, and J.L.D. Alves, Clustered edge-by-edge preconditioners for non-symmetric finite element equations. In: *Proceedings of the 4th World Congress on Computational Mechanics*. Buenos Aires, Argentina, 1998 (CD-ROM). (Cited on p. 253)

[33] K.-Y. Chien, Predictions of channel and boundary-layer flows with a low-Reynolds number turbulence model. *AIAA J.* 20 (1982) 33–38. (Cited on pp. 286, 287)

[34] A.J. Chorin, Numerical solution of the Navier-Stokes equations. *Math. Comp.* **22** (1968) 745–762. (Cited on p. 273)

[35] A.J. Chorin, On the convergence of discrete approximations to the Navier-Stokes equations. *Math. Comp.* **23** (1969) 341–353. (Cited on p. 273)

[36] R. Codina, A discontinuity-capturing crosswind-dissipation for the finite element solution of the convection-diffusion equation. *Comput. Methods Appl. Mech. Engrg.* **110** (1993) 325–342. (Cited on pp. 225, 226)

[37] R. Codina, Comparison of some finite element methods for solving the diffusion-convection-reaction equation. *Comput. Methods Appl. Mech. Engrg.* **156** (1998) 185–210. (Cited on pp. 221, 222, 263)

[38] R. Codina and J. Blasco, Stabilized finite element method for the transient Navier-Stokes equations based on a pressure gradient projection. *Comput. Methods Appl. Mech. Engrg.* **182** (2000) 277–300. (Cited on pp. 269, 270)

[39] R. Codina, J. Blasco, G.C. Buscaglia and A. Huerta, Implementation of a stabilized finite element formulation for the incompressible Navier-Stokes equations based on a pressure gradient projection. *Int. J. Numer. Methods Fluids* **37** (2001) 419–444. (Cited on p. 270)

[40] G. Cohen, P. Joly, J.E. Roberts, and N. Tordjman, Higher order triangular finite elements with mass lumping for the wave equation. *SIAM J. Numer. Anal.* **38**:6 (2001) 2047–2078. (Cited on p. 217)

[41] G. Comini, M. Manzan, and C. Nonino, Analysis of finite element schemes for convection-type problems. *Int. J. Numer. Methods Fluids* **20** (1995) 443–458. (Cited on p. 222)

[42] M. Crouzeix and P.A. Raviart, Conforming and non-conforming finite element methods for solving the stationary Stokes equations. *R.A.I.R.O.* **R-3** (1973) 77–104. (Cited on p. 269)

[43] C.T. Crowe, J.D. Schwarzkopf, M. Sommerfeld, and Y. Tsuji, *Multiphase Flows with Droplets and Particles*. CRC Press, 2nd edition, 2011. (Cited on p. 293)

[44] C. Cuvelier, A. Segal, and A.A. van Steenhoven, *Finite Element Methods and Navier-Stokes Equations*. Kluwer, 1986. (Cited on pp. 24, 93, 102, 122, 127, 130, 147, 156, 179, 292, 293)

[45] L. Demkowicz, *Computing with Hp-Adaptive Finite Elements. Vol. 1: One and Two Dimensional Elliptic and Maxwell Problems. Vol. 2: Frontiers: Three Dimensional Elliptic and Maxwell Problems with Applications*. Chapman & Hall/CRC, 2007, 2008. (Cited on p. 293)

[46] C.R. DeVore, An improved limiter for multidimensional flux-corrected transport, *NASA Technical Report* **AD-A360122** (1998). (Cited on p. 243)

[47] G.S. Dietachmayer, A comparison and evaluation of some positive definite advection schemes. In: J. Noyle and R. May (eds.), *Computational Techniques and Applications*, Elsevier, 1986, pp. 217–232. (Cited on p. 240)

[48] J. Donea, A Taylor-Galerkin method for convective transport problems. *Int. J. Numer. Meth. Engrg.* **20** (1984) 101–120. (Cited on pp. 197, 221)

[49] J. Donea, S. Giuliani, H. Laval, and L. Quartapelle, Time-accurate solution of advection-diffusion equations by finite elements. *Comput. Methods Appl. Mech. Engrg.* **193** (1984) 123–145. (Cited on p. 228)

[50] J. Donea and A. Huerta, *Finite Element Methods for Flow Problems.* John Wiley & Sons, Chichester, 2003. (Cited on pp. 18, 30, 147, 184, 187, 195, 198, 220, 221, 224, 225, 263, 268, 269, 292, 293, 294)

[51] J. Donea and L. Quartapelle, An introduction to finite element methods for transient advection problems. *Comput. Methods Appl. Mech. Engrg.* **95** (1992) 169–203. (Cited on p. 263)

[52] J. Donea, L. Quartapelle, and V. Selmin, An analysis of time discretization in the finite element solution of hyperbolic problems. *J. Comput. Phys.* **70** (1987) 463–499. (Cited on pp. 197, 198, 222, 223)

[53] J. Donea, B. Roig, and A. Huerta, *High-Order Accurate Time-Stepping Schemes for Convection-Diffusion Problems.* Monograph CIMNE **42**, Barcelona, 1998. (Cited on pp. 198, 221)

[54] J. Donea, V. Selmin, and L. Quartapelle, Recent developments of the Taylor-Galerkin method for the numerical solution of hyperbolic problems. In: *Numerical Methods for Fluid Dynamics III*, Oxford, 1988, 171–185. (Cited on pp. 223, 225, 227, 229)

[55] J. Douglas, Jr. and T. Dupont, Interior penalty procedures for elliptic and parabolic Galerkin methods. In: R. Glowinski and J.-L. Lions (eds.), *Computing Methods in Applied Sciences*, Lecture Notes in Physics, **58**. Springer, Berlin, 1976, pp. 207–216. (Cited on p. 227)

[56] J. Douglas, Jr. and J. Wang, An absolutely stabilized finite element method for the Stokes problem. *Math. Comp.* **52** (1989) 495–508. (Cited on pp. 221, 270)

[57] D.A. Drew and S.L. Passman, *Theory of Multicomponent Fluids.* Springer, 1999. (Cited on p. 293)

[58] D. Drikakis and W. Rider, *High-Resolution Methods for Incompressible and Low-Speed Flows: Fundamentals and Applications.* Springer, 2004. (Cited on p. 293)

[59] J.-J. Droux and T.J.R. Hughes, A boundary integral modification of the Galerkin least squares formulation for the Stokes problem. *Comput. Methods Appl. Mech. Engrg.* **113** (1994) 173–182. (Cited on p. 223)

[60] D.R. Durran, *Numerical Methods for Fluid Dynamics.* Springer, 2nd edition, 2010. (Cited on p. 293)

[61] H.C. Elman, D.J. Silvester, and A.J. Wathen, *Finite Elements and Fast Iterative Solvers*, Oxford University Press, 2005. (Cited on pp. 75, 147, 280, 292, 293)

[62] ELMER: Open Source Finite Element Software for Multiphysical Problems, `http://www.csc.fi/english/pages/elmer` (Cited on pp. 36, 139, 173)

[63] ELMER documentation, `http:sourceforge.net/projects/elmerfem` (Cited on pp. 35, 139)

[64] M.S. Engelman, R.L. Sani, and P.M. Gresho, The implementation of normal and/or tangential boundary conditions in finite element codes for incompressible fluid flow. *Int. J. Numer. Meth. Fluids* **2** (1982) 225–238. (Cited on p. 284)

[65] V. Eyert, A comparative study on methods for convergence acceleration of iterative vector sequences. *J. Comput. Phys.* **124** (1996) 271–285. (Cited on p. 251)

[66] H. Fang and Y. Saad, Two classes of multisecant methods for nonlinear acceleration. *Numer. Linear Algebra Appl.* **16** (2009) 197–221. (Cited on p. 251)

[67] P.E. Farrell, M.D. Piggott, C.C. Pain, G.J. Gorman, and C.R. Wilson, Conservative interpolation between unstructured meshes via supermesh construction. *Comput. Methods Appl. Mech. Engrg.* **198** (2009) 2632–2642. (Cited on p. 231)

[68] M. Feistauer, J. Felcman, and I. Straškraba, *Mathematical and Computational Methods for Compressible Flow*. Clarendon Press, Oxford, 2003. (Cited on p. 293)

[69] J.H. Ferziger and M. Perić, *Computational Methods for Fluid Dynamics*. Springer, 1996. (Cited on pp. 18, 22, 27, 29, 31, 71, 114, 293)

[70] B.A. Finlayson. *Numerical Methods for Problems with Moving Fronts*. Ravenna Park Publishing, Seattle, 1992. (Cited on pp. 263, 293)

[71] C.A.J. Fletcher, *Computational Techniques for Fluid Dynamics*. Springer, 1988. (Cited on pp. 213, 293)

[72] C.A.J. Fletcher, The group finite element formulation. *Comput. Methods Appl. Mech. Engrg.* **37** (1983) 225–243. (Cited on p. 213)

[73] C.A.J. Fletcher, A comparison of finite element and finite difference solutions of the one- and two-dimensional Burgers' equations. *J. Comput. Phys.* **51** (1983) 159–188. (Cited on p. 213)

[74] L.P. Franca and S.L. Frey, Stabilized finite element methods: II. The incompressible Navier-Stokes equations. *Comput. Methods Appl. Mech. Engrg.* **99** (1992) 209–222. (Cited on pp. 158, 269, 270)

[75] L.P. Franca, S.L. Frey, and T.J.R. Hughes, Stabilized finite element methods: I. Application to the advective-diffusive model. *Comput. Methods Appl. Mech. Engrg.* **95** (1992) 253–276. (Cited on pp. 221, 263)

[76] P.J. Frey and P.L. George, *Automatic Mesh Generation, Applications to Finite Methods*. Hermes Science Publishing, Oxford, Paris, 2000. (Cited on pp. 16, 293)

[77] A.C. Galeão and E.G.D. do Carmo, A consistent approximate upwind Petrov-Galerkin Method for convection-dominated problems. *Comput. Methods Appl. Mech. Engrg.* **68**:1 (1988) 83–95. (Cited on p. 226)

[78] P.L. George, *Automatic Mesh Generation, Applications to Finite Methods*. John Wiley & Sons, New York, 1991. (Cited on pp. 16, 293)

[79] V. Girault and P.A. Raviart, *Finite Element Methods for Navier–Stokes Equations*. Springer, Berlin, Heidelberg, 1986. (Cited on pp. 147, 269, 292)

[80] R. Glowinski, *Finite Element Methods for Incompressible Viscous Flow*. In: P.G. Ciarlet and J.L. Lions (eds.), *Handbook of Numerical Analysis, Vol. IX: Numerical Methods for Fluids* (Part 3). North-Holland, 2003. (Cited on pp. 147, 201, 272, 273, 276, 292, 293)

[81] R. Glowinski, Viscous flow simulation by finite element methods and related numerical techniques. In: E.M. Murman and S.S. Abarbanel (eds.), *Progress and Supercomputing in Computational Fluid Dynamics*. Birkhäuser, Boston MA, 1985, pp. 173–210 (Cited on p. 272)

[82] S.K. Godunov, Finite difference method for numerical computation of discontinuous solutions of the equations of fluid dynamics. *Mat. Sb.* **47** (1959) 271–306. (Cited on pp. 192, 206, 233)

[83] G.H. Golub and C.F. Van Loan, *Matrix Computations*. Johns Hopkins University Press, 3rd edition, 1996. (Cited on p. 293)

[84] S. Gottlieb and C.W. Shu, Total Variation Diminishing Runge-Kutta schemes. *Math. Comp.* **67** (1998) 73–85. (Cited on p. 241)

[85] S. Gottlieb, C.-W. Shu, and E. Tadmor, Strong stability-preserving high-order time discretization methods. *SIAM Review* **43** (2001) 89–112. (Cited on p. 241)

[86] P.M. Gresho and R.L. Sani, On pressure boundary conditions for the incompressible Navier-Stokes equations. *Int. J. Numer. Methods Fluids* **7** (1987) 1111–1145. (Cited on p. 266)

[87] P.M. Gresho and R.L. Sani, *Incompressible Flow and the Finite Element Method*. John Wiley & Sons, 2000. (Cited on pp. 147, 266, 278, 292, 293)

[88] F. Grinstein, L. Margolin, and W. Rider, *Implicit Large Eddy Simulation*. Cambridge University Press, 2007. (Cited on pp. 71, 293)

[89] S. Gross and A. Reusken, *Numerical Methods for Two-Phase Incompressible Flows*. Springer, 2011. (Cited on p. 293)

[90] H. Grotjans and F. Menter, Wall functions for general application CFD codes. In: *ECCOMAS 98, Proceedings of the 4th Computational Fluid Dynamics Conference*. John Wiley & Sons, 1998, pp. 1112–1117. (Cited on pp. 285, 286, 289)

[91] J.-L. Guermond and A. Ern, *Theory and Practice of Finite Elements*. Springer, 2005. (Cited on p. 293)

[92] J.-L. Guermond and L. Quartapelle, On stability and convergence of projection methods based on pressure Poisson equation. *Int. J. Numer. Methods Fluids* **26** (1998) 1039–1053. (Cited on p. 274)

[93] W. Hackbusch, *Multigrid Methods and Applications*. Springer, Berlin, 1985. (Cited on pp. 28, 293)

[94] J. Hämäläinen, *Mathematical Modelling and Simulation of Fluid Flows in the Headbox of a Paper Machine*. Universität Jyväskylä, Mathematisches Institut, Bericht **57**, Doctoral thesis, 1993. (Cited on p. 62)

[95] J. Hämäläinen and J. Järvinen, *Elementtimenetelmä virtauslaskennassa* (The Finite Element Method in Computational Fluid Dynamics). CSC Press, 1994. (Cited on pp. vii, viii, 79, 158)

[96] P. Hansbo, Aspects of conservation in finite element flow computations. *Comput. Methods Appl. Mech. Engrg.* **117** (1994) 423–437. (Cited on pp. 216, 228)

[97] P. Hansbo and C. Johnson, Adaptive streamline diffusion methods for compressible flow using conservation variables. *Comput. Methods Appl. Mech. Engrg.* **87** (1991) 267–280. (Cited on p. 226)

[98] A. Harten, High resolution schemes for hyperbolic conservation laws. *J. Comput. Phys.* **49** (1983) 357–393. (Cited on pp. 206, 240, 248)

[99] A. Harten, On a class of high resolution total-variation-stable finite-difference-schemes. *SIAM J. Numer. Anal.* **21** (1984) 1–23. (Cited on pp. 206, 240, 248)

[100] A. Harten, J. Hyman, and P. Lax, On finite-difference approximations and entropy conditions for shocks. *Comm. Pure Appl. Math.* **29** (1976) 297–322. (Cited on p. 228)

[101] J.S. Hesthaven and T. Warburton, *Nodal Discontinuous Galerkin Methods: Algorithms, Analysis, and Applications*. Springer, 2007. (Cited on p. 293)

[102] K. Hiltunen, A. Jäsberg, S. Kallio, H. Karema, M. Kataja, A. Koponen, M. Manninen, and V. Taivassalo, *Multiphase Flow Dynamics. Theory and Numerics*. VTT Publications **722**, 2009. (Cited on p. 293)

[103] J.O. Hinze, *Turbulence*. McGraw-Hill, New York, 1975. (Cited on pp. 66, 293)

[104] C. Hirsch, *Numerical Computation of Internal and External Flows*. John Wiley & Sons, Chichester, 1990. 2nd edition: Butterworth-Heinemann, 2007. (Cited on pp. 30, 210, 293)

[105] G.M. Homsy et al. (eds.), *Multimedia Fluid Mechanics*. Cambridge University Press, 2nd edition, 2008 (DVD-ROM). (Cited on pp. 59, 293)

[106] M.E. Hubbard, Non-oscillatory third order fluctuation splitting schemes for steady scalar conservation laws. *J. Comput. Phys.* **222** (2007) 740–768. (Cited on p. 260)

[107] T.J.R. Hughes and A. Brooks, A multidimensional upwind scheme with no crosswind diffusion. In: T.J.R. Hughes (ed.), *Finite Element Methods for Convection Dominated Flows*, AMD **34**. AMSE, New York, 1979, pp. 19–35. (Cited on p. 220)

[108] T.J.R. Hughes, G. Engel, L. Mazzei, and M.G. Larson, The continuous Galerkin method is locally conservative. *J. Comput. Phys.* **163**:2 (2000) 467–488. (Cited on pp. 31, 216)

[109] T.J.R. Hughes, L.P. Franca, and M. Balestra, A new finite element formulation for computational fluid mechanics: V. Circumventing the Babuska–Brezzi condition: A stable Petrov–Galerkin formulation of the Stokes problem accommodating equal order interpolation. *Comp. Meth. Appl. Mech. Engrg.* **59** (1986) 85–99. (Cited on p. 269)

[110] T.J.R. Hughes, L. Franca, and G. Hulbert, A new finite element formulation for fluid dynamics: VIII. The Galerkin least-squares method for advective-diffusive equations. *Comput. Methods Appl. Mech. Engrg.* **73** (1989) 173–189. (Cited on p. 221)

[111] T.J.R. Hughes and M. Mallet, A new finite element formulation for computational fluid dynamics. IV. A discontinuity capturing operator for multidimensional advective-diffusive systems. *Comput. Methods Appl. Mech. Engrg.* **58** (1986) 329–336. (Cited on p. 225)

[112] T.J.R. Hughes, M. Mallet, and A. Mizukami, A new finite element formulation for computational fluid dynamics. II. Beyond SUPG. *Comput. Methods Appl. Mech. Engrg.* **54**:3 (1986) 341–355. (Cited on p. 226)

[113] T.J.R. Hughes and G.N. Wells, Conservation properties for the Galerkin and stabilised forms of the advection-diffusion and incompressible Navier-Stokes equations. *Comput. Methods Appl. Mech. Engrg.* **194**:9-11 (2005) 1141–1159. Erratum: *Comput. Methods Appl. Mech. Engrg.* **195**:9–12 (2006) 1277–1278. (Cited on p. 31)

[114] W. Hundsdorfer and J.G. Verwer, *Numerical Solution of Time-Dependent Advection-Diffusion-Reaction Equations*. Springer, 2003. (Cited on pp. 192, 201, 210, 233, 272, 293)

[115] F. Ilinca, J.-F. Hétu, and D. Pelletier. A unified finite element algorithm for two-equation models of turbulence. *Comp. & Fluids* **27** (1998) 291–310. (Cited on pp. 289, 290)

[116] A. Jameson, Analysis and design of numerical schemes for gas dynamics 1. Artificial diffusion, upwind biasing, limiters and their effect on accuracy and multigrid convergence. *Int. J. CFD* **4** (1995) 171–218. (Cited on pp. 208, 209)

[117] A. Jameson, Computational algorithms for aerodynamic analysis and design. *Appl. Numer. Math.* **13** (1993) 383–422. (Cited on pp. 219, 233, 236, 246, 247)

[118] A. Jameson, Positive schemes and shock modelling for compressible flows. *Int. J. Numer. Meth. Fluids* **20** (1995) 743–776. (Cited on p. 233)

[119] V. John, *Large Eddy Simulation of Turbulent Incompressible Flows. Analytical and Numerical Results for a Class of LES Models*. Springer, 2004. (Cited on pp. 71, 293)

[120] V. John and P. Knobloch, On spurious oscillations at layers diminishing (SOLD) methods for convection-diffusion equations: Part I. A review. *Comput. Methods Appl. Mech. Engrg.* 196:17–20 (2007) 2197–2215. (Cited on pp. 219, 220, 221, 226)

[121] V. John and E. Schmeyer, On finite element methods for 3D time-dependent convection-diffusion-reaction equations with small diffusion. *Comput. Meth. Appl. Mech. Engrg.* **198** (2008) 475–494. (Cited on pp. 226, 244, 260)

[122] C. Johnson, *Numerical Solution of Partial Differential Equations by the Finite Element Method.* Studentlitteratur, Lund, 1987. (Cited on pp. 122, 220, 263, 293)

[123] C. Johnson, A. Szepessy, and P. Hansbo, On the convergence of shock-capturing streamline diffusion finite element methods for hyperbolic conservation laws. *Math. Comput.* **54** (1990) 107–129. (Cited on p. 226)

[124] C.T. Kelley, *Iterative Methods for Linear and Nonlinear Equations.* Frontiers in Applied Mathematics **16**, SIAM, Philadelphia, 1995. (Cited on p. 293)

[125] D.W. Kelly, S. Nakazawa, O.C. Zienkiewicz, and J.C. Heinrich, A note on anisotropic balancing dissipation in finite element approximation to convection diffusion problems, *Int. J. Numer. Methods Engrg.* **15** (1980) 1705–1711. (Cited on p. 220)

[126] J. Kim, *Investigation of Separation and Reattachment of a Turbulent Shear Layer: Flow over a Backward Facing Step.* PhD thesis, Stanford University, 1978. (Cited on p. 290)

[127] J. Kim, P. Moin, and R.D. Moser, Turbulence statistics in fully developed channel flow at low Reynolds number. *J. Fluid Mech.* **177** (1987) 133–166. (Cited on pp. 287, 288)

[128] N.I. Kolev, *Multiphase Flow Dynamics*, Vols. 1–3. Springer, 2nd edition, 2007. (Cited on p. 293)

[129] M. Křížek and P. Neittaanmäki, *Finite Element Approximation of Variational Problems and Applications.* John Wiley & Sons, New York, 1990. (Cited on pp. 85, 293)

[130] D. Kuzmin, *Numerical Simulation of Reactive Bubbly Flows.* PhD thesis, Jyväskylä Studies in Computing **2**, University of Jyväskylä, 1999. (Cited on p. 277)

[131] D. Kuzmin, Positive finite element schemes based on the flux-corrected transport procedure. In: K.J. Bathe (ed.), *Computational Fluid and Solid Mechanics*, Elsevier, 2001, pp. 887–888. (Cited on pp. 236, 240, 242)

[132] D. Kuzmin, On the design of general-purpose flux limiters for implicit FEM with a consistent mass matrix. I. Scalar convection. *J. Comput. Phys.* **219** (2006) 513–531. (Cited on p. 245)

[133] D. Kuzmin, Algebraic flux correction for finite element discretizations of coupled systems. In: E. Oñate, M. Papadrakakis, B. Schrefler (eds.), *Computational Methods for Coupled Problems in Science and Engineering II.* CIMNE, Barcelona, 2007, 653–656. (Cited on pp. 240, 246)

[134] D. Kuzmin, Explicit and implicit FEM-FCT algorithms with flux linearization. *J. Comput. Phys.* **228** (2009) 2517-2534. (Cited on pp. 240, 241, 242, 258)

[135] D. Kuzmin, *A Guide to Numerical Methods for Transport Equations.* `http://www.mathematik.uni-dortmund.de/~kuzmin/Transport.eps`. (Cited on pp. 32, 240)

[136] D. Kuzmin, Linearity-preserving flux correction and convergence acceleration for constrained Galerkin schemes. *J. Comput. Appl. Math.* **236** (2012) 2317–2337. (Cited on pp. 236, 240, 246, 248, 258)

[137] D. Kuzmin, Algebraic flux correction I. Scalar conservation laws. In: D. Kuzmin, R. Löhner, S. Turek (eds.), *Flux-Corrected Transport: Principles, Algorithms, and Applications*. Springer, 2nd edition, 2012, pp. 145–192. (Cited on pp. 32, 233, 234, 236, 238, 240, 242, 243, 244, 245, 246, 248, 259, 260, 261, 262)

[138] D. Kuzmin, R. Löhner, and S. Turek (eds.), *Flux-Corrected Transport: Principles, Algorithms, and Applications*. Springer: Scientific Computation, 1st edition, 2005; 2nd edition, 2012. (Cited on p. 293)

[139] D. Kuzmin, O. Mierka, and S. Turek, On the implementation of the $k-\varepsilon$ turbulence model in incompressible flow solvers based on a finite element discretization. *Int. J. Comp. Sci. Math.* **1** (2007) 193–206. (Cited on pp. 282, 283, 284, 287, 288, 289, 290, 291)

[140] D. Kuzmin and M. Möller, Goal-oriented mesh adaptation for flux-limited approximations to steady hyperbolic problems. *J. Comput. Appl. Math.* **233** (2010) 3113–3120. (Cited on p. 264)

[141] D. Kuzmin, M. Möller, J.N. Shadid, and M. Shashkov, Failsafe flux limiting and constrained data projections for equations of gas dynamics. *J. Comput. Phys.* **229** (2010) 8766–8779. (Cited on pp. 231, 262, 263)

[142] D. Kuzmin, M. Möller, and S. Turek, High-resolution FEM-FCT schemes for multidimensional conservation laws, *Comput. Methods Appl. Mech. Engrg.* **193** (2004) 4915-4946. (Cited on p. 240)

[143] D. Kuzmin, M.J. Shashkov, and D. Svyatskiy, A constrained finite element method satisfying the discrete maximum principle for anisotropic diffusion problems. *J. Comput. Phys.* **228** (2009) 3448–3463. (Cited on p. 236)

[144] D. Kuzmin and S. Turek, Flux correction tools for finite elements. *J. Comput. Phys.* **175** (2002) 525–558. (Cited on pp. 236, 238, 240, 256, 257, 258)

[145] D. Kuzmin and S. Turek, High-resolution FEM-TVD schemes based on a fully multidimensional flux limiter. *J. Comput. Phys.* **198** (2004) 131–158. (Cited on pp. 240, 245)

[146] D. Kuzmin and S. Turek, Multidimensional FEM-TVD paradigm for convection-dominated flows. In: *Proceedings of the IV European Congress on Computational Methods in Applied Sciences and Engineering (ECCOMAS 2004)*, Volume II, ISBN 951-39-1869-6. (Cited on pp. 282, 283)

[147] O.A. Ladyzhenskaya, *The Mathematical Theory of Viscous Incompressible Flow*. Gordon and Breach Science Publishers, New York, 1969. (Cited on p. 150)

[148] L.D. Landau and E.M. Lifshitz, *Fluid Mechanics*. Pergamon Press, New York, 2nd edition, 1987. (Cited on pp. 54, 293)

[149] C.B. Laney, *Computational Gasdynamics*. Cambridge University Press, 1998. (Cited on p. 293)

[150] A. Lapidus, A detached shock calculation by second order finite differences. *J. Comput. Phys* **2** (1967) 154–177. (Cited on p. 199)

[151] S. Larsson and V. Thomee, *Partial Differential Equations with Numerical Methods*. Springer, 2003. (Cited on p. 293)

[152] B.E. Launder and D.B. Spalding, The numerical computation of turbulent flows. *Comput. Methods Appl. Mech. Engrg.* **3** (1974) 269–289. (Cited on pp. 66, 69, 282)

[153] B.E. Launder and D.B. Spalding, *Lectures in Mathematical Models of Turbulence*. Academic Press, London, 1972. (Cited on pp. 70, 293)

[154] H. Laval and L. Quartapelle, A fractional-step Taylor-Galerkin method for unsteady incompressible flows. *Int. J. Numer. Meth. Fluids* **11** (1990) 501–513. (Cited on p. 276)

[155] W. Layton, *Introduction to the Numerical Analysis of Incompressible Viscous Flows*. SIAM, Philadelphia, 2008. (Cited on pp. 147, 150, 292, 293)

[156] R.J. LeVeque, *Numerical Methods for Conservation Laws*. Birkhäuser, 1992. (Cited on pp. 20, 31, 185, 192, 202, 207, 208, 209, 210, 293)

[157] R.J. LeVeque, *Finite Volume Methods for Hyperbolic Problems*. Cambridge University Press, 2002. (Cited on pp. 210, 293)

[158] R.J. LeVeque, High-resolution conservative algorithms for advection in incompressible flow. *SIAM J. Numer. Anal.* **33** (1996) 627–665. (Cited on pp. 31, 258)

[159] A.J. Lew, G.C. Buscaglia, and P.M. Carrica, A note on the numerical treatment of the *k*-epsilon turbulence model. *Int. J. Comput. Fluid Dyn.* **14** (2001) 201–209. (Cited on pp. 283, 285)

[160] R.W. Lewis, P. Nithiarasu, and K.N. Seetharamu, *Fundamentals of the Finite Element Method for Heat and Fluid Flow*. John Wiley & Sons, U.K., 2005. (Cited on pp. 147, 293)

[161] R. Liska, M. Shashkov, P. Váchal, and B. Wendroff, Optimization-based synchronized flux-corrected conservative interpolation (remapping) of mass and momentum for arbitrary Lagrangian-Eulerian methods. *J. Comput. Phys.* **229** (2010) 1467–1497. (Cited on p. 231)

[162] R. Löhner, Edges, stars, superedges and chains. *Comput. Methods Appl. Mech. Engrg.* **111** (1994) 255–263. (Cited on pp. 252, 254)

[163] R. Löhner, *Applied CFD Techniques: An Introduction Based on Finite Element Methods*. John Wiley & Sons, 2nd edition, 2008. (Cited on pp. 16, 18, 25, 33, 217, 219, 223, 225, 229, 230, 231, 246, 247, 251, 252, 263, 293, 294)

[164] R. Löhner and M. Galle, Minimization of indirect addressing for edge-based field solvers. *Commun. Numer. Methods Eng.* **18**:5 (2002) 335–343. (Cited on p. 252)

[165] R. Löhner, H. Luo, and J.D. Baum, Selective edge removal for unstructured grids with Cartesian cores. *J. Comput. Phys.* **206**:1 (2005) 208–226. (Cited on p. 18)

[166] R. Löhner, K. Morgan, J. Peraire, and M. Vahdati, Finite element flux-corrected transport (FEM-FCT) for the Euler and Navier-Stokes equations. *Int. J. Numer. Meth. Fluids* **7** (1987) 1093–1109. (Cited on pp. 224, 225, 229, 240, 256, 257)

[167] R. Löhner, K. Morgan, M. Vahdati, J.P. Boris, and D.L. Book, FEM-FCT: Combining unstructured grids with high resolution. *Commun. Appl. Numer. Methods* **4** (1988) 717–729. (Cited on pp. 224, 240, 256, 257)

[168] R. Löhner, K. Morgan, and O.C. Zienkiewicz, The solution of non-linear hyperbolic equation systems by the finite element method. *Int. J. Numer. Methods Fluids* **4** (1984) 1043–1063. (Cited on p. 223)

[169] R. Löhner, K. Morgan, and O.C. Zienkiewicz, An adaptive finite element procedure for compressible high speed flows. *Comput. Methods Appl. Mech. Engrg.* **51** (1985) 441–465. (Cited on pp. 224, 225)

[170] P.R.M. Lyra, *Unstructured Grid Adaptive Algorithms for Fluid Dynamics and Heat Conduction*. PhD thesis, University of Wales, Swansea, 1994. (Cited on pp. 217, 246, 247)

[171] P.R.M. Lyra, R.B. Willmersdorf, M.A.D. Martins, and A.L.G.A. Coutinho, Parallel implementation of edge-based finite element schemes for compressible flow on unstructured grids. In: *Proceedings of the 3rd International Meeting on Vector and Parallel Processing*. Faculdade de Engenharia da Universidade do Porto, Porto, Portugal, 1998. (Cited on p. 252)

[172] M. Malinen, Boundary conditions in the Schur complement preconditioning of dissipative acoustic equations. *SIAM J. Sci. Comput.* **29** (2007) 1567–1592. (Cited on p. 42)

[173] M. Malinen, P. Råback, and J. Ruokolainen, The parallel solution of discrete linearized Navier–Stokes equations in the frequency domain. *Acta Acustica* united with *Acustica* **100** (2014) 102–112. (Cited on p. 42)

[174] T.P.A. Mathew, *Domain Decomposition Methods for the Numerical Solution of Partial Differential Equations*. LNSCE **61**, Springer, 2008. (Cited on p. 294)

[175] Millennium Prize Problems, `http://en.wikipedia.org/wiki/Millennium_Prize_Problems`. (Cited on p. 54)

[176] B. Mohammadi and O. Pironneau, *Analysis of the k-epsilon Turbulence Model*. John Wiley & Sons, 1994. (Cited on pp. 68, 69, 282, 292, 293)

[177] M. Möller, *Adaptive High-Resolution Finite Element Schemes*. PhD thesis, Dortmund University of Technology, 2008. (Cited on pp. 252, 255)

[178] K. Morgan and J. Peraire, Unstructured grid finite element methods for fluid mechanics. *Reports on Progress in Physics* **61**:6 (1998) 569–638. (Cited on p. 217)

[179] K. Morgan, J. Periaux, and F. Thomasset (eds.), *Analysis of Laminar Flow over a Backward Facing Step, A GAMM Workshop*. Notes on Numerical Fluid Mechanics **9**, Vieweg & Sohn, Braunschweig/Wiesbaden, 1984. (Cited on p. 75)

[180] K.W. Morton, *Numerical Solution of Convection-Diffusion Problems*. Chapman & Hall, 1996. (Cited on pp. 263, 293)

[181] P. Ni, *Anderson Acceleration of Fixed-point Iteration with Applications to Electronic Structure Computations*. PhD thesis, Worcester Polytechnic Institute, 2009. (Cited on p. 251)

[182] E.S. Oran and J.P. Boris, *Numerical Simulation of Reactive Flow*. Cambridge University Press, 2nd edition, 2001. (Cited on pp. 32, 293)

[183] S. Osher and R. Fedkiw, *Level Set Methods and Dynamic Implicit Surfaces*. Springer, New York, 2003. (Cited on p. 293)

[184] A.K. Parrott and M.A. Christie, FCT applied to the 2-D finite element solution of tracer transport by single phase flow in a porous medium. In: *Numerical Methods for Fluid Dynamics*, Oxford Univ. Press, London, 1986, pp. 609–619. (Cited on p. 240)

[185] S.V. Patankar, *Numerical Heat Transfer and Fluid Flow*. McGraw-Hill, New York, 1980. (Cited on pp. 29, 234, 282, 287, 293)

[186] J. Peraire, M. Vahdati, J. Peiro, and K. Morgan, The construction and behaviour of some unstructured grid algorithms for compressible flows. *Numerical Methods for Fluid Dynamics* **IV**, Oxford University Press, 1993, pp. 221–239. (Cited on pp. 246, 252)

[187] J.B. Perot, An analysis of the fractional step method. *J. Comput. Phys.* **108** (1993) 51–58. (Cited on p. 280)

[188] O. Pironneau, *Finite Element Methods for Fluids*. John Wiley & Sons, 1989. (Cited on pp. 147, 293)

[189] S.B. Pope, *Turbulent Flows*. Cambridge University Press, Cambridge, 2000. (Cited on p. 66)

[190] A. Prohl, *Projection and Quasi-Compressibility Methods for Solving the Incompressible Navier-Stokes Equations.* Advances in Numerical Mathematics, B. G. Teubner, Stuttgart, 1997. (Cited on pp. 274, 292)

[191] J.N. Reddy and D.K. Gartling, *The Finite Element Method in Heat Transfer and Fluid Dynamics*. CSC Press, Taylor and Francis, London, 3rd edition, 2010. (Cited on p. 147)

[192] L. Quartapelle, *Numerical Solution of the Incompressible Navier-Stokes Equations*. ISNM **113**, Birkhäuser, Basel, 1993. (Cited on pp. 147, 198, 221, 228, 274, 292, 293)

[193] L. Quartapelle and V. Selmin, High order Taylor–Galerkin methods for nonlinear multidimensional problems. In: *Finite Elements in Fluids*, Pineridge, Swansea, 1993, pp. 1374–1384. (Cited on p. 198)

[194] A. Quarteroni, F. Saleri, and A. Veneziani, Factorization methods for the numerical approximation of Navier–Stokes equations. *Comput. Methods Appl. Mech. Engrg.* **188** (2000) 505–526. (Cited on p. 281)

[195] A. Quarteroni and A. Valli, *Numerical Approximation of Partial Differential Equations*. Springer, Berlin, 1994. (Cited on pp. 147, 156, 201, 221, 269, 272, 292, 293)

[196] A. Quarteroni and A. Valli, *Domain Decomposition Methods for Partial Differential Equations*. Oxford University Press, Oxford, 1999. (Cited on pp. 18, 294)

[197] V.V. Ranade, *Computational Flow Modeling for Chemical Reactor Engineering*. Academic Press, 2002. (Cited on p. 293)

[198] R. Rannacher, On Chorin's projection method for the incompressible Navier-Stokes equations. In: *The Navier-Stokes Equations II: Theory and Numerical Methods*. Lecture Notes in Math. **1530**, Springer, 1992, pp. 167–183. (Cited on p. 274)

[199] R. Rannacher and S. Turek, A simple nonconforming quadrilateral Stokes element. *Numer. Meth. PDEs* **8** (1992) 97–111. (Cited on p. 269)

[200] H.G. Roos, M. Stynes, and L. Tobiska, *Numerical Methods for Singularly Perturbed Differential Equations*. Springer, 1996. 2nd edition, 2008. (Cited on p. 293)

[201] Y. Saad, *Iterative Methods for Sparse Linear Systems*. SIAM, 2nd edition, 2003. (Cited on p. 293)

[202] A. Safjan and J.T. Oden, High-order Taylor-Galerkin and adaptive h-p methods for second-order hyperbolic systems: Application to elastodynamics. *Comput. Methods Appl. Mech. Engrg.* **103** (1993) 187–230. (Cited on p. 198)

[203] A. Safjan and J.T. Oden, High-order Taylor-Galerkin methods for linear hyperbolic systems. *J. Comput. Phys.* **120** (1995) 206–230. (Cited on p. 198)

[204] P. Sagaut, *Large Eddy Simulation for Incompressible Flows*. Springer, 3rd edition, 2006. (Cited on pp. 71, 293)

[205] A. Samarskii and P. Vabishchevich, *Numerical Methods for Solving Convection-Diffusion Problems* (in Russian). Editorial URSS, Moscow, 1999. (Cited on p. 293)

[206] P.A.B. de Sampaio and A.L.G.A. Coutinho, A natural derivation of discontinuity capturing operator for convection-diffusion problems. *Comput. Methods Appl. Mech. Engrg.* **190** (2001) 6291–6308. (Cited on pp. 222, 225, 226)

[207] R.L. Sani, J. Shen, O. Pironneau and P.M. Gresho, Pressure boundary condition for the time-dependent incompressible Navier–Stokes equations. *Int. J. Numer. Methods Fluids* 50 (2006) 673–682. (Cited on p. 266)

[208] M. Schäfer and S. Turek (with support of F. Durst, E. Krause, and R. Rannacher), Benchmark computations of laminar flow around cylinder. In E.H. Hirschel (ed.), *Flow Simulation with High-Performance Computers II.* Notes on Numerical Fluid Mechanics 52, Vieweg, 1996, pp. 547–566. (Cited on p. 165)

[209] F. Schieweck, On the role of boundary conditions for CIP stabilization of higher order finite elements. *Electronic Transactions on Numerical Analysis* **32** (2008) 1–16. (Cited on p. 227)

[210] V. Selmin, Finite element solution of hyperbolic equations. I. One-dimensional case. *INRIA Research Report* **655**, 1987. (Cited on pp. 227, 228, 229)

[211] V. Selmin, Finite element solution of hyperbolic equations. II. Two-dimensional case. *INRIA Research Report* **708**, 1987. (Cited on pp. 227, 229)

[212] V. Selmin, Third-order finite element schemes for the solution of hyperbolic problems. *INRIA Research Report* **707**, 1987. (Cited on p. 198)

[213] V. Selmin, The node-centred finite volume approach: bridge between finite differences and finite elements. *Comput. Methods Appl. Mech. Engrg.* **102** (1993) 107–138. (Cited on pp. 25, 217, 218)

[214] V. Selmin and L. Formaggia, Unified construction of finite element and finite volume discretizations for compressible flows. *Int. J. Numer. Methods Engrg.* **39** (1996) 1–32. (Cited on pp. 25, 217, 236, 252)

[215] J.A. Sethian, *Level Set Methods and Fast Marching Methods: Evolving Interfaces in Computational Geometry, Fluid mechanics, Computer Vision, and Materials Science.* Cambridge University Press, 1999. (Cited on p. 293)

[216] P.K. Smolarkiewicz and G.A. Grell, A class of monotone interpolation schemes. *J. Comput. Phys.* **101** (1992) 431–440. (Cited on p. 231)

[217] P. Solin, *Partial Differential Equations and the Finite Element Method.* John Wiley & Sons, 2005. (Cited on p. 293)

[218] P. Solin, K. Segeth, and I. Dolezel, *Higher-Order Finite Element Methods.* Chapman & Hall/CRC Press, 2003. (Cited on p. 293)

[219] P.K. Sweby, High resolution schemes using flux limiters for hyperbolic conservation laws. *SIAM J. Numer. Anal.* **21** (1984) 995–1011. (Cited on pp. 206, 208, 240, 248)

[220] R. Temam, *Navier-Stokes Equations*, North-Holland, 1977. (Cited on p. 150)

[221] R. Temam, Sur l'approximation de la solution des équations de Navier-Stokes par la méthode des pas fractionnaires (I). *Arch. Rational Mech. Anal.* **32** (1969) 135–153. (Cited on p. 273)

[222] R. Temam, Sur l'approximation de la solution des équations de Navier-Stokes par la méthode des pas fractionnaires (II). *Arch. Rational Mech. Anal.* **33** (1969) 377–385. (Cited on p. 273)

[223] H. Tennekes and J. L. Lumley, *A First Course in Turbulence.* MIT Press, 1972. (Cited on p. 66)

[224] S. Thangam and C.G. Speziale, Turbulent flow past a backward-facing step: a critical evaluation of two-equation models. *AIAA Journal* 30 (1992) 1314–1320. (Cited on p. 289)

[225] J.F. Thompson, Z.U.A. Warsi, and C.W. Mastin, *Numerical Grid Generation: Foundations and Applications.* North-Holland, Amsterdam, 1985. (Cited on pp. 16, 18, 293)

[226] E.F. Toro, *Riemann Solvers and Numerical Methods for Fluid Dynamics*. Springer, 1999. (Cited on p. 293)

[227] A. Toselli and O. Widlund, *Domain Decomposition Methods – Algorithms and Theory*. Springer, 2004. (Cited on p. 294)

[228] D.A. Trangenstein, *Numerical Solution of Hyperbolic Partial Differential Equations*. Cambridge University Press, 2009. (Cited on pp. 195, 210, 293)

[229] S. Turek, On discrete projection methods for the incompressible Navier-Stokes equations: An algorithmical approach. *Comput. Methods Appl. Mech. Engrg.* **143** (1997) 271–288. (Cited on p. 278)

[230] S. Turek, *Efficient Solvers for Incompressible Flow Problems: An Algorithmic and Computational Approach*. LNCSE **6**, Springer, 1999. (Cited on pp. 28, 147, 220, 249, 273, 278, 280, 292, 293)

[231] S. Turek, C. Becker, and S. Kilian, Hardware-oriented numerics and concepts for PDE software, In: Special Journal Issue for *PDE Software, Elsevier, International Conference on Computational Science ICCS'2002*, Amsterdam, 2002. *FUTURE* **1095** (2003), 1–23. (Cited on p. 33)

[232] S. Turek, C. Becker, and S. Kilian, Some concepts of the software package FEAST. In: J. M. Palma, J. Dongarra, and V. Hernandes (eds.), *VECPAR'98 - Third International Conference for Vector and Parallel Processing*. Lecture Notes in Computer Science, Springer, Berlin, 1999. (Cited on p. 33)

[233] S. Turek and D. Kuzmin, Algebraic flux correction III. Incompressible flow problems. In: D. Kuzmin, R. Löhner, and S. Turek (eds.), *Flux-Corrected Transport: Principles, Algorithms, and Applications*. Springer, 2nd edition, 2012, pp. 239–298. (Cited on pp. 150, 278, 280, 287, 291)

[234] S. Turek and A. Ouazzi, Unified edge-oriented stabilization of nonconforming FEM for incompressible flow problems: Numerical investigation. *J. Numer. Math.* **15**:4 (2007) 299–322. (Cited on p. 227)

[235] UMFPACK web site, `http://www.cise.ufl.edu/research/sparse/umfpack/`. (Cited on pp. 27, 138)

[236] P. Váchal and R. Liska, Sequential Flux-Corrected Remapping for ALE Methods. In: A. Bermudez de Castro, D. Gomez, P. Quintela, and P. Salgado (eds.), *Numerical Mathematics and Advanced Applications* (ENUMATH 2005). Springer, 2006, pp. 671–679. (Cited on p. 231)

[237] J. van Kan, A second-order accurate pressure-correction scheme for viscous incompressible flow. *SIAM J. Sci. Statist. Comput.* **7** (1986) 870–891. (Cited on p. 274)

[238] R.S. Varga, *Matrix Iterative Analysis*. Prentice-Hall, Englewood Cliffs, NJ, 1962. (Cited on pp. 32, 234, 235)

[239] H.F. Walker and P. Ni, Anderson acceleration for fixed-point iterations. *SIAM J. Numer. Anal.* **49** (2011) 1715–1735. (Cited on pp. 29, 150, 251)

[240] R.F. Warming and B.J. Hyett, The modified equation approach to the stability and accuracy of finite difference methods. *J. Comput. Phys.* **14** (1974) 159–179. (Cited on pp. 195, 196)

[241] P. Wesseling, *Principles of Computational Fluid Dynamics*. Springer, Berlin, Heidelberg, 2001. (Cited on pp. 210, 293)

[242] P. Wesseling, *An Introduction to Multigrid Methods*. John Wiley & Sons, Chichester, 1992. Corrected Reprint: R.T. Edwards, Inc., Philadelphia, 2004. (Cited on pp. 28, 293)

[243] D.C. Wilcox, *Turbulence Modeling for CFD*. DCW Industries, Inc., La Canada CA, 3rd edition, 2006. (Cited on pp. 66, 293)

[244] B. Wohlmuth, *Discretization Techniques and Iterative Solvers Based on Domain Decomposition*. LNSCE **17**, Springer, 2001. (Cited on p. 294)

[245] S.T. Zalesak, Fully multidimensional flux-corrected transport algorithms for fluids. *J. Comput. Phys.* **31** (1979) 335–362. (Cited on pp. 203, 205, 230, 240, 243, 257, 258)

[246] S.T. Zalesak, A preliminary comparison of modern shock-capturing schemes: Linear advection. In: R. Vichnevetsky and R. Stepleman (eds.), *Advances in Computer Methods for PDEs*. Publ. IMACS, 1987, pp. 15–22. (Cited on p. 203)

[247] S.T. Zalesak, The design of Flux-Corrected Transport (FCT) algorithms for structured grids. In: D. Kuzmin, R. Löhner, and S. Turek (eds.), *Flux-Corrected Transport: Principles, Algorithms, and Applications*. Springer, 2005, pp. 29–78. (Cited on pp. 203, 205, 230)

[248] O.C. Zienkiewicz and K. Morgan, *Finite Elements and Approximation*. University of Wales, Swansea, John Wiley & Sons, 1983. (Cited on p. 80)

[249] O.C. Zienkiewicz and R.L. Taylor, *The Finite Element Method*. Butterworth-Heinemann, Oxford, 2000. (Cited on pp. 263, 293)

[250] O.C. Zienkiewicz, R.L. Taylor, and P. Nithiarasu, *The Finite Element Method for Fluid Dynamics*. University of Wales, Swansea, Elsevier, 6th edition, 2009. (Cited on p. 293)

Index